地方重塑
PLACE REMAKING

首届国际公共艺术研究员会议文集

总策划 汪大伟　主编 章莉莉

上海大学出版社

目 录
CONTENTS

- 5 序言 FOREWORD
- 7 编者语 EDITOR'S COMMENTS
- 9 会议宗旨 AIMS
- 11 会议背景 BACKGROUND

- 13 嘉宾简介 GUESTS
- 33 会议综述 OVERVIEW

- 77 研究员访谈 INTERVIEW
- 78 公共艺术要让人触手可及——研究员伊芙·莱米斯尔访谈
  Public Art within Reach : An Interview with Eve Lesmele

- 88 艺术的内涵根植于艺术品之中——研究员沃恩·赛迪访谈
  Explore the Content from the Public Art Works :
  An Interview with Vaughn Sadie

- 98 公共艺术作为一种社会治疗——研究员里奥·谭访谈
  Public Art as a Social Therapy : An Interview with Leon Tan

- 108 公共艺术与城市身份——研究员茱茜·乔克拉访谈
  Public Art as an Urban Identity :
  An Interview with Giusy Checola

- 123 案例研究 CASE STUDY
- 124 基于子区域的公共艺术研究分区方案
- 129 大洋洲 Oceania
- 173 南亚 中西亚 South Asia, Central and West Asia
- 213 中南美洲 Central and South America
- 245 欧洲 Europe
- 285 东亚 东南亚 East Asia, Southeast Asia
- 337 北美 Northern America
- 389 非洲 Africa

- 428 感悟 PERSONAL UNDERSTANDING
- 433 后记 AFTERWORD

序 言
FOREWORD

在全球化背景下，人类文化多样性问题日益受到关注。文化生态的健康与否，既是衡量一个地区文化潜力的重要指标，也是保护地域文化遗产的关键要素。"地方重塑"不仅是局部的、实时性的地方再造，也是整体的、历时性的地区文化生态工程。重视推进文化生态研究，把文化置于生态之中，用历史视角解读地域文化变迁，研究文化演变与这一地区其他生态系统如自然、经济、社会等的关系，这些研究既是了解一个地区的重要资源，也为艺术家制定地区可持续发展方略提供借鉴和启示。

虽然世界不同地区存在不同的文化差异性，但公共艺术是文化认同的载体，一方面以艺术的形式表达群体的文化本质，另一方面公众又以艺术的方式参与其中并形成文化认同。因此，在"地方重塑"的过程中，公共艺术对地区的介入，就文化学意义而言，正是一种认同与再造的结合，"地方重塑"的理念也在这种认同中具有了普遍意义。

"地方重塑"将艺术引入大众日常生活，引导大众重新认识、界定自己的生存环境，赋予地方建设以更大的开放性。加入艺术元素的社区环境生发出一种双向机制：日常生活既因为艺术的到来而获得新的层级结构和发展潜质，导致新的文化价值思考，同时，生活中的许多元素又对艺术的存在方式及发生逻辑产生冲撞，从而启示出更加具有社会属性和公众意义的艺术样式，在交叉中使艺术类型更加多元化并产生新的动力。

我相信，"地方重塑"将是公共艺术的永恒主题。

上海大学美术学院 院长、教授
2014 年 10 月

编者语
EDITOR'S COMMENTS

围绕"地方重塑"的价值观和理念，我们看到城市和乡村生活中的空间环境、文化脉络、日常生活等公共领域，当公共艺术介入后发生的变化。艺术的作用非常奇妙，其呈现方式从社区改造、空间转换、壁画雕塑到艺术活动等非常多元，引发了人们的思考和认知，促进了社会关系，改变了空间感受，让人类共同生活的空间充满文化的光芒。

我们关注公共艺术，其实是在关注人类自身的生存空间和生活常态。重塑地方人文，艺术介入生活，此次研究员会议邀请国际公共艺术研究员们各抒己见，头脑风暴，共同探讨"什么是好的公共艺术"的标准和理念。

这本《地方重塑——首届国际公共艺术研究员会议文集》围绕构建国际公共艺术智库网络的目标，组织来自全球的近30位国际公共艺术研究员，在已成功举办的两届国际公共艺术论坛及评奖活动的基础上，进一步推动公共艺术的国际学术研究和交流发展。

书籍收录了公共艺术研究员们推荐分享的全球最新公共艺术案例，从公共艺术项目的发起动因、组织运作、资金保障、公众反馈、时间评价等角度进行分析，并提供了不同思考角度的解读，也期待读者能用自己的方式去解读公共艺术。

章莉莉
2016年9月

会议宗旨
AIMS

由上海大学美术学院与国际公共艺术协会（IPA）共同主办的 2014 国际公共艺术研究员会议，于 2014 年 11 月 21—22 日在上海大学美术学院举行。

上海大学美术学院通过创办"国际公共艺术奖"（IAPA）评选活动、国际公共艺术论坛，利用国际专家研究团队资源，构建了一个国际公共艺术智库网络。在 2012 年和 2014 年两届评奖活动中，该智库网络已经起到了积极的作用。首先，来自该智库网络的全球公共艺术研究员，通过考察、研究共提交了来自全球的 267 个优秀公共艺术案例的研究分析报告，目前已出版成册。其次，2013 国际公共艺术论坛的成功举办，进一步丰富智库网络的专家资源。最后，基于智库的咨询报告研究，上海大学美术学院为后世博园改造、地铁公共艺术规划、浙江美丽乡村建设提出咨询方案，其中部分方案已被采纳，该智库在中国公共艺术项目中发挥了实质性的作用。

本次研究员会议旨在进一步发挥国际公共艺术智库网络的积极作用，推动公共艺术学科的发展。来自全球的近 30 位国际公共艺术协会研究员参加了本次会议。会议旨在总结首届和本届"国际公共艺术奖"评审经验，进一步推进公共艺术研究，建立国际性公共艺术研究与交流平台，以及推动国际公共艺术专家智库网络的构建。

会议的主要内容是：

1、探索公共艺术在未来的总体发展趋势，确定今后"国际公共艺术奖"评选活动的主题。

2、集中规划网络发展及覆盖力，在现有联合国统计署全球地理区域划分基础上，兼顾世界文化区、世界人口因素，并参照前两届公共艺术案例增减地区的变化，设定合理的划分方案，使公共艺术案例筛选工作的覆盖面更加广泛，让全球各文化区的优秀公共艺术案例得以呈现。

3、探索最佳合作与工作方式，实现工作平台以及技术语言的统一，构建一流的国际公共艺术研究与交流平台。

会议背景
BACKGROUND

希望通过研究员会议，扩大"国际公共艺术奖"活动的影响，加强国际交流，达成共识，共同推进国际公共艺术的发展。

国际公共艺术奖（IAPA）由中国《公共艺术》杂志和美国《公共艺术评论》杂志于 2011 年共同创立。该奖项是国际公共艺术领域最高成就的象征，为世界各地区正在开发中的城市提供公共艺术建设范例，引领公共艺术发展潮流。

IAPA 评奖活动每两年举办一次，首届评选会于 2012 年 5 月 10—11 日在上海召开，来自全球各大洲的公共艺术研究员，从 2006 年 1 月 1 日—2011 年 9 月 30 日完成的作品中提名 141 件杰出作品，并由评审委员会评选出 25 个最佳案例以及 6 个获奖案例。2013 年 4 月 12 日—15 日在上海大学举行首届"国际公共艺术奖"颁奖仪式，同时举办国际公共艺术论坛。

首届 IAPA 评奖活动首届 IAPA 评奖活动过程中，为了建立全球性的公共艺术智库网络，促进国际公共艺术研究以及分享世界各地公共艺术信息，于 2013 年 2 月由上海大学资助成立了"国际公共艺术协会"（IPA）。该协会主张通过评选、出版物、网站、媒体等方式，倡导公共艺术理念和优秀的公共艺术实践，促使世界各地的决策者关注公共艺术。

在上海大学美术学院的支持下，今年又迎来了第二届"国际公共艺术奖"评选活动。截止到 2014 年 8 月，评选会共收到 30 位全球专家组成的研究团队推荐的 124 件作品案例，涵盖装置、建筑、雕塑、壁画、行为、活动、景观设计、空间规划等多种类型，均为 2007 年 1 月—2012 年 12 月完成。本届评选更加侧重于体现出以艺术家为主导的、卓有成效的地方重塑、社区建设或社会实践艺术，以及高水平的执行效果和对所在地的长远（或潜在长期）的积极影响。代表全球七大洲的评委经过多轮的评审，选出 32 件入围作品、7 件优秀作品及 1 个大奖。奖项将于 2015 年在新西兰奥克兰举行的第二届"国际公共艺术奖"颁奖仪式暨国际公共艺术论坛中揭晓。

嘉宾简介
GUESTS

Lewis Biggs	路易斯·比格斯
Wang Dawei	汪大伟
Jack Becker	杰克·贝克尔
Jin Jiangbo	金江波
Peter Morales	皮特·莫拉莱斯
Bruce Phillips	布鲁斯·菲利普斯
Lesya Prokopenko	莱西亚·普洛科朋科
Megan Guerber	梅根·古尔伯
Raphael Chikukwa	拉斐尔·奇古瓦
Jun Kitazawa	北泽润
Leon Tan	里奥·谭
Sara Black	萨拉·布莱克
Kelly Carmichael	凯利·卡迈克尔
Vaughn Sadie	沃恩·赛迪
Gabriela Ribeiro	加布里埃拉·里贝罗
Hsiung Peng Chu	熊鹏翥
Eve Lemesle	伊芙·莱米斯尔
Giusy Checola	茱茜·乔克拉
Peter Shand	彼得·尚德
Stella Prasetya	史黛拉·普拉瑟塔赫
Jessica Fiala	杰西卡·费亚拉
Nahla Al Tabbaa	拉·阿尔·塔瓦

路易斯·比格斯
Lewis Biggs

国际公共艺术奖（IAPA）组委会主席。1979—1984 年担任阿尔诺芬尼·布里斯托尔展览协调员，并任英国文化协会视觉艺术部展览馆员至 1987 年。1986 年担任圣保罗双年展英国专员。1987 年加入英国利物浦泰特美术馆，1990—2000 年担任泰特美术馆主管。1998 年与利物浦双年展的詹姆斯·摩尔和珍妮·蓝晶·里德一起成为创始受托人，2000—2011 年担任双年展行政长官。

Chairman of the Organising Committee of IAPA. Lewis Biggs was Exhibitions Coordinator at Arnolfini, Bristol, 1979—1984, and Exhibitions Officer with the British Council Visual Arts Department until 1987. He was British Commissioner for the Sao Paulo biennale in 1986. In 1987 he joined Tate, and was Director of Tate Liverpool from 1990 to 2000. In 1998 he became a founding Trustee, with James Moores and Jane Rankin Reid, of Liverpool Biennial, and has been Chief Executive of the Biennial from 2000 to 2011.

汪大伟
Wang Dawei

上海大学美术学院院长、教授、博导,《公共艺术》杂志主编;现任中国美术家协会理事、中国美术家协会平面设计艺委会副主任、上海市文联委员、上海创意设计工作者协会主席、上海美术家协会副主席、教育部高等院校艺术设计专业指导委员会委员、教育部艺术硕士专业指导委员会委员、上海地铁建设环境艺术委员会委员、上海双年展艺术委员会委员、上海艺术博览会艺术委员会委员。

Professor, Advisor of Institute for Public Art (IPA), Dean of Fine Arts College, Shanghai University, Chief Editor of *Public Art*, the Standing Council Member of Chinese artists, Deputy Director of the Arts Council of Graphic Design, member of Chinese Artists Association, Chairman of Shanghai Creative Design Workers Association, Vice Chairman of Shanghai Artists Association, member of Master of Fine Arts Steering Committee of Education Ministry, member of Environmental Art of Shanghai Metro Construction Committee, member of Art Committee of Shanghai Art Fair, Vice Director of Art Committee of Shanghai Art and Design Exhibition.

杰克·贝克尔
Jack Becker

始创于1978年的美国非营利性艺术机构"预测公共艺术"的创始人和执行董事,《公共艺术评论》杂志主编。"预测公共艺术"的总部设在明尼苏达州圣保罗。作为一名公共艺术家及项目管理者,擅长参与项目,架起连接艺术家创意与地方需求的桥梁,结合社区与机会创作有意义的公共艺术。组织过70多个展览、50份出版物和大量有特色的项目。

Jack Becker is founder and executive director of Forecast Public Art, which was established in 1978. He is also the publisher of *Public Art Review*. Forecast Public Art is a nonprofit arts or ganization based in St. Paul, Minnesota. As a public artist and program administrator, he specializes in projects that connect the ideas and energies of artists with the needs and the opportunities of communities to create meaningful public art. Jack has organized more than 70 exhibitions, 50 publications, and numerous special events.

金江波
Jlin Jiangbo

国际公共艺术协会(IPA)副主席,发起人。上海公共艺术协同创新中心执行主任,上海大学美术学院院长助理、教授、硕士生导师;中国工业设计协会信息与交互设计专业委员会副主任;上海青年创意人才协会理事;上海创意设计工作者协会理事及多媒体设计专业委员会主任;首届国际公共艺术奖与公共艺术论坛秘书长;首届北京国际设计三年展"可能的世界"策展人;上海艺术设计展艺术委员会委员;2011"上海市十大青年高端创意人才"。

Jiangbo Jin graduated and gained PhD in Art from Academy of Fine Arts, Tsinghua University. He is Secretary-General of the first International Award for Public Art Presentation Ceremony and Public Art Forum, director of Shanghai Designer Association , director of Shanghai Young Creative Talent Association and director of Multimedia Design Committee. In 2011 he was invited to be curator of the 1st Beijing International Design Triennial. In 2013 he became the member of Shanghai Art Design committee. Currently, he is dean assistant of Fine Art College of Shanghai University. In 2011 he won the Shanghai Top Ten Youth Top-level Creative Talents.

皮特·莫拉雷斯
Peter Morales

国际公共艺术协会 (IPA) 研究员，雕塑家，来自美国明尼苏达，长期从事大型户外雕塑作品创作，作品多次参加国际艺术展，精通英语、西班牙语。

IPA researcher, sculptor, he created a large number of Large-Scale Outdoor Sculpture, his works take part in numerous international exhibitions. And he is proficient in English and Spanish.

布鲁斯·菲利普斯
Bruce Phillips

国际公共艺术协会 (IPA) 研究员，新西兰惠灵顿维多利亚大学在读硕士，国际策展人，策划多起大型国际及个人展览。

IPA researcher, curator, a current student of Victoria University Wellington, New Zealand.

莱西亚·普洛科朋科
Lesya Prokopenko

国际公共艺术协会(IPA)研究员,国立基辅莫希拉学院文化研究专业学士,国际策展人。

IPA researcher, Culture Studies BA of the National University Kyiv-Mohyla Academy, International Curator.

梅根·古尔伯
Megan Guerber

国际公共艺术协会(IPA)研究员策展人,《公共艺术评论》编辑,2013—2014 国际公共艺术奖研究员。

IPA Researcher, curator, editor of *Public Art Review Magazine*, researcher of IAPA 2013—2014.

拉斐尔·奇古瓦
Raphael Chikukwa

国际公共艺术协会 (IPA) 研究员，伦敦金斯顿大学策展当代设计硕士，津巴布韦国家美术馆主策展人，2010—2011年津巴布韦国家馆策展创始人。成功策划了2011年首届津巴布韦馆第54届威尼斯双年展和2013年第二届津巴布韦馆第55届威尼斯双年展。

Raphael Chikukwa was awarded the 2006—2007 Chevening Scholar and he holds an MA Curating Contemporary Design from Kingston University, London. He is an independent curator for many years before joining the National Gallery of Zimbabwe in the mid-2010 as it's Chief Curator. He is the founding Zimbabwe Pavilion curator in 2010—2011 and he has curated the 1st Zimbabwe Pavilion in 2011 at the 54th Venice Biennale and the 2nd Zimbabwe Pavilion at the 55th Venice Biennale in 2013. He is also the founding coordinator of the 1st Zimbabwe curatorial workshop and forum that saw regional emerging and established curators exchange at the National Gallery of Zimbabwe in Bulawayo in 2013.

北泽润
Jun Kitazawa

国际公共艺术协会 (IPA) 研究员,当代艺术家,北泽君八云工作室代表,策展人,艺术项目负责人。组织策划多起艺术项目,如 2014 年在德岛县佐那河内农园内的《满月聚餐》,2013 年在冲绳县荣町市场内的《荣町市场客厅》,舞鹤市京都府《时光旅行博物馆》等。

IPA researcher, Contemporary artist, representative of Jun Kitazawa, Yakumo office, curator. Organizing and planning many art projects. Such as 2014 *Full Moon Dining* at Sanagochi Village, Tokushima pref, 2013 *LIVING ROOM SAKAEMACHI-ICHIBA* at Sakaemachi-market, Naha city, Okinawa pref and *TIME TRAVEL MUSEUM* at Maizuru city, Kyoto pref, etc.

里奥·谭
Leon Tan

国际公共艺术协会 (IPA) 研究员，来自奥克兰的艺术与文化批评家，策展人，教育工，作者，心理治疗师。奥克兰大学艺术史博士，出版多部专著，参与 2012 台北双年展，苏黎世数字艺术周等多项国际大展。

IPA Researcher, an art and culture critic, curator, educator and psychotherapist, with a PhD in Art History from the University of Auckland. His doctoral research was based on the analysis and theorization of the psychology, aesthetics and politics of networked expression. He is also a senior lecturer in the Department of Design and Contemporary Arts at Unitec Institute of Technology. He is a professional member of the International Association of ArtCritics (AICA, Hong Kong Section).

萨拉·布莱克
Sara Black

国际公共艺术协会 (IPA) 研究员，福尔茅斯艺术学院文学硕士管理实践专业的创建人之一，英国福尔茅斯大学的客座讲师。英国康沃尔策展人和制作人。曾为亨利摩尔基金会、利物浦双年展、第二届福克斯顿三年展和工程基地工作，组织策划多项关于场所艺术的会议和演讲系列活动。

IPA researcher, Sara was a founding partner in the MA Curatorial Practice at Falmouth School of Art, and is a visiting lecturer at Falmouth University, UK. And she is acurator and producer based in Cornwall, UK. Since 1998 Sara has worked for Henry Moore Foundation, Liverpool Biennial, Folkestone Triennial and Project Base and independently, producing 33 place-based projects with artists in rural locations, festivals and urban centres.

凯利·卡迈克尔
Kelly Carmichael

国际公共艺术协会(IPA)研究员,城市画廊举办的"居住在丽塔格斯"展览策展人。曾在洛杉矶现代艺术博物馆实习。系列纪录片《英国雕塑史》的专业研究员。

IPA researcher, and she has recently joined asCurator, Special Projects "The Rita Angus Residency at City Gallery". She began with a curatorial internship at the Museum of Contemporary Art, Los Angeles; worked in a Parisian dealer gallery and dabbled intelevision production as the specialist researcher fora six part documentary series on the history of British sculpture.

沃恩·赛迪
Vaughn Sadie

国际公共艺术协会 (IPA) 研究员，德班理工学院公共管理技术博士在读，成功策划了多次个人与团体展。入围 MTN 新当代决赛，2012 年，当选《邮卫报》南非人 200 强，2011 年于伊斯坦布尔被授予 iDANS 奋进奖。

IPA researcher, he currently is the Doctorate of Technology in Public Management of Durban Institute of Technology , he curated and participated in many person shows, Collaborations and group shows. He was awarded Finalist in MTN New Contemporaries and *Mail and Guardian* Top 200 South Africans in 2012, also iDANS Critical Endeavour Award in Istanbul in 2011.

加布埃拉·里贝罗
Gabriela Ribeiro

国际公共艺术协会 (IPA) 研究员，巴西圣保罗大学美学和艺术史硕士，巴西 Vale do Paraíba 专业学士，在视觉艺术、媒体与设计方面有丰富的经验。

IPA researcher, a Currently student in Aesthetics and Art History at the University of São Paulo (USP-PGEHA) program. Experience in Visual Arts with an emphasis on design, communication and production of exhibitions. Served in non-formal education with exhibits and workshops on mediation and teaches art in formal education.

熊鹏翥
Hsiung PengChu

国际公共艺术协会 (IPA) 研究员，毕业于美国纽约大学 (NYU) 人类学研究所与博物馆学研究所，现为帝门艺术教育基金会执行长，曾任世界宗教博物馆筹备处助理研究员与顺益台湾原住民博物馆研究组主任。

IPA researcher, graduated from Anthropology Institute / Museum Institute in the year of 1989. Currently is the CEO of Emperor art education foundation. Was a Preparatory office assistant researcher of the museum of world religions, and the director of the research team of Shun yi Taiwan aboriginal museum.

伊芙·莱米斯尔
Eve Lemesle

国际公共艺术协会(IPA)研究员,策展人,巴黎索邦大学艺术管理专业在加拿大、欧洲、印度从事策展和艺术管理,2009年定居孟买,印度what about art(WAA)创始人,关注南亚艺术家和艺术案例,并从事南亚艺术品收藏。

IPA researcher, curator. She is a graduate of La Sorbonne in arts management and holds a diplomain South-Asia studies from INALCO (Paris). She settled in Mumbai in 2009, and started What about art (WAA), a first-of-its-kind arts management agency in India.

茱茜·乔克拉
Giusy Checola

国际公共艺术协会(IPA)研究员,意大利作家,研究国际艺术项目策展人,关注艺术与公众的关系、城乡公共空间、公共艺术的影响力等问题。

IPA researcher, author and research curator of international artistic projects oriented both to the research and production. Currently she focuses her work on the placemaking, currently she is member of the curatorial board of the SoutHeritage Foundation for Contemporary Art based in Matera.

彼得·尚德
Peter Shand

国际公共艺术协会(IPA)研究员，奥克兰大学国家创意艺术与工业研究院副院长、美术副教授，策展人，策划了多起大型国际艺术展览。

IPA researcher, Deputy Dean, National Institute for Creative Arts and Industries, Associate Professor of Fine Arts, Curator.

史黛拉·普拉瑟塔
Stella Prasetya

国际公共艺术协会(IPA)研究员，英国创意艺术大学硕士，曾为马拉塔纳基督大学讲师，参与过多次艺术策划与艺术活动。

IPA researcher, MA graduated from the University for the Creative Arts (Canterbury, Kent, UK), and she was a Lecturer of Undergraduate Arts and Design Programme of Maranatha Christian University, Bandung, Indonesia.

杰西卡·费亚拉
Jessica Fiala

国际公共艺术协会(IPA)研究员，来自美国的跨文化学者，艺术管理，舞蹈家，美国《公共艺术评论》作者，明尼苏达州大学教授。

IPA researcher, Interdisciplinary scholar, dancer, writer of *Public Art Review*.

纳赫拉·阿尔·塔瓦
Nahla Al Tabbaa

国际公共艺术协会(IPA)研究员，英国Bath Spa大学艺术与设计专业硕士，沙迦艺术基金会社区宣传项目协调员，玛干艺术空间项目与空间协调员。

IPA researcher, graduated from Bath Spa University, Community & Outreach Projects Coordinator at The Sharjah Art Foundation/Sharjah Biennial, Project & Space Coordinator at Makan Art Space.

会议综述
OVERVIEW

周娴 / 整理

宋晓伟 滕健俊 / 摄影

2014年11月21—22日，由上海大学美术学院与国际公共艺术协会（IPA）共同主办的2014国际公共艺术研究员会议在上海大学美术学院举行。

来自全球近30位国际公共艺术协会研究员参加了本次会议。会议旨在积极发挥上海大学美术学院国际公共艺术智库网络的作用，推动公共艺术学科发展，进一步推进公共艺术研究，建立国际性公共艺术研究与交流平台。

致 辞

在国际公共艺术研究员会议召开之际,谨以上海大学美术学院的名义向出席会议的研究员表示热忱的欢迎,并对你们在公共艺术领域做出卓有成效的研究和努力表示崇高的敬意。

国际公共艺术协会(IPA)与上海大学美术学院共同主办的本次研究员会议是继前两届国际公共艺术奖的评审之后的重要学术活动,它为国际公共艺术研究提供了一个研究、交流和共享的平台,它汇聚了全球五大洲七个地区当下公共艺术的研究成果,更可喜的是来自 14 个国家和地区 22 位研究员其乐融融、汇聚一堂共商公共艺术研究之路,共议公共艺术发展趋势,共建公共艺术智库网络,为国际公共艺术奠定学术研究基础,为探索公共艺术多元发展提供各种可能。这也是国际公共艺术协会(IPA)为研究员搭建的学术研究网络。我相信本次研究员会议将对推动国际公共艺术发展具有实质性的意义和深远地影响。

预祝会议圆满成功,祝各位研究员度过一个愉快而值得回忆的上海之行。

汪大伟 教授
国际公共艺术协会 顾问
上海大学美术学院 院长

As Institute for Public Art—Research Network Meeting is to be held in Shanghai, I would like to extend our warmest welcome to all researchers on behalf of Fine Arts College, Shanghai University, and express our sincere thanks to you for your hard work and achievements in the area of public art.

The meeting, co-organized by Institute for Public Art (IPA) and Fine Arts College, Shanghai University, is an important academic activity after the two judging processes for the International Award for Public Art. It provides a research, exchange and sharing platform for international public art. It has attracted fruitful research results of contemporary public art from the five continents, seven regions in the world. Even better, 22 researchers from 14 countries and regions are here to discuss the research and development of public art, as well as the establishment of intelligence network for public art. It will lay a foundation for academic research and explore various development for international public art. This is also an academic research network for researchers established by the Institute for Public Art (IPA). I am sure that the meeting will have a far-reaching impact and substantive significance for international public art.

Wish the conference success! Wish all of you have an enjoyable and worthwhile trip in Shanghai.

Professor Wang Dawei
Advisor of Institute for Public Art
Dean of Fine Art College, Shanghai University

致 辞

我谨代表上海大学美术学院的合作伙伴——公共艺术协会,欢迎各位前来参加 2014 年 11 月 20—23 日举办的研讨会。

公共艺术协会的目标是研究公共艺术形式和宣传相关方面的知识;建立并维护致力于公共艺术的专业人才网;向开发商和市政部门等潜在的客户倡导公共艺术。

不同国家和地区对公共艺术的认识存在很大的差异,不同的文化背景对"公共空间"甚至"艺术"的理解也可能迥然不同。通过对比,有助于公共艺术研究人员更为深刻地理解不同的文化背景和其多样性。

这次研讨会聚集了来自世界各国关注和致力于研究公共艺术的学者与艺术家,尤其是以地方重塑为研究方向的艺术家。

研讨会期间,我们将会看到一些对公共艺术案例的研究。我们将会讨论已经在"国际公共艺术奖"中形成的公共艺术的研究准则和方法,并探讨研究报告的格式。我们还会讨论公共艺术协会根据地域的文化特征,为"国际公共艺术奖"而提出区域划分。如果我们有足够的时间,我们将会和艺术家与专业机构就市政开发进行信息沟通。

我们首先考虑的工作是进一步深化和阐明对 2017 年"国际公共艺术奖"的研究指导。

我们深信,由上海大学慷慨资助举办的本次研讨会,将会鼓励我们对所讨论问题的兴趣和扩大研究员的规模。

路易斯 · 比格斯
公共艺术协会主席

On behalf of The Institute for Public Art (IPA) in partnership with Shanghai University College of Fine Arts, I would like to welcome you to this seminar 20—23 November 2014.

The aims of the IPA are to research the forms of public art and propagate knowledge about them; to create and maintain a network of professionals concerned for public art; and to advocate public art to potential clients such as developers and city halls.

Art in public space differs greatly from one region and country to another—the conditions of "public space", and even the convention of "art" itself being understood very differently in different cultures. This variety creates a challenge to researchers who would like to propagate knowledge about public art practices, compare them with each other, and advocate those that are most successful in their own cultural contexts.

This seminar brings together people, from many countries around the world, who are involved in research about art in public space, and in particular artist-led place-making.

During the seminar, we will look at some case studies of art in public space. We shall discuss the research criteria and methods that have been developed for the International Award for Public Art (IAPA), and the format in which the research is presented. We shall review the geographic / cultural "regions" developed by the IPA in relation to research for the IAPA. If we have time, we shall discuss advocacy to city halls and developers and how best to exchange information among artists and professional agencies. Our priority is to further develop and clarify the guidelines for research towards the 2017 IAPA.

But we also confidently expect that during these days, so generously hosted by Shanghai University, our discussions about issues and concerns of mutual interest will augment and nourish our existing network of researchers.

Lewis Biggs
Chairman, IPA

第一部分 "国际公共艺术协会"以及"国际公共艺术奖"回顾与介绍

《公共艺术评论》(Public Art Review)杂志主编、IPA 副主席杰克·贝克尔(Jack Becker)担任会议主持人。

他首先介绍了设立"国际公共艺术奖"的背景。2011 年,由路易斯·比格斯(Lewis Biggs)、上海大学美术学院汪大伟院长、新西兰艺术基金会总督学院主席约翰·麦科马克(John McCormack)以及我本人共同讨论设立该奖项,通过《公共艺术评论》(美国)与《公共艺术》(中国)两本杂志的合作,提高大家对公共艺术的认知,让大家更好地了解什么是公共艺术。他指出《公共艺术评论》杂志已出版 25 年了,但因为语言和文化上的障碍,他们并不了解非洲、南美洲以及中美洲和中东地区公共艺术的发展。在上海大学美术学院的支持下,IPA 寻找到来自世界各地的公共艺术研究员,他们在各个领域搜集相关信息,通过研究与分享,使人们能够了解世界各地公共艺术的发展情况。他提出 IPA 的目标是,促进公众对公共艺术的理解、欣赏与支持,让人们了解全球的公共艺术。实现这个目标的第一步就是要进行分享,但如何进行研究并将研究数据和信息让更多的人获得和分享,对于 IPA 来说是一个挑战。从教育的角度来说,IPA 今天所做的就是要帮助建立一个更大的研究网络,收集、归纳和分享更多的公共艺术信息,使更多的人能够理解和欣赏公共艺术。

上海大学美术学院院长、IPA 顾问汪大伟介绍了上海大学美术学院公共艺术学科的发展情况。

他指出,1998 年他曾在《装饰》杂志上发表题为《公共艺术将是 21 世纪的朝阳学科》的论文,在国内率先提出"公共艺术"的概念,并把公共艺术作为学科来建设,同时也列入"国家 211 工程"建设的重点学科。现在,公共艺术已经被中国广泛接受。伴随着上海城市发展,

01 2014 国际公共艺术研究员会议与会嘉宾合影

上海大学美术学院与时俱进，尤其是在 20 世纪城市大规模建设时期，参与了上海的轨道交通建设以及上海的环境设计，有几个数字可以反映参与的力度：2010 年上海整个轨道交通建设通车的 112 个车站中，有 66 个车站的总体设计是由上海大学美术学院担当的。到目前为止，将近 360 个车站中 85% 的壁画和环境艺术，以及上海 50% 以上的城市雕塑都是由学院设计、创作的。另外，2010 上海世博会，在上海很多主题公园中，都能看到上海大学美术学院公共艺术学科发展的足迹。汪大伟指出，随着大量的艺术实践，更需要理论和学科的凝练和提升。于是，他与美国《公共艺术评论》杂志主编杰克·贝克尔、路易斯·比格斯一起共同策划设立"国际公共艺术奖"，后又成立"国际公共艺术协会"，推动公共艺术研究网络建设。现在，公共艺术的话题带来了更多朋友的关注。今天正值上海大学美术学院建院 55 周年，在这 55 周年的历程中，重点学科的建设推进了整个学院发展。学院目前有 3 个一级博士点，有博士生 80 位，硕士生 450 位，本科生 150 位左右，教职工将近 200 人。最后，汪院长还介绍了上海大学美术学院的博士后流动站，这是推进公共艺术深入研究，加强国际合作的一个工作站。在学校政策的规定下，希望吸纳全世界有志于公共艺术发展的研究员进入这个工作站工作，并提供一定的生活补贴和科研经费。

IPA 主席路易斯·比格斯向大家介绍了创办"国际公共艺术协会"的背景。

他指出，在他作为利物浦美术馆馆长以及利物浦双年展的策展人和总监时，利物浦城市开始衰退，城市环境非常糟糕。他希望人们通过自己的双手创造周围的环境和条件，让生活变得美好。因为上海和利物浦是友好城市，2000 年来参加上海双年展时见到汪大伟院长，两人对于公共艺术有着同样的热情，都认为公共艺术能够解决城市环境中的一些问题。但他认为公共艺术应该成为城市规划和城市设计中的一部分，需要在规划之初就预防问题的出现，而不是等问题出现了再解决。他提出IPA的主要目标是，通过对话实现一种有机的成长，主要包括三个方面：一是做研究，了解全世界公共艺术领域发生的事情。二是打造一个网络，让对公共艺术有热情的研究人员能够一起做研究、分享思想，并把研究成果付诸实践。成员还应该包括开发者、市政厅的城市规划人员。三是积极推动所做的事情，获得政府的支持。中国政府对于公共艺术也非常重视，这太好了！给世界其他地方树立了一个非常好的典范。而要实现这些目标，首要的就是设立"国际公共艺术奖"奖项。这个奖项不仅是一个竞赛，而且是推介我们的研究成果，并得到媒体关注的一个契机。颁奖活动每两年一届，2013年4月在上海举行了首届颁奖仪式；2015年6月底将在奥克兰举办第二届颁奖仪式，这一次是由山东工艺美术学院共同承办的。评奖是一个很好的机制，而研究员网络又是另一个机制，这是我们所做的核心工作。如果人们之间能够相互理解，并且分享大家的热情，我们就可以实现目标。除了颁奖仪式外，还计划举办国际论坛，研究人员将会参与这个论坛，也可以请开发者来参加，他们虽然不做研究，但可以了解研究员们所做的工作。由于颁奖和论坛都是双年的，在两年中我们将在非洲、中国香港、圣保罗等地举办一些小的论坛，不仅要构建全球性的平台，同时也鼓励打造一些重要的区域性平台和论坛。

汪大伟院长回顾了2013"国际公共艺术奖"、国际公共艺术论坛的情况，并介绍了2014"国际公共艺术奖"评审相关情况。

"国际公共艺术奖"由美国《公共艺术评论》和中国《公共艺术》杂志作为发起单位，国际公共艺术协会与上海大学美术学院共同承办了第一届、第二届"国际公共艺术奖"的评选活动。经过一年的征集，2011年10月31日共收到全球专家组成的研究团队推荐的142个优秀公共艺术案例，涵盖了装置、建筑、雕塑、壁画、行为表演、活动等类型。其中很大的一部分是国外具有代表意义的作品，在当地具有相当的影响力，也获得了其他的各种奖项。第一届评委会主席就是现在IPA的主席路易斯，由杰克·贝克尔、荷兰阿姆斯特丹艺术和公共空间基金会主管弗尔雅·厄尔德姆奇（Fulya Erdemci）、日本东京当代艺术馆总馆长长谷川佑子（Yuko Hasegawa）、新西兰艺术基金会总督学院主席约翰·麦科马克、巴西圣保罗大学当代艺术教授卡提亚·坎顿（Katia Canton），还有我本人组成的评审团对案例进行评审。评审的标准是：①案例必须是在2006年1月—2011年4月完成的作品；②必须对当地产生过积极影响，而不是以艺术家个人的成就影响作为评价标准；③无论是永久的还是临时作品或项目，都必须以塑造空间为主；④案例要体现艺术家的主导作用，从创新、影响力等方面进行评选；⑤案例通过艺术家的创意与协调，能够启发人们，让人们参与其中。2012年5月10—11日两天，评委们按照以上标准，通过认真仔细的阅读和推介，以及激烈的争论和讨论，最后以投票的方式在142个案例里选出了25个最佳案例，其中6个案例入围最后大奖的评选。2013年4月12日举办了"首届国际公共艺术颁奖仪式暨公共艺术论坛"，国内公共艺术研究专家、学者以及国际研究员汇聚一堂，做了"关于地方重塑对于公共艺术的作用"专题论坛。上海市教委副主任袁雯教授、上海大学校长罗宏杰教授、奥克兰大学美术学院院长拉娜·达文波特（Rhana Devenport）以及几位获奖艺术家和策展人等都做了专题演讲。论坛还包括三个具有针对性的分论坛，分别就"公共艺术如何进入上海地铁公共空间"、"世博后公共艺术如何介入和激活世博园区"以及"公共艺术如何参与中国浙江乡村城镇化建设"三个主题进行讨论。其中第三个分论坛是在浙江乡镇开的，当地村长向专家提出了"公共艺术如何改变生活"、"公共艺术是否能增加收入"等实际问题。乡村百姓们最关心的是如何提高他们的收入问题，而我认为公共艺术对于提高收入的作用往往是望尘莫及的。但是我告诉他们，尽管收入不高，但是公共艺术可以带来精神方面的愉悦，也是一种幸福。会议得到了当地镇政府、企业和村民的支持。当时美术学院跟他们签订了一个"公共艺术实验基地"的合作项目。我希望以后的"国际公共艺术奖"一定要为举办地解决一些实际问题。我们IPA集聚全球的研究网络，来对接这些问题，给当地政府、当地举办者提供解决问题方案，以此来增强评奖活动的魅力。换句话说，哪里的公共空间有问题，就会来争取举办"国际公共艺术奖"，因为背后有一个国际的研究团队和研究网络，至少能够针对所在地方的具体问题，提供多种、多元的解决思路。

接下来介绍一下2014国际公共艺术奖评审的情况。本届评审在2014年9月刚刚结束，评委会主席由大洋洲评委拉娜·达文波特担任，她是新西兰戈维布鲁斯特艺术馆馆长、作家、策展人。评委会成员组成如下：欧亚大陆区评委是新加坡当代艺术中心创始董事、南洋科技大学教授尤特·米塔·鲍尔（Ute Meta Bauer）；非洲区评委是南非开普敦大学教授、策展人、导演杰·佩瑟（Jay Pather）；南美洲区评委是美国独立策展人比尔·凯利（Bill Kelley）；北美洲区评委是美国独立策展人、作家切尔西·海恩斯（Chelsea Haines）；贝鲁特到孟加拉区评委是印度控杰（Khoj）国际艺术家联盟创始成员兼董事、策展人普加·苏德（Pooja Sood）；东亚区评委是我。评审中要从内容丰富、形式各异、各具特色的案例中进行遴选，这给评委们带来了很大的挑战。同时，IPA主席路易斯担任本届评委会的监审长，他对整个评审过程的公正性、有效性进行监审。经过评委的评述、案例讨论和公共投票，评委们从全球19位研究员推荐的125个公共艺术案例中评选出32个入围作品，同时选出了7个优秀奖和1个大奖。获奖情况暂未公开，将于2015年6月在奥克兰的颁奖大会和论坛上公布。评审结束后，还举办了一场"什么是好的公共艺术?"评委对话会,评委们从"作品与空间"、"公众的审美导向"、"社会关注的问题"以及"流行的趋势"等不同的角度阐述了各自的见解,体现了当今世界跨地域的公共价值观。大家认为:好的公共艺术首先要考虑到公众。公共艺术立项应该反映公众需求,项目公示需要真实采纳民意,而不能被架空。而我们IPA的职责,就是要建立一个广泛的全球公共艺术研究网络来推动世界公共艺术的发展,同时也希望它能够促进艺术家、策展人、公众以及政府更好地理解公共艺术概念。不仅欣赏它,而且能够积极地参与公共艺术重塑地方的活动。

杰克·贝克尔对第一部分会议进行了总结:

汪大伟院长提出希望我们能够打造一个系统来支持研究网络做更多研究,我们是需要这样的一个支持系统推动未来IPA的工作以及研究、分享和颁奖工作。我们可以与城市、大学建立合作,不仅能够获得最新的研究成果,而且可以帮助大学的教学,也可以促进城市社区、文化和经济方面的发展。如果我们能够把最好的艺术家带到城市、大学里,我想也是符合城市与大学的发展需要,因为他们也面临着一些挑战。

第二部分 公共艺术案例研究

路易斯·比格斯介绍了英国福克斯通三年展的情况:

如何在本土和国际之间寻求一个很好的平衡?这是我一直在思考的问题。福克斯通是离法国最近的一个英国小镇,与法国隔海相望。这里的地理和历史问题,对于公共艺术来说是一个非常重要的基础和支撑。福克斯通虽然只有5万人,却有着悠久的历史。欧洲海底隧道通车后,福克斯通发生了很大的变化。对于这样一个小镇来说,是非常激动人心的,从巴黎来的人和伦敦来的人都可以在这里相遇。20世纪,曾有很多英格兰家庭希望到福克斯通度假,当时这里有很多海边设施。但在20世纪60年代后,因为大家都愿意坐飞机到其他有阳光的地方去度假,福克斯通的度假功能也逐渐丧失了。我担任了福克斯通三年展的策展人。因为小镇没有任何的美术馆,所以想把街道作为展示艺术作品的空间。这次三年展共展出了18件艺术作品,我们希望艺术展之后能够达到25—30件。

在这张地图(图02)上,蓝点代表以前的作品,红点代表2014年在我的领导下创作的一些新作品。从图上可以看到,蓝点是沿着海岸线自

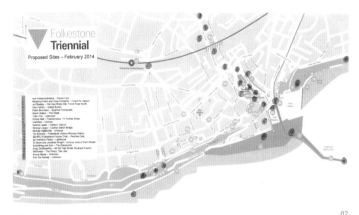

02 福克斯通三年展作品地图

东向西排列的。海岸线面朝法国,可以说是福克斯通的脸面,和度假紧密联系在一起。而从南到北的大部分是红点,这些点代表了福克斯通的历史,因为福克斯通有一条纵贯的河延伸到了海岸线。我策展的目的,就是要充分挖掘那些被人们忽视的福克斯通的历史。这也是三年展想要达到的一些目标。整个展览的作品都是为当地居民创作的,同时也面向外来的游客和居民。

《水箱下的暗河》(Pent River Water Tanks)(图03/图04/图05/图06)是由本土艺术家戴安娜·德弗(Diane Dever)和乔纳森·莱特(Jonathan Wright)创作的。他们向我提出了方案,想要发掘那些失去的河流。因为房地产开发,很多人不知道福克斯通地底下有一条河流的存在。河流是艺术的源泉,也是精神的源泉,它非常重要。我想不通为什么开发商要把这条河覆盖起来,因为通常河流能够使房地产升值。艺术家在这条失去的河流所流经的地方建造了一系列的贮水箱。当人们沿着这个路线走,就会发现很多这种水箱,引发人们去思考这里为什么会有这种水箱,是不是地底下有水在流动呢?放置水箱的地点都是经过选择的,与河流的历史是相关联的。

我们在努力发掘福克斯通过去的地理和历史,以及与现代福克斯通居民生活之间的联系。另外一个案例是《绿/光》(Green/light)(图07/图08),由当地的艺术家乔·布瑞德林(Jyll Bradley)创作。作品设置的地点以前是一个居民区的煤气厂,1966年建成,现在已经完全破败了,需要更多人来关注和管理。很多人都希望这里不要像一个废墟那样存在,而是能为当地创造一些价值。作品基于当地的历史,也与乡间景色相联系。也许这个作品最终是需要拆掉的,因为土地所有者只给了一个短期租约,我们现在正在与他们谈判。因为这个地区太糟糕,开发商不愿意来开发,但如果不开发的话,这里仍然会很糟糕,说来说去就是利润的问题。如果一些人需要地方住,他到这儿来自己造点房子

03

04 05

06

03/04/05/06 《水箱下的暗河》（Pent River Water Tanks）
04 2014 英国福克斯通三年展作品
07/08 2014 英国福克斯通三年展作品《绿/光》（Green/light）
09 2014 英国福克斯通三年展作品《帕亚斯公园》(Payers Park) 项目之挖地活动

不就行了吗？我们正在请当地人到这里建房子。所以我们现在也在讨论，希望将来能够把这块地接管过来。这件艺术品就像一个特洛伊木马，希望在 50 年后能够启动这个进程，能够把这块地重新打造起来。

第三个案例是《帕亚斯公园》(Payers Park)，它是艺术家和建筑师一起合作的项目，艺术机构 Muf architecture/art 在这里组织了多个社区活动。公园位于老城中心位置，因其特殊的地理环境，这里一直是荒地，不仅位于陡坡之下，且地底下就是河流，地基不稳，人们一般不愿意在这里建造东西，所以一直是没有被利用起来。这个项目中，艺术家和建筑师与居民合作，邀请他们参加这个活动，并了解到当地居民对这里有什么愿望。这个区域的居民多为移民家庭，居住环境非常拥挤，所以喜欢在户外活动。正因为如此，大家常会闹意见，有时也打架，甚至还有一些犯罪活动。从不同的利益出发，大家都想利用这块地方，却持有不同的观点和想法。有一个活动是发动人们挖地（图 09），看在这片荒地上能挖出什么来，并利用挖出来的东西，来说明这个地方将来可以用作什么用途。另一个活动是做了一些临时烤箱（图 10），教人们烤披萨，尝试一些不同的菜谱，包括从其他国家带来的一些菜谱。还有一个活动是"跳跳舞"（图 11），就是在一定的区域跟着音乐跳舞。这些活动都反映了当地居民的需求。但设计者希望这里不要被过度设计，使空间还能有进一步开发新用途的可能性，以激励人们做更多不同的事情。还有一个是当地艺术家做的陶瓷作品（图 12），放在一棵树上。这个项目有意思的地方是，我们通过谈判将这个私人空间变为公共空间。这里本是开发商的地，因为利润薄不愿意开发而荒废。通过引进艺术，人们可以在这里对话，使得这里变成一个共享的公共区域，创造出一种积极的用途。

第四个案例是加布里埃尔·莱斯特（Gabriel Lester）创作的《港口平台》（Harbour Steps）（图 14），位于铁路与港口的交汇处。因为当地已经不再从这个码头运输货物，所以铁路方面也不跑火车了。这个城镇能否像纽约的《高线公园》项目一样从废弃的基础设施上获益？当地人认为这个铁路上应该再跑火车，但这纯粹是一种情感上的诉求。是否应该考虑一下，这里能否被保护并且利用这个历史遗迹？艺术家在上海也做过项目，发现在中国用竹子做的脚手架非常有意思。竹子虽然是很好的材料，但在英国利用得并不多。所以艺术家就用竹子做了一个装置，人们不仅可以在上面攀爬，而且装置本身也是一个观景台，在上面能够看到整个铁路码头。作品非常成功，很多人都到这里参观，还有人晚上在这里睡觉。

10 烤披萨活动

11 跳跳舞活动

12 当地艺术陶瓷作品放在树上

13 新西兰奥克兰的《维多利亚女王雕像》

上海市慈善基金会、上海市慈善教育培训中心主任徐佩莉向大家介绍了"牵手艺术彩虹，献给未来的地球"慈善公共艺术项目：

20年前，上海市慈善基金会和上海第二工业大学共同创办了上海市慈善教育培训中心，其宗旨是知识扶贫，授人以渔，助人发展。"牵手艺术彩虹，献给未来地球"是我们中心做的一个少年儿童慈善艺术项目，旨在帮助上海的3—17岁、不同国籍的困难家庭或困境中的热爱艺术的少年儿童来发掘艺术灵感，感恩五彩生活，热心回报社会。经过我们的培训，使孩子们能具有更多热爱世界、善待地球、崇尚环保、温暖人间的品德和素养，以心灵和双手来描绘心中未来地球的模样。活动由上海市慈善基金会、上海市儿童基金会主办，由"艺术花园"、我们中心和上海大学公共艺术创意中心承办。2013年11月至今，不仅得到了高校的支持，还得到了部分企业、社区和社会组织的支持。在项目进行中开展了一批校舍、校企结合的课程，制作了食品，也印制了宣传册，并组织了两次少儿创意作品成果展示和多次义卖活动。至今已经有5000多人次参与，取得了少年儿童慈善创意艺术培训的一些经验。项目开展一年多来，社会反响很好，也得到了政府、社会的称赞，上海市教委也愿意为项目做些指导。

"艺术花园"是上海一批年轻白领志愿者组成的机构，通过组织艺术展览、艺术作品义卖募集善款。善款主要用于两个项目：一是"启智通艺"项目，主要帮助困难家庭和困境中的孩子学艺术；二是"小小创意家"项目，主要针对热爱艺术、有爱心的健康孩子，进一步提高他们的创意能力和艺术追求。我们开设了剪纸、水粉画、折纸等课程。此外，与高校合作，让孩子们到上海大学公共艺术创意中心学习，章莉莉老师还带领孩子到地铁站讲解她设计地铁导视系统的思路。与创意产业机构合作，让孩子们走进M50、南翔创意园学习；还让孩子们到都市农庄体验，到上海市佘山少儿活动营地、上海艺术博物馆等地学习。项目主要筹资的途径是：大学、中小学、幼儿园艺术的老师们捐画义卖；来自上海高校的、企业的、社会的艺术大师作品捐赠义卖；有爱心的孩子们捐赠作品义卖；企业捐款以及慈善机构的资金支持，比如上海市慈善基金会、儿童基金会。

希望能够在国际公共艺术协会的指导或主办下，能有更多国家、地区的艺术家共同来关注少年儿童公益慈善事业的发展，携手推进世界公共艺术文化的发展，如果对我们这个项目感兴趣，可以通过我们研究员的研究和参与，共同制定一些在各个国家、各个地区推进的计划。通过我们的理念，我们的行动来影响当地公益组织，推动企业、高校、市民对少年儿童公共艺术和公益慈善事业的关注，支持扶贫项目的开展。总的来说，希望我们项目能够实现艺术研究与社会服务相结合，艺术实践与公益慈善的行动相结合，在这个方面能够多出一些力，以我们的研究和行动献上我们"蓝天下的挚爱"。

14 《港口平台》（Harbour Steps）
15 新西兰时轮公园的项目《Evening Echo》
16 奥克兰大学的雕像
17 新西兰奥克兰的项目《无题》

IPA 研究员、奥克兰大学国家创意艺术与工业研究院副院长彼得·尚德（Peter Shand）向大家介绍了新西兰公共艺术的历史、发展的现状以及新西兰的案例：

首先，给大家介绍新西兰以及大英帝国殖民地一个非常普遍的公共艺术项目《维多利亚女王雕像》（图 13）。奥克兰的维多利亚女王雕像比惠灵顿的高很多。在新加坡、澳大利亚等地都可以看到维多利亚女王的雕像，可以说这是在公共领域里一个非常独特的符号，充分显示了和空间、地域、人口之间的相互关系。这个雕像非常平静，双手在胸前合在一起，高高耸立，俯瞰众生。新西兰并没有《宪法》，女王是国家的象征，是宪法的化身，具有权威性，树立在公众领域由公众来敬仰。她出现在新西兰的货币上，是我们信念的捍卫者，在很多方面我们都是她的臣民。可以说她是我们公共艺术的一个核心，很多公共空间都充满了这样的国家象征。维多利亚女王雕像在新西兰树立了以后，她成为了象征性的符号，代表着很多意义和价值观，比如：英勇、权威以及意识形态。另一座位于奥克兰大学的雕像（图 16），双手抱臂，也同样具有权威感。我希望大家不要误解我的意思，我举这些例子，并不是说公共艺术只是展示国家力量这样一个象征。任何一个公共艺术都有它自己的方法，都有代表着某种公众力量的方式，包括公众的钱、公众的愤怒，公众的控制以及公众的一切。

IPA 研究员、惠灵顿维多利亚大学硕士、国际策展人布鲁斯·菲利普斯（Bruce Phillips）也介绍了几个新西兰的公共艺术案例：

第一个案例是《Evening Echo》（图 15），这个项目是在时轮公园（Shalom Park）做的，位于盖斯步道（Gas works road）和艾尔伯特路（Albert road）。1990 年开幕，是跟当地犹太社区紧密联系的一个项目。这个烛台灯柱每一年的这个时间点燃，并通过当地的报纸进行宣传。这个活动还要取决于当地的居民有什么样的记忆。这是一个永久装置，表现生命是一个漫长的、缓慢的、消亡的过程。因为大家对于城市社区的记忆正在逐渐减少，如果大家认为公共艺术目的就是要造成这种永久性的社会变化的话，那可能就是很大的误读。作为一个社会的项目，它只是建立了一种可能性，并不能确保所提出的这些东西能够永久留存。在这里，设计师试图建立起一种积极的公共区域，建立起概念性的框架使得有一种物理性的参与方式，使得艺术家自己的取向被公众所忽视或者是所抛弃。这样的一个方式，使得这种社会项目拥有了不同程度的参与度。

第二个是 2012 年奥克兰的一个项目，名为《无题》（图 17），作者是卢克·威尔斯·汤普森（音），这是一个非常大胆的尝试。项目引起了非常多的争论。2008 年 1 月 28 号凌晨，一个年轻人和他的亲戚正

18 颜名宏《玉米田的时间——家 X 记忆的渡口》中国台湾地区高雄市驳 2 艺术特区
19 颜名宏《玉米田的时间——家 X 记忆的渡口》人们参加玉米采收活动
20 颜名宏《玉米田的时间——家 X 记忆的渡口》"稻草人"计划

在一个白人的车库门上涂鸦，当时被白人发现后将其杀害。由于种族偏见，杀人者埃默里（音）只在监狱里待了 11 个月就被放出来。车库门上面保留了受害者留下的痕迹，艺术家把这些门搜集来，以纪念那个逝去的生命。当时围绕这件事有很多新闻报道，而被涂鸦的门就成为了报道中的热门话题。涂鸦者并没有被当做一个悲剧的暴力事件受害者，很多年后，这个车库大门被人收购，并放入了奥克兰国家美术馆作为一个永久的公共纪念，以此来消除人们心理的创伤。

我们如何来评判一个我们没有参与或仔细观看的作品？如何更好地充分了解很多介入社会的作品？这些作品对社会、社区产生了什么影响？有哪些区域没有得到我们足够的关注？有哪些偏见是我们必须要克服和消除的？或是我们还没有意识到的？我觉得这是我们这一次研讨会可以讨论的。这是从我个人的一些角度触发所能想到的问题，也是根据自己介绍的案例所能想到的问题。

IPA 副主席、上海大学公共艺术协同创新中心执行主任金江波介绍了他在新西兰做的一个艺术项目：

2010 年，我拍摄了新西兰和全球殖民者有关的一些社会图像，希望以此来了解全球的殖民运动以及后殖民方式。欧洲殖民者经过 50 年左右的发展之后，欧洲的工业化标准发生了改变。例如罐头制作的标准变了，这里的罐头厂的生产跟不上欧洲的行业标准，导致整个行业的衰退。这个美丽的港口小镇，因为工厂的发展、产业链的营造和社会生态的构建，变成了有 30 000 人左右的小规模城市。但是由于行业标准改变，投资者离开了这里，小镇慢慢衰退，到 20 世纪 80 年代，这里只留下 200 户人家。工厂倒闭造成了巨大的公共建筑资源浪费。因为很多材料的堆积，这里变成了影响当地生活的一种污染源。后来毛利人和英国殖民者签订了一个《停战协议》，那一天可能有一些毛利人或者说其他人把这个工厂烧成了废墟，使得工厂变成了一个巨大无比的空旷障碍物，成为了一个具有强烈反思意义的，可以称之为"社会雕塑"的"作品"保留在那里，以此来纪念或者让我们反思这段历史。我想此时艺术家的介入其实显得微不足道，我所做的是用图像保留这个历史。这种公共事件，改变了我们的生活，改变着一个地方的历史，它对这里有着很大的作用力。而我们公共艺术实际上就是为了关注这些东西、发掘这些东西。我想研究时，是不是可以关注这类公共事件，这个历史本身是具有艺术价值的。

IPA 研究员、台湾帝门艺术教育基金会执行长熊鹏翥介绍了中国台湾地区公共艺术发展的情况，并推荐了四个案例：

21 《海市蜃楼——台湾闲置公共设施摄影计划》姚瑞中和失落社会档案室共同完成
22 中国台湾地区《绿色美学》项目

公共艺术不只是一个名词,而一个动词,它所指的是一个公共化的过程。下面介绍的四个案例,我将分别讨论这个过程。我先介绍一下中国台湾地区公共艺术发展的状况,有两个脉络:第一,1992年开始在法规上规定,政府机构在建筑以及相关的建设必须有一定的百分比的费用来执行公共艺术。第二,有公共精神的艺术创作与艺术计划。这个部分的经费,不一定完全来自政府,但是政府也有计划地支持这个艺术计划。所以说,公共艺术在中国台湾地区是由上而下的,即法规规定必须编预算做的,另外一个是跟社区营造有关的,是由下而上所发生的具有公共精神的艺术计划。下面通过四个案例分别介绍这些不同性质的计划。

第一个计划,是由艺术家执行的。它不是法令规定下的公共艺术,但项目是位于一个政府所规划出来的公共空间,费用来自其他的展览预算。这个案例名为《玉米田的时间——家X记忆的渡口》(2013年5月—8月)(图18),位于中国台湾地区高雄市驳2艺术特区,艺术家是颜名宏。驳2艺术特区是利用旧仓库、旧厂房改造成的艺术空间。艺术家辟出了一块田地,通过在城市空间中种植玉米,表现对环境改变、都市情境改变的一种思考。这个项目包括装置作品,即一个不锈钢的方盒和一些断裂的楼梯。几张空置的楼梯放置在田间,象征着被破坏、瓦解。而这个断裂的空间象征一个无法回到的家。艺术家通过玉米的成长过程,使得植物的生长连接了被断裂的楼梯。艺术家在玉米成长的过程中还举办了相关的活动,邀请民众来协助采收玉米(图19),之后把采收完毕的植物再回归土壤。另外,还有一些其他的艺术计划,像"稻草人"(图20)等。

第二个计划叫《海市蜃楼——台湾闲置公共设施摄影计划》(2010年至今)(图21),由艺术家姚瑞中和失落社会档案室共同完成。姚瑞中常做社会相关的纪实摄影,喜欢拍摄废弃的空间、场景等。2010年在学校带摄影课程时,他就请学生回到家乡去拍摄废弃的建筑物。后来,他惊讶地发现中国台湾地区各处有非常多利用公共资金设置的无用建筑物,包括被废弃的学校、军营,也包括为了好大喜功所建设的科学园区、港口设施等。这个作品不只记录了废弃的空间,更重要的是它让台湾当局必须采取新的措施去解决这些空置、被浪费等问题。这些建筑被称为"蚊子馆",因为它空旷无人,只有蚊子在里面。这个计划执行至今,已有约4年的时间,我相信这个计划还会继续执行下去,它对于中国台湾地区的政治,还有社会的影响也会逐渐地发酵。

第三个计划是《绿色美学》(图22)(2013年8月23日—2014年2月26日)。这是一个大学的项目,属于百分比计划。在过去的"百分比计划"中,大家都希望看到实体的作品,但这个项目中,策划团队成功说服了学校,使得项目中过程性的作品比重超过了实体作品。其中一个作品(图23)是一位日本艺术家邀请民众采集区域不同的土壤夯土成墙,随着时间流逝,这个墙将瓦解,植物从这里生长起来。

23

24

25

26

23 中国台湾地区《绿色美学》项目中,日本艺术家邀请民众采集区域不同的土壤夯土成墙,随着时间流逝,这个墙将瓦解,植物从这里成长起来

24 中国台湾地区《树梅坑溪环境计划》,举办让民众了解过去溪流生态环境的社区说明会

25 中国台湾地区《树梅坑溪环境计划》,带领民众顺溪而上认识溪流与人的关系的活动

26 墨西哥华雷斯城的《红鞋子》项目

27

28

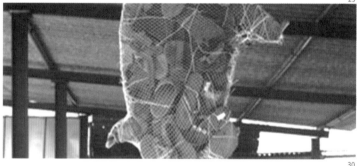

29

30

27 巴西圣保罗《城市自然》项目
28 巴西双胞胎街头艺术家 Os Gemeos 壁画项目。
29 印度孟买的项目《相遇》
30 印度孟买的艺术家，收集了很多海边的废弃物做成了一个城市形状的装置最后一个叫《树梅坑溪环境计划》（2009—2012 年），是由吴玛俐与竹围地区联合做的。树梅坑溪其实已被污染，人与溪水的关系也已断裂，艺术家希望通过水来连接破碎的土地与社区关系。这个计划包括很多子计划，如让民众了解过去溪流生态环境的社区说明会（图 24）；带领民众顺溪而上认识溪流与人的关系的活动（图 25）；此外还有与周边学校合作的校内教学——"我们的溪流故事书"计划，让学生来描述他们的环境或者想象溪流的未来，并且在不同季节举办早餐会。这个计划至今已有 5 年时间，目前只局限于下游 2.5 公里左右的流域。我期待未来这个计划可以覆盖整个 10 公里的流域，相信一定能够达到更好的效果。

IPA 研究员、美国雕塑家皮特·莫拉莱斯（Peter M Morales）介绍了南美的项目：

第一个是墨西哥华雷斯城的《红鞋子》（Red Shoes，2009 年）（图 26）。这个装置艺术源于艺术家伊琳娜·肖维特（Elina Chauvet）个人悲剧性的经历。她在妹妹被杀害后就开始研究墨西哥北部女性被害的情况。后来发现很多女性遭遇过绑架或谋杀。这个项目中，她邀请 33 位女性捐献自己的鞋子，这些鞋子都是红色的或者被涂成了红色。展览吸引了很多媒体的关注。该项目也在世界各地做过展出。

第二个项目是《城市自然》（图 27）。2007 年，巴西圣保罗市推出了"清洁城市"的政策，以清除城市空间中的广告、LOGO 等视觉污染。因为有 3% ~ 7% 的广告牌现在不能够再登广告了，所以艺术家 Grupo Bijari 和 José Subero 决定使用广告牌的基础设施来做艺术。他们使用一种很特别的沟通方式来阐释尊重自然的理念。

第三个是壁画项目（图 28），也与"清洁城市"政策有关。在这项政策颁布后，空出来很多墙体空间。艺术家就利用这些墙体来画壁画。项目由好几个艺术家协同完成。巴西双胞胎街头艺术家 Os Gemeos 已经做艺术很长时间了，让我感到吃惊的是，这些壁画非常具有启发意义，能够启发我们进行反思，反思我们是谁，我们在做什么。

第四个项目是智利的壁画。它是由涂鸦艺术家 INTI 创作的。作品尺寸非常大，覆盖了好几面墙。壁画中的图像来自玻利维亚与节日有关的图像，艺术家把不同地区、不同文化背景的元素融合在一起了。

IPA 研究员、印度 WAA 创始人、策展人伊芙·莱米斯尔（Eve Lemesle）介绍了 5 个南亚项目：

印度和孟加拉国没有公共艺术政策，政府也不提供资金支持。从这个角度来说，这里公共艺术的发展是具有挑战性的。公共艺术项目很难获得许可。如果没有许可，项目在任何时候都可能被叫停。政府要求在做项目之前申请许可，但这又面临腐败的问题。这些都是挑战。我关注的是由艺术家引领的以及由不同组织来帮助的公共艺术案例。

第一个是印度孟买的项目《相遇》（Encounters）（图29），是由几位艺术家组成的团体"Art Oxygen"做的。当地有一个公共艺术节，通常在一些热闹的如市场、车站以及城市中心地带等区域举行活动。当地人一般没有机会看到当代艺术作品，所以艺术家把当代艺术在活动中展示给人们，往往会去探索一些能够对人们日常生活产生影响的问题。为艺术家们与市场商贩合作的作品。市场很多摊位都放着装有动物的笼子。艺术家将这些动物画下来，再把它们从笼子里解放出来，作品完成后被挂在了屋檐下展示。还有一件作品位于孟买海滩地区（图30）。由于城市化的进程，孟买一些小渔村正在渐渐消失，来自当地渔民村落的艺术家收集了很多海边的废弃物做成了一个城市形状的装置。另外一件作品是艺术家在海滩上用盐写了一些文字（图31）。盐，除了可以让一些食物储存更久外，它在印度历史文化中还有着非常重要的意义。这件作品让盐的性质和意义与海滩冲刷的瞬间性进行对比。

第二个项目是由非政府机构发起的项目。他们邀请来自不同国家的艺术家共同在德里开展一个项目。我推荐的这个项目是两个德国艺术家做的。在印度，自行车是一种传统的出行方式，但现在城市中骑自行车的人越来越少。图32中乘坐在自行车后面的人可以与司机互相交换来骑车，分享骑车的体验。项目旨在倡导和宣传绿色的出行方式。

第三个是今年在达卡艺术峰会上开展的项目。达卡艺术峰会每两年举行一次。这个项目名为《与此同时在别处》（Meanwhile Elsewhere）（图33），由Summit艺术基金会进行运作。这个项目使用了城市中的一些广告牌，广告牌上有两个钟，钟面词语是有着某种联系的。当你在城市中穿梭的时候，你会在广告牌上看到一组组有着某种联系的词，在你获得了各种各样的信息时，大脑可能就会思考不同概念之间的关联。

第四个项目是印度钦奈的凤凰城市场公共艺术项目（Phoenix Market City Public Art Program）（图34），由ArtC基金会组织。项目邀请策展人来进行策展，在大型商场里展出当代艺术作品，有的是永久性的，有的是临时性的，还有一些是通过商场屏幕来介绍国际性或本土项目。很多印度艺术家的作品都是为，国外展览或博物馆所创作的，这些作品回到印度后很难找到合适的展示场地。利用这样一个大型的商场空间，能够向当地公众展示印度艺术家在国外展示的作品。商场实际上并不能完全算作公共空间，现在随着各方的努力，我们可以看到很多这样的商业空间在举办艺术展。

31 印度孟买海边的公共艺术项目,艺术家在海滩上用盐写了一些文字
32 由德国艺术家在印度做的一个倡导绿色出行的项目
33 印度达卡艺术峰会的项目《与此同时在别处》
34 印度钦奈的凤凰城市场公共艺术项目

35

36

37

38

35 印度2014年新开通的孟买国际机场新航站楼的公共艺术项目
36 南非开普敦项目《感知自由》,迈克尔·埃利恩
37 2009—2011年间在南非约翰内斯堡的火车上做的试验项目,让大学生们到火车上做服务工作
38 南非开普敦东部的涂鸦项目图,将曾经被一些吸毒者和罪犯所占据的空间清理出来做涂鸦

最后一个案例是在 2014 年新开通的孟买国际机场新航站楼的公共艺术项目。这里汇集了 7000 多件艺术品和工艺品（图 35），数量远超在印度其他任何博物馆或其他地方所能看到的。这个项目规模非常大，而且是由私人出资的。这个公司的所有人自己开发了机场，而且是一个收藏家。

IPA 研究员、德班理工学院博士沃恩·赛迪（Vaughn Sadie）介绍了南非的案例：

南非约翰内斯堡公共艺术发展已经有 6 年了，但一些项目并不成功，一些基础设施因为没有发挥作用而被废弃。德班和开普敦现在也因选择过程中不透明以及缺少一些公平的筛选条件而引起了争议。

第一个是开普敦的项目，名为《感知自由》（Perceiving Freedom，2013）（图 36），艺术家是迈克尔·埃利恩（Michael Elion）。这件作品是一副巨型的眼镜朝向曼德拉被关押了 30 年的岛上，充分体现了一个被囚禁人的孤独感，同时表现了曼德拉的坚强意志。这个作品让很多人意识到公共艺术的存在。它充分显示了，如何针对不同观众以及公众空间中复杂的关系来进行公共艺术创作是一个非常复杂的问题。

另一个案例是实验性项目。它是 2009—2011 年在约翰内斯堡的火车上做的，目的是让大学生们到火车上做服务工作，通过人与人之间的互动来增进对当地的了解。由于种族隔离的历史原因，城市被分成很多白人区和黑人区。很多黑人被迫搬离了原来居住的地方，这列火车主要是把黑人又运回到他们原先离开的地方。图 37 中学生手里拿着《圣经》，从一个车厢走向另一个车厢，在车厢里布道。这个活动也有一些非洲建筑和文学方面的专家参与。

最后一个是开普敦东部的涂鸦项目（图 38）。在 20 个世纪 30 年代，这里实行了种族隔离政策，建立了一些黑人居住区。这些空间后来曾经被一些吸毒者和罪犯所占据，现在艺术家把这个空间清理出来做涂鸦。

IPA 研究员、巴西圣保罗大学美学和艺术史硕士布里埃拉·里贝罗（Gabriela Ribeiro）介绍了南美案例：

第一个是由 EDUARDO SRUR 在圣保罗做的装置项目（图 39）。因为人们经常往这条河里扔塑料瓶而导致河流污染，艺术家将很多巨大的塑料瓶装置放在河边，以唤醒当地居民意识到河流被污染，反思生态问题。这些瓶子在夜晚还能发光。同时，这也是一个教育项目，展览结束之后，塑料瓶使用的材料被做成双肩背包赠送给当地孩子。

第二个是 MUNDANO 跟其他艺术家一起合作的在圣保罗的项目，强调的是生态保护和可持续发展理念。这是一个"社会艺术"的真实案例。如图 40，他们在车上涂鸦，写着"我的这辆车被污染了"。

第三个是永久性的装置，它是一个八层的床。作品想要反映社会中一些人的居住环境非常拥挤、狭窄，甚至是无家可归的状况。装置安装好的第二天，一些人就已经将其作为日常起居的地方了。

IPA 研究员、英国巴斯泉大学艺术与设计专业硕士纳赫拉·阿尔·塔瓦（Nahla Al Tabbaa）介绍了中东地区的案例：

第一个是黎巴嫩贝鲁特的项目，名为《吉卜兰·图韦尼纪念馆》（Gebran Tueni Memorial，2008）（图 41）。这件作品是为纪念黎巴嫩著名的新闻记者吉卜兰·图韦尼而作，位于贝鲁特市中心的一个直线走道上。纪念碑上用三种语言（法语、黎巴嫩语、英语）写的文字传递了爱国主义的价值。艺术家 Vladimir Djurovic 来自黎巴嫩，我认为这样一个作品不需要引入国际艺术家来创作。这里种植的树是黎巴嫩土生土长的，使人联想起黎巴嫩的历史，鹅卵石上面刻有名字，公众可以随便拿走。

第二个是埃及的项目。项目名称译作《非埃及化的餐厅》（El Matam El Mish Masry）（图 42），是位于开罗的贫民窟中的一个当地小餐馆。它是在废弃的农田上重建的。埃及很多农田、庄稼和水都受到了严重的污染，很多当地人无法获得清洁的水源和食物。艺术家 Asunción Molinos Gordo 在这个餐馆里为人们提供当地最好食材做出的食物。但随着预算越来越小，只好在能力范围内寻找好的食材，最后艺术家只能找到一些垃圾食材呈给食客。在穆巴拉克执政的时候，很多这样的问题都被忽视了。艺术家希望通过这个项目反映当地的社会问题，促使民众关注自己的生活，引发思考。政府不太喜欢这个项目，最终这些餐馆被关闭了。但幸运的是，我们拍了视频，很好地记录下了整个过程。

第三个也是在埃及开展的项目，名为《Bakabooza》（图 43），是由同一个艺术团队做的。"Bakabooza"是一个木偶，它参与了总统选举，并在马路上唱歌、进行表演、发表演说。艺术家进入一些贫民窟中，让那里的居民对于总统竞选时的政治主张提出自己的看法，比如：种一些芒果树、提供免费的医疗、在公共场所播放音乐等。这些可以说也是埃及人的梦想。所以说，"Bakabooza"并不是一个乌托邦式的人物，而是充分吸取人民意见而创作出的人物。

第四个项目叫《从海湾到海湾到海湾》（图 44），是一个印度媒体合作机构在阿联酋设立的项目。在海湾地区的历史中，劳动力都是从印度、巴基斯坦和孟加拉国引进的。在当地，阿拉伯人与南亚移民在语言和文化上有障碍，所以经常发生冲突。项目是艺术家与说不同国家语言的船

39 EDUARDO SRUR 在圣保罗做的装置项目,将很多巨大的塑料瓶装置放在河边,以唤醒当地居民意识到河流被污染,反思生态问题

40 MUNDANO 跟其他艺术家一起合作在圣保罗做的项目,强调的生态保护和可持续发展理念。他们在车上涂鸦,写着"我的这辆车被污染了"

41 黎巴嫩贝鲁特的项目《吉卜兰·图韦尼纪念馆》

42 埃及的项目《非埃及化的餐厅》

43 埃及项目《Bakabooza》
44 《从海湾到海湾到海湾》项目
45 黎巴嫩难民区的壁画项目
46 日本当代艺术家北泽润的《起居室》项目

员,乘坐一艘印度制造的船航行到索马里,艺术家和船员通过沟通结成亲密的关系。整个过程用视频记录下来,然后进行展示,使人们能够了解船员和水手的真实生活。

最后是一个黎巴嫩难民区的壁画项目(图45),由Sabra和Shatila创作。黎巴嫩有大量的难民区域,这个地区曾经在1980年代遭受过大屠杀。人们带着大屠杀之后的心理创伤和苦难记忆仍然生活在这片区域,而且条件非常差。为了纪念大屠杀35周年,来自美国的艺术家邀请了很多艺术家与当地居民一起创作壁画,以纪念这一历史事件,唤醒人们进行反思。很多壁画内容都来自当年拍摄的视频。

IPA研究员、日本当代艺术家北泽润(Jun Kitazawa)介绍了他在日本做的几个公共艺术项目:

第一个项目叫《客厅》(Living Room)(图46),是将商店转变成起居室,地点位于东京附近的一个有4 000居民的住宅区。住宅区附近有一个空着的小商店,艺术家设计一个规则,鼓励居民把自己不用的家具放在这个空间中,让其变成一个起居室。居民积极地参与,他们拿来很多不同东西,还在这里举行了不同的活动。有人搬来了钢琴,还举办过演唱会,吸引了150名观众。还有人把播放电影的设备拿过来,这里就变成了一个放映厅。艺术家也拿了一些厨具,让大家可以在这里互相交换厨艺。项目已经持续了5年,目前还在继续中,由当地社区的主妇管理,现在有些人居住在这里。在这个自主空间,所有人都可以发挥自己的想像力,通过各种各样的家具把这里变成他们想要的样子,可以说这个项目是由居民来共同创作的。在冲绳、北海道以及日本其他区域都有类似的项目,居住在这些地区的人民创造了这个艺术本身。这个起居室里的物品都是社区所特有的,带上了地域的特殊烙印。

第二个项目叫《我的小镇集市》(图47)。2011年福岛核泄漏事件后,很多居民搬到了一些临时居住地。在这三年中,我多次去福岛的临时居住地,并一直在思考如何来重建这个废墟之城。这个项目是用一天的时间将临时居住区域变成一个集市。让孩子可以在这里实现对未来城镇的设想。这个集市里有银行、医院、学校、警察局、剧院图书馆……,孩子们可以扮演不同的社会角色。如果你来到这个集市,必须要使用集市上特有的货币。项目在过去的3年中共开展了11次,目前还在继续进行中。通过这样的小镇市场,能够建立起新的社区文化,而且是由当地的居民来共同创建的。

最后一个项目叫《太阳自助旅馆》(Sun Self Hotel)(图48),是将一些空房间转变为自助旅馆。酒店的大厅、房间和装饰都是由居民自己动手打造的。我们还用移动的太阳能电池来收集电能。很多来自日本的其他地区旅客参加了这个活动。房间服务是人们自己想出来的，例如和客人进行交流或者进行游戏。每个人都从客人的角度出发，想出一些好点子。晚餐后我们做了一些像灯笼一样的东西来点亮夜空，就像创造出了夜空中的太阳。这个项目由当地的居民继续推进，已经举办了好多次活动。这个项目也在台湾做过。我认为艺术的根源是要由人们创作，所以我非常愿意继续做一些以社区为基础、以社区为特色的项目。

IPA 研究员、乌克兰国立基辅莫希拉学院文化研究专业学士、策展人莱西亚·普洛科朋科（Lesya Prokopenko）介绍了东欧的案例：

第一个是波兰艺术家 Miroslaw Balka 做的，叫《OTWOCK》（图49）。OTWOCK 小镇是艺术家的家乡，欧经济萧条时期这个小镇很荒凉。图 49 上这个作品是在一幢被拆掉的历史建筑的位置上放置了一张当年这幢建筑的照片，超过 10 个艺术家给这个项目提供了他们的作品。

第二个项目是著名艺术家 Luc Tuymans 做的，他在一个老式家庭住宅窗子外做了一个装置。窗户上的图像与当地历史文化有关，窗外还有各种表演活动。艺术家想要打破窗内和窗外的边界，打破私人和公共空间的边界，用艺术重新诠释空间。

第三个项目也是在乌克兰，叫《从一个纪念碑到另一个纪念碑》（2009年）（图 50），艺术家是 Zhanan Kadyrova。Shargorod 小镇是一个非常典型的乌克兰小镇，镇上的人习惯政府的管制和限制，很多人不参与公共活动或者无法自己提出一些好想法。图 50 中是艺术家创作的纪念碑，它以非常创新的方法来提醒人们积极参与公共活动。她向市政府要求必须要有一个地方安放这个纪念碑。很多的志愿者自愿建起了这个广场放置纪念碑，从无到有，自己创造了一个新的公共空间。

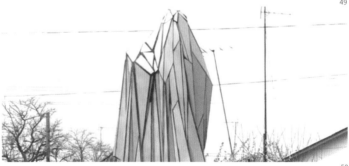

47 日本当代艺术家北泽润的《我的小镇集市》项目
48 日本当代艺术家北泽润的《太阳自助旅馆》项目
49 波兰艺术家 Miroslaw Balka 做的《OTWOCK》项目
50 艺术家 Zhanan Kadyrova 在乌克兰做的项目《从一个纪念碑到另一个纪念碑》

第三部分　　公共艺术案例研究

金江波：我生活的浙江渔村曾在改革开放初期展开社会大生产，仅有10万人的小城里生产出提供给全球约80%家庭的阀门，但这样的社会大生产对社会公共资源造成了浪费。公共艺术能为我们带来幸福感和振兴社区精神，而这些作用是社会生产无法替代的。去年，我的家乡邀请我回去，希望我们用公共艺术给他们的生活方式以及产业形态进行一些调整，提升人们生活质量，使城镇变得充满温情、更具社区精神。我们今天聚焦城市、城乡的发展以及未来的生活方式，下面有请来自中国美术学院的卓旻教授谈一谈，怎么把我们的文化力量注入到城市、空间、人、艺术和我们的生活。

IPA研究员、中国美术学院城市设计系主任卓旻发表了题为"公共艺术：一种当代的意识形态"的主题演讲。

公共艺术到底能为我们带来什么？上一届公共艺术论坛聚焦于"地方重塑"，我认为"地方重塑"是一个非常精准的目标，也是一个多层面的目标。从这个目标出发，我想从三个方面来讨论一下公共艺术可能的议题。

1、重塑地方的物质空间和场所感

（1）公共艺术和现代主义城市空间。现代主义大城市中形成了无尽的、索然无味的失落空间。这些失落的空间，在我看来不仅是需要治疗的病态，还需要去填充，重新赋予社会学功能和象征性意义。对于这个在规划学和建筑学被证明失效的场所，公共艺术往往显示出一种神奇的效果。因为公共艺术可以是一种微创手术，更可以是一种动态的干预，它往往不是永久性的，可能是某种临时的存在。但对于城市而言，公共艺术所需要承担的失败的责任和规划或者建筑相比，微乎其微，而其成功所达到的效果，则远远超过别的任何形式。在上届公共艺术奖案例中，艺术家在阿联酋Sharyah Museu广场上做的作品，以一种颠覆式的干预注入到艺术场地当中，最终达到场地存在的意义。看似和艺术毫无关联的足球场，又恰恰是最为艺术性的改造城市空间的手法。

（2）公共艺术作为城市的图像符号存在。对于认识一个城市而言，大众不再依赖传统的建筑或者是城市空间,转而依赖于某种连续的图像和符号,所以公共艺术在图像符号方面的生产,是其重塑城市印象的一个持续议题。而广告、标牌、涂鸦这些能够快速产生的并和大众审美品位相结合的艺术形式,在我看来远比博物馆内的艺术作品更有力量。但是需要强调的是,在消费主义的时代,"消费"和"大众活动"在启发艺术家的同时,艺术也可能轻易地成为消费主义的同谋,公共艺术尤其处在一个危险的境地,因为公共艺术的公众属性显然很难去拒绝大众的一种审美倾向,从而丧失自身的某种独立和敏锐,这是所有公共艺术家需要警惕的。

(3)公共艺术抵抗当代城市的同质化。我们的日常生活在商业和资本的推动下已经日益简单化和同质化。"购物"和"消费"似乎已被证明为我们现存公共活动的唯一形式。购物中心就像吸引教徒的教堂,飞机场正广泛地将旅客转化为顾客而获利。对于由"信息"和"消费"的快速传播所引发的同质化,我认为地域性或许是消除这个倾向的不二选择,而这提供了公共艺术的一个导向,即公共艺术如果根植于地域,那么地域的历史、地理、文化,乃至传统意义的民风民俗,都是公共艺术生存的现实基础,而这些都是可以帮助公共艺术来重新塑造地域特征的。在上届公共艺术奖案例中,我发现了萨尔瓦多街头壁画的例子,不仅在题材上展现了中美洲的传统,还表达了对当下中美洲社会问题的关注。

2、重塑地方的人群关系

(1)公共艺术和自治式的社区再造。不管是在西方还是东方,社区问题都是人群关系中的一个重要问题。它不仅关系到整个社会的稳定,也能帮助人们寻找到归属感。一个运转良好的社区,一定也是具有相当自治能力的社区。但要让市民投身到自己的自治管理当中,需要一定的激励因素,对此,显然没有比具有某种艺术气质的集体活动更好的性质了。正如前几年上海大学美术学院所做的曹杨新村的社区再造案例,就是获得了相类似的效果。在这之前,在大众还没有过多关注公共艺术的时候,我们已经有了大量衰败的社区因艺术家而获得重生的案例,比如纽约的SOHO。

(2)公共艺术和族群的融合。地球上人们有着不同的肤色、不同的信仰、不同的政治倾向、不同的地域文化,但是对于差异和分歧,人们往往更愿意诉诸暴力来解决问题。艺术不可能避免战争,但的确也是医治战争创伤的良药。旷日持久的巴以冲突,用协议换来的和平并不能消除人们心中的仇恨,而在《乌鸦巢》案例中,通过引入自然界的生物加快改造军事要塞,以鸟类的和平来隐喻逐渐取代要塞带给人们的关于冲突的联想,使得政治问题在大自然当中解体,从而尽快消弭不同种族之间曾经的心理对抗。这些公众参与的艺术项目,应该能够将和平的希望撒播到不同的族群之中。而面对移民问题、弱势群体等问题，都应该是公共艺术所指向的议题。不仅是社会学家，艺术家也应该有义务对他们进行帮助，并通过公共艺术来唤醒全社会的良知。在这方面公共艺术应该是公众的领导者，致力于消除对少数群体的歧视，加强不同人群之间的融合。

3、重塑地方的发展

（1）公共艺术和地方经济发展。公共艺术从项目的资助和推动来看，离不开公共资金的支持，而公共艺术也需要对其资助者创造某种价值。但公共艺术所带来的收益，有的时候很难评估，有时的确会带来真正的"金钱"方面的资本收益，但有时只是提高了某个地区的知名度。我认为就地方发展的收益评估而言，公共艺术的评价体系可以参照慈善事业模式，很多慈善事业所追求的目标不一定是直接创造钱方面的价值，而是触发更多的赞助者或公众投入到改善某个地区或者族群的生活标准上来。公共艺术在很多第三世界国家有着更为广阔的潜力，贫穷的地区往往缺乏除人力以外的可能的自然、社会资源以及社会关注，而公共艺术相对来说是一种性价比非常高的社会促进手段，例如在上届公共艺术奖案例中的 Nalpar 项目。

（2）公共艺术和公民教育。公共艺术已经跨出了传统艺术的界定，在教育、医疗等领域都存在着公共艺术介入的可能。对于提高欠发达地区的教育和医疗水准，不可能永远依靠外界的援助，更多的是需要唤醒当地社会和民众自身对于教育和医疗的重视，而这就为公共艺术提供了一个绝佳的介入机会。欠发达地区的教育设施和教育水平都比较落后，为那里营造像大城市那样的教育环境并不现实，这就要求艺术家们以一种具有创意性的方式去介入，公共艺术的实施过程就是教育当地儿童或居民的过程。上海大学在乡村做的美丽乡村公共艺术实践，在我看来很难用资本价值衡量。这种有限的活动当然不可能就此改变乡村的面貌，但是我相信艺术可能已经永久地镌刻在那些还从未离开过家乡的孩童脑海里。而这种艺术的冲击，必定在无形中为他们将来的成长打开无数的可能性。当新一代集体审美意识提高，公众所迸发出来的对于生活质量和产品美学的渴求，将是推动地区发展的永久动力。

以上这些角度、空间、人、发展是紧紧缠绕在一起的，只是在不同的地区或是族群间可能有不同的侧重，这恰恰构成了公共艺术的多元性。当汪院长第一次给我介绍第一届公共艺术研究主题是"地方重塑"的时候，我就联想到了包豪斯第一任院长在包豪斯成立时发表的宣言："建筑师们，画家们，雕塑家们，我们必须回到工艺，因为没有所谓的'职业艺术'，艺术家和工匠之间没有本质的区别，当然艺术家们是更高层次的工匠。"这个宣言也同样回答了我对于当代公共艺术的设问。我们这个时代不再是一个绘画和雕塑的时代，在当代的全球化语境当中，艺术家还需要博物馆吗？我想公共艺术家是不需要的，艺术家们更需要去拥抱大众，参与到社会当中去。我想汪院长和 IPA 以及这里的公共艺术论坛正在发出这样的战斗檄文,正在吹响公共艺术的号角。当代的艺术家要走到公众当中去,因为不存在所谓的职业艺术。艺术和意识形态没有根本的区别。意识形态在政治性层面不能解决问题,却往往可以在艺术性层面得到解决。我最后一句话就是:当代的艺术家应该是一个"更高层次的社会改革者"。

路易斯·比格斯:非常感谢卓教授对我们去年评奖的一些作品做出了清晰的分类。我们需要更加清晰的思路,因为这次研讨会主要的目的就是要寻找到一些更好的研究方法,寻找更好的机遇,让我们能够更好地对这些入围作品进行评选。我想邀请各位研究员提出一些好的想法,我鼓励各位能够提出一些评选标准,可以帮助我们这些评委在评选时进行参考,比如:这件作品是不是和公众有紧密的联系?艺术是不是和社区有紧密的联系?

IPA研究员、意大利作家,研究国际艺术项目策展人茱茜·乔克拉(Giusy Checola):《乌鸦巢》这个案例的第一层意义是鸟的隐喻,鸟象征着自由。我们必须亲眼看到作品,并且和当地人进行交流,才真正能够得到启发。比如:要让这个作品真正融入到自然中;要让建筑充满神圣感,事实上这个项目有很多隐含的意义,有一些意义并不是一眼就能够看到的。因为这个项目装置是非常精致的,人们很多时候如果第一眼看,只能看到第一层意义。

路易斯·比格斯:从实际的角度来说,你对于我们研究的过程提出一些实用的建议,比如:这个作品不仅仅要有一层意义,要有好几层意义。

茱茜·乔克拉:是的。艺术家本人的目的以及他的出发点可能与这个作品实际传达出的效果之间会存在差距。我们要去探寻这个作品背后深层的意义。

路易斯·比格斯:艺术家的想法必须非常清晰地体现在作品中,否则我们无法判断它是好作品。

IPA研究员、《公共艺术评论》编辑梅根·古尔伯(Megan Guerber):卓旻教授的发言非常有意思,但有一点我想自己的看法,就是你所提到的城市空间以及商业化。比如:像SOHO这样的项目与商业化之间的关系,实际上它是非常同质化的一个地方。我觉得对于研究者真正的挑战是研究如何通过艺术激发它?

IPA研究员、策展人凯利·卡迈克尔(Kelly Carmichael):我认为任何评委或者任何研究人员写案例的时候,必须要亲眼看一下这个项目,了解作品所在地的文化及其周围的社会环境,看到它的问题才能够解决问题。

纳赫拉·阿尔·塔瓦:我觉得有的作品本身也可以被操纵,可以找一些人摆拍，然后到网上一发。这些东西都是很容易弄出来的。

沃恩·赛迪：我们研究员在研究时撰写文案，要做好文本的记录，这样才能对这个项目进行很好的说明，同时要考虑项目的价值取向，还有道德取向的问题，文本中要体现某一种理想或者某种价值观。"公共艺术"到底是什么？我们所要传达的价值观是什么？艺术家和艺术传达过程要达到的内在价值观是什么？我认为研究时要把地掀开，好好地找一找，把这些东西找出来。我们要确保向公众推荐的这些公共艺术，是他们所需要的。

路易斯·比格斯：评奖标准中所谓"艺术家引领的项目"这个说法，是指艺术家应该在任何的设计团队、项目团队或者教育团队中都是处于非常显著的位置。

沃恩·赛迪：如果考虑到公众的话，你必须要考虑到好的领导力与坏的领导力。有好的领导力，可能不会让任何事情都随便发生。在公共艺术领域，到底领导力如何发挥，就是指这种话语权的制衡应该是怎样一种关系。我们研究人员就要对这些项目进行筛选研究，不让有些项目随便就能够溜进来。

杰克·贝克尔：艺术家在公共艺术创作当中能学到什么？我认为是通过教育制度把公共艺术家培养出来，但是目前的培养方法往往缺乏对于现在最佳的前沿实践的认识。很多教育体系中培养的艺术家都希望把作品送进美术馆、画廊。当然，艺术家需要掌握一些技能，才能在公众领域成为好的领导者，必须善于去倾听社会对你的反应，并把你的创意用在建设性的目的上，这些都是非常有用的技能。现在的制度中，艺术家与他人合作还有着非常大的鸿沟。对于研究者而言，要了解公共艺术创作的全部过程是一个很大的挑战。

汪大伟：我突然想到"盲人摸象"的寓言。一群盲人从不同的角度、不同的感觉对这个大象做出自己的判断和定义。我觉得刚才的讨论就像盲人摸象的过程。卓旻刚才牵出一头大象给我们，抛出一连串的从物质空间到城市问题的话题给我们。大家的发言都从亲身体验出发，但却没能说出大象的本质，即大象到底是什么东西？有意思的是，围绕这个说不清楚的话题，大家努力想把它说清楚，这就是我们会议一个很有趣的地方。能不能假设每个人说的都对。例如"什么是好的公共艺术"这个问题，从评委的角度，希望挑选一些使艺术家、政府、受众都能够得益的典型案例推荐给大家；从研究员的角度，希望能够推荐自己最熟悉的案例，揭示出真相，探索其规律，提供给大家去参考；从艺术家的角度，希望充分发挥自己的才能，把激情和才华最大化地释放出来。所以从每个人的角度来定义"公共艺术"都对，但是背后有一个核心问题，即公共艺术到底是什么？

我们把"地方重塑"作为公共艺术的一个价值所在,实际上"地方"有着多重含义。第一,它是物理空间,可以指向各种类型的空间。空间中出现了问题,需要把它改过来,正如卓旻提到的,当城市建筑师、规划师的设计出了问题以后,希望艺术能做一些调整与改进。我们应该反思,艺术虽然可以解决问题,但它不是一种辅助品,它应该成为主角,艺术的价值不在于装饰,其真正存在的意义是起到融合,甚至是去弥补一些缺陷的特殊功能。在城市化进程当中,社区里可能会存在一些问题需要艺术去弥补和解决,这种"弥补"、"解决"就是一个重塑过程。所以我认为公共艺术的核心价值就是地方重塑。因为我们涉及到一个很大的空间概念,包括物质空间、行为空间、社会空间,尤其是在社会空间中,可以用艺术的方式去解决信仰、种族、文化方面的一些冲突。这些都是从不同的层面,对于地方重塑的一种智慧贡献。作为评委,作为研究员,作为艺术家,要站在自身的职责去挑选和思考。有时候我们会错位,为什么瞎子相互之间会争起来?是因为他坚定地相信自己的感受才是最真实的。所以我认为艺术家不应该想评委的事,研究员不应该去替代艺术家职责,评委更不能替代研究员或艺术家,这种定位将有助于我们正确地判断公共艺术。而且这种分层次的思考和自己对公共艺术的价值判断,就形成了一个"我们提倡什么,不要什么"的标准。这种最终的价值判断,也就成了每个人在做每一项工作的选择标准。

路易斯·比格斯:如果研究员有更多的时间,他可以从更多的角度、更多的层次、更多的资源来分析解读,也许会发现大象身上的其他部位,从而扩展他对于这个大象的定义。

IPA研究员、英国福尔茅斯大学的客座教师、策展人萨拉·布莱克 (Sara Black):我认为公共艺术是一种艺术的形式,它是和公众相连的。颁布这个奖项应该要明确公共艺术的定义,对公共艺术进行分类,如果这样的话,我们能够更好来进行研究,能更深入地研究,能使我们的研究变得更具有价值。首先要了解公共艺术的不同分类,公共艺术涵盖了很多方面,很难有一个普遍的价值涵盖到所有的价值方面,我很难区分到底什么是好的公共艺术,什么是坏的公共艺术。所以我觉得还是要有一个明确的分类,明确的定义。我在想是不是可以将奖项进行细分,分成具体的奖项。

IPA研究员、津巴布韦国家美术馆主策展人拉斐尔·奇古瓦 (Raphael Chikukwa):南非有做公共艺术的传统,而从津巴布韦、西非到尼日利亚以及北非有一些相反,因为大部分国家没有博物馆或者艺术馆来展现艺术作品。在南非当然还有一个非常重要的问题,那就是作品到底是归功于艺术家还是组织者?

路易斯·比格斯:我很难回答你的问题,因为取决于具体的项目、具体的时间。事实上我们已经有过很多的争论了,很多艺术节被认为是公共艺术。那么这个艺术节到底是由艺术家主导还是由其他人主导?如果组织方主办了艺术节,那么他们是不是也成为艺术家了呢?事实上我们最好尽可能来避免回答这个问题。在第一次颁奖的时候我们已经讨论决定,只要对公共空间有所贡献,我们就把它看作是公共艺术。我觉得我们是不是还要继续更好地来定义到底什么是地方重塑?或者是要在公共艺术里面找到一个比较明确的领域,能够让评奖的过程更加有针对性、更加清晰?我们不想讨论什么是公共艺术,因为这个话题是讨论不完的。这就跟你问什么是公共,什么是艺术一样。在不同的文化中,大家的定义是不一样的。

杰克·贝克尔:公共艺术如何帮助我们来学习?如何来解读空间?如何通过艺术体验来获取意义?在公共艺术领域,艺术家是有机会来教育和开发目标受众的,增加受众对于环境的敏感度,增加大家对于这些事件的敏感度。艺术家应该从传统的视域中跳出来,创作一些以前不存在的东西,也许能够改变环境、改变我们的生活,甚至改变政策或者改变我们的视角。

汪大伟:关于"地方重塑"的概念,我想再重新表述一下。"塑造"与"重塑",实际上是有不同含义的。"塑造"是从"无"到"有",而"重塑",实际上是在原有的基础上重新赋予一种新的含义,换句话说,就是去解决问题。对于公共艺术的职能,我认为它是用一种艺术语言和方式去介入社会、介入生活,去解决"地方",即各种形态的空间包括物质空间、社会空间、行为空间等出现的一些问题。对于公共艺术的主题,我提出三种思路:一是多角度地提出主题。二是看它对于空间的积极意义。三是一个主题和多个分主题,我比较赞成"一个主题"与"多个分主题"的概念。在"地方重塑"这个主题下,希望公共艺术能够对地方公共空间起到积极的意义。尽管重塑在方法上、思路上、针对的对象上、形式上各不相同,但都可以切入到讨论的方向和主题中去。

北泽润:我认为公共艺术是社会的问题,艺术项目应该对社会提出一些问题,应该对社会和生活提出一些问题。虽然艺术作品本身并不是一个社会问题的解决方案,但可以从不同的角度来讨论社会的问题,提供一些崭新的角度。我的创作方法是制定一个基础由其他人共同参与,我认为这也是公共艺术创作的一个好方法,即艺术家来主导这个项目、这个体系,但是会请其他的人来参与。艺术家提出一些问题,引发其他人来思考和讨论。

路易斯·比格斯:我一直有这样的观点,我经常和其他人辩论,我认为我们的目的是要提出问题,而不是回答问题。艺术家到底是要提出问题,还是要提出一个思考的方式?我们可以邀请大家共同讨论,但是我觉得艺术家并不应该指手划脚告诉人家做什么,要给问题提出解决方案。

本次会议中，上海大学美院公共艺术中心教师刘景明还向大家介绍了上海大学美术学院公共艺术数据库构建的情况，并介绍了如何使用网站数据库进行研究以及提交研究成果。

会议最后，上海大学美术学院博士研究生冯莉向大家汇报了她对于"公共艺术区域划分方案"的研究。她对两届公共艺术奖提交案例进行分析，并从自然地理学、政治地理学、文化地理学以及人口地理学四个方面进行深入分析研究，最终划分出非洲、东亚和东南亚、大洋洲、拉丁美洲、欧洲、北美、南亚和中西亚七大区域。报告结束后，与会嘉宾探讨了国际公共艺术奖作品征集的区域划分方案，在激烈的讨论后，最终决定以后按照这个方案中的区域划分来征集作品。

结 语

汪大伟院长最后对会议做出了总结：本次会议不仅回顾了我们之前的工作，还分享了很多优秀的案例。大家在非常融洽的氛围中探讨公共艺术的土题，并对今后研究工作的开展给出了具体的建议。IPA建立这样一个"国际公共艺术研究员"网络是非常重要的，在座诸位是我们IPA的重要研究力量，希望凭借我们的研究网络，能够更加深化我们的研究内容和层次，把这个研究平台打造成一个提升国际影响力的窗口。这一次会议给了我很大的信心，我们学院做这件事情，不仅是为了能够推动学科建设，另外还希望能够对中国的公共艺术有所影响，同时也非常希望积极地参与到国际公共艺术的发展潮流中去。

研究员访谈
INTERVIEW

公共艺术要让人触手可及
——研究员伊芙·莱米斯尔访谈
Public Art within Reach:
An Interview with Eve Lesmele

艺术的内涵根植于艺术品之中
——研究员沃恩·赛迪访谈
Explore the Content from the Public Art Works:
An Interview with Vaughn Sadie

公共艺术作为一种社会治疗
——研究员里奥·谭访谈
Public Art as a Social Therapy: An Interview with Leon Tan

公共艺术与城市身份
——研究员荼茜·乔克拉访谈
Public Art as an Urban Identity:
An Interview with Giusy Checola

01 伊芙·莱米斯尔

公共艺术要让人触手可及
——研究员伊芙·莱米斯尔访谈

Public Art within Reach:
An Interview with Eve Lesmele

研究员：伊芙·莱米斯尔
采访：陈文佳

Researcher: Eve Lesmele
Interviewer: Chen Wenjia

伊芙·莱米斯尔（Eve Lesmele）是国际公共艺术协会（IPA）研究员。她在艺术领域有着 15 年的管理与策划经验，经常和来自不同专业背景的艺术家及艺术机构一起策划公共艺术项目。她的个人经历横跨了传统艺术的故乡——法国，当代艺术的新兴地——北美区域，以及传统手工艺和当代艺术新贵——印度，在旅居过多个不同城市，感受不同文化的碰撞带来的新意之后，伊芙在 2009 年选择了定居孟买——印度的经济中心和南亚新兴现代城市。很多到访过印度的人会把对孟买这座城市的印象总结为"强烈的对比"，古老的历史带来丰富的传统意蕴，和现代设施为城市带来快节奏的生活产生出巨大落差。同样，在这座充满强烈对比和诸多机遇的城市里，公共艺术也有着非常广泛的灵感和基础。伊芙·莱米斯尔在印度创建了"什么是艺术"（What About Art，简称 WAA）的非盈利机构，作为公共艺术项目策划和商业运作的平台，致力于在印度推广跨越文化的公共艺术项目。

本文就公共艺术项目的运营采访了伊芙·莱米斯尔女士，通过她了解在印度及北美的策划和运作公共艺术项目的社会和政策环境，以及项目运作和策划的方式。

请介绍一下您在公共艺术领域的工作。

伊芙·莱米斯尔：我从事与艺术领域相关的工作有 15 年之久，工作经历遍及加拿大、印度以及欧洲。我毕业于巴黎索邦大学艺术管理专业，并在巴黎东方文化语言学院（INALCO，Paris）获得了南亚文化学的文凭。之后我在加拿大蒙特利尔的一个非盈利艺术机构中开始了我的职业生涯。我的第一份工作是在蒙特利尔的达林方德利视觉艺术中心担任艺术品经理人和策展人。后来，我移居到我的故乡巴黎，并在巴黎的 Art Centre Point Ephemère 策划了一个国际艺术家驻地项目。

2003 年起，我开始了第一次印度之行。此后，印度变成了我出行的常规目的地。在 2009 年，我决定定居在孟买并开始了一次创业，这也是"什么是艺术"（以下简称 WAA）非盈利组织诞生的背景。WAA 是在印度同类型的艺术管理机构中的先驱，它为成长中的南亚艺术圈以及艺术家们提供了富有创造力和培养潜力的艺术项目。自创立以来，WAA 与印度的许多文化机构和画廊合作过很多项目，并为诸如 Shilpa Gupta, William Kentridge, Rashid Rana, Nikhil Chopra, Tejal Shah 等印度和国际艺术家们策划画展和进行艺术品经纪管理。此外，WAA 也在印度策划公共艺术项目，譬如印度第四大城市——金奈（Chennai）的"艺术 C"项目（Art C），以及加拿大摄影艺术家雷蒙德·阿普瑞尔（Raymonde April）位于孟买浩南环形公园的公共展览。WAA 是 2016 年达卡艺术峰会的会员，并在 2013 年开始策划了非盈利性的艺术家驻地项目，以此来吸引更多国际艺术家到访印度，与印度文化碰撞，产生火花。

PA: Could you please give us a brief introduction about what you have done in the field of public art?

Eve Lesmele: I have been working in the arts sector since 15 years in Canada, Europe and India. I am a graduate of La Sorbonne in arts management and hold a diploma in South-Asia studies from INALCO (Paris). I started my career in the non-for-profit sector as an arts manager and curator at the Darling Foundry in Montreal. I moved back to Paris and curated international residency programs at the art centre Point Ephemère. After several visits to India starting in 2003, I settled in Mumbai in 2009, and startedWhat about art? (WAA), a first-of-its-kind arts management agency in India. WAA produces ambitious art projects within the growing South-Asian art scene. Since its creation, WAA has collaborated with numerous galleries and cultural institutions in India, and handled projects for artists such as Shilpa Gupta, William Kentridge, Rashid Rana, Nikhil Chopra, Tejal Shah to name a few. Among WAA recent public art projects is the production of the Public art program ArtC (Chennai), the exhibition of Canadian photographer Raymonde April in Horniam Circle Park (Mumbai) and the agency will be the associate producers of the Dhaka Art Summit 2016. In 2013 we also started a not-for-profit artists' residency.

请介绍一下在您的居住国中的公共艺术发展以及相关的公共艺术政策。

伊芙·莱米斯尔：当前的印度并没有公众基金或公共政策来支持公共艺术的实施。而在印度要取得当地政府对艺术项目的许可，其过程是非常冗长的，部分机构存在的官僚主义使得这一过程更加困难，这为当地的公共艺术项目带来很大的挑战。但是，有时当艺术家、独立机构以及私人基金共同策划出一个非常有意思的公共艺术项目计划时，即便没有政府的许可，这样的项目也会被落地实施。这样的项目所选择的地点或在街道上成为流动的风景线，或在所谓的具有私人所有权的"半公共空间"如办公楼或商场等。例如，我们最近策划的《艺术C》项目，就是在金奈一个商场里放置许多大型雕塑、装置艺术品以及影像和声音艺术品。这一公共艺术项目在周末的时候一天就有约5万名观众参观。

最近一段时间，作为独立艺术机构，我们也通过参加一些公共艺术项目来寻求政府的资助，比如印度柯钦双年展（Kochi Biennale）以及孟买国际机场新建的二号航站楼中的公共艺术计划。在我看来，这样的项目为当地政府对公共艺术的态度和政策传递了一个非常积极的信号。

我们非常希望藉此能看到支持公共艺术项目的有关公共政策能够在印度正式出台和运行。

PA: Please introduce the development/policy of public art in your home country /resident country?

Eve Lesmele: In India there is a lack of public funding and no policy for public art. Difficulties in getting permissions and a complex bureaucracy makes public art projects challenging.

However few artists, independent organisations and private foundations have been putting together very interesting public art projects, sometimes without official permission, either in the streets, or in "private-public" places like office complexes and malls. For example, we have recently produced the public art program ArtC in a mall in Chennai with lots of large scale sculptures, installations, video art and sound art. This is an initiative of the art patron Mr Vijay Choraria and the mall has 50,000 visitors per day on the week-ends.

In the recent past there have been few initiatives that have found the support of the government, such as the Kochi Biennale and the art program at the new Terminal 2 of the Mumbai international airport. This is of course very positive and we hope that public policy will soon contribute to the development of public art projects in the country.

通常公共艺术的产生需要多方面的支持，这一创造过程离不开诸如出资方、策展人、艺术家以及公众。在您看来，什么是公共艺术创造过程中最重要的一方？

伊芙·莱米斯尔：因为我现在生活在印度，所以想以印度为假设的环境来回答这个问题。在印度，包括公共艺术在内的当代艺术是一个非常狭

02 参观者在印度金奈商场中观看《艺术 C》展览的艺术品
摄影：Thukral & Tagra

窄的圈子，所有的流程通常在各个私人画廊之间策划或流通，普通大众是不会参与到这些展览中的。但是，因为在这里没有专门的当代艺术博物馆对公众开放，所以有能力将当代艺术带到公共范围内的私人艺术机构就在公共艺术策划"创作"实施的整个过程中显得尤为重要。在我看来，印度的公共艺术能够将艺术的门槛降低，降到普通大众触手可及的维度，同时激发年轻一代对于美与艺术的追求和对于创造力的培养。从这一点而言，将公共艺术带给公众的意义是很深远的。公共艺术的另外一个重要功能是增进城市空间规划的合理利用。在城市发展中，最重要的莫过于理解城市中的空间、地方特征和社区精神。公共艺术恰恰能在这一点上很好地展现和帮助人们理解空间特征和地区精神。同时，优秀的公共艺术作品也能增进人们生活品质，结合公众的同时将城市的精神和社区的文明提升到更深的层次。

PA: What is the most important part in the creation of public art, in your opinion?

Eve Lesmele: In India, contemporary art is mostly confined and circulated within private art galleries, and the general public doesn't attend these exhibitions. Hence bringing contemporary art to the public sphere is relevant as there aren't too many public museums for contemporary art. Public art can help democratise access to the art and develop interest and creativity of the younger generations.

Another important function of public art is linked to its integration to urban planning. In urban development, the most important thing is to understand the space, the place and the community. Public art can enhance the experience of living and engaging with the city and its concerns in a deeper manner.

您认为在当前社会中，公共艺术在人们的生活中所承担的社会功能是什么？

伊芙·莱米斯尔：公共艺术的一个重要社会功能就是增强人们在社会生活中的参与感。区别于仅仅限于少数人的传统形式，公共艺术将艺术的形式放在较大的规模中带给普通民众。从空间环境而言，公共艺术重新定义空间并将美的元素和人的参与带到整个空间中，让空间与人互动，这是为现代城市和城市空间带来灵魂的方式。而艺术家在公共艺术创作的过程中也完成了与社会大众的互动，在把与社区特质相关的艺术品带到社区和公共环境的同时，也为社区居民和公共环境带来活力和关注；而这点对于某些不具有能力和资源去拥有艺术品的社区和公共区域而言，是极为重要的一种与社会其他端口对话的方式，这就是公共艺术非常重要的社会功能。

03/04/05 《领奖台 / 立方体》孩子们与作品互动
摄影：Reena Kallat

PA: what is the social function of public art, in your opinion.

Eve Lesmele: Public art plays an important social function to engage people, to bring art to the common people and to make art inclusive to the public at large and not exclusive to a few people. It creates beauty, brings soul to concrete urban spaces and modern cities. Artists who are socially engaged create art projects that are relevant and specific to certain communities that otherwise might not have the privilege or resources to engage or experience arts, this is a very vital function of public art.

虽然公共艺术是属于公众的，但正像您之前所述，其策划和创作的过程中仍然离不开资方和相关的支持，您认为当前的艺术市场是以什么样的方式为公共艺术的发展提供推动力的呢？

伊芙·莱米斯尔：在印度，有许多艺术品收藏家长时间以来在当地的艺术品市场表现活跃，这部分艺术品收藏家已经开始成立有关的艺术基金为公共艺术项目提供持续的资助和支持，可以把这种行为看做是一种观念的转变。以往，当人们收购一件艺术品后，往往只是为了在家中独自欣赏这件艺术品，而现在，收藏家们更愿意通过系列展览和支持有关的公共艺术项目来和公众分享他们的艺术收藏。这样一种和公众分享的过程可以被看做是一种传播爱的慈善行为，同时也是分享了对艺术品的欣赏力。

PA: Although Public art is supposed to contribute to the public, however, public art cannot develop without fund and essential supports from organizations. Currently, management and trade in Art markets provided fund and continence for contemporary art. Do you think art market can also provide encouragements for public art? And by what means?

Eve Lesmele: I would say that many art collectors in India have been involved for a long time in the art market, and consequently started forming art foundations and supporting public art projects. People who enjoy art in their home through their own collection are now open to sharing their collection with the public by way of opening art museums, putting exhibitions and public art projects etc. They want to share with the public through a philanthropic way of sharing their love and what they have experienced with artwith others.

在您看来，什么样的公共艺术作品能够激发观众的参与性？请您列举一些印度公共艺术项目中的案例。

伊芙·莱米斯尔：我认为能引发公众兴趣并激发他们参与到整个作品之中的公共艺术才是优秀的作品。互动的艺术作品往往通过公众的共同参与来完成整件作品，这样的作品往往能激发公众的参与性。同样，以未完成的形式感出现的实验性艺术品也能让观众更乐于思考作品的意义和与作品有关的意境。我想用以下两个案例来说明我的观点：

06 / 07 / 08 达拉维艺术之家艺术品创造过程
来自达哈维社区的孩子们和他们完成的作品

第一个是《领奖台 / 立方体———为儿童创作的公共艺术项目》(Podium/Cube—A public Art Project for Children)，2012 年，实木上漆，122cm x 122 cm，展示尺寸根据现场调整。这件互动艺术装置是定居印度孟买的艺术家瑞娜·卡拉特（Reena Kallat）在海滩上以儿童为参与对象进行创作的。这件装置由 20 个木质的方块组成，这些木块是利用体育比赛上的领奖台改建而成的，因此每个单独的方块上都有一个对应的数字。20 个木块可以组合成一个大型的立方体。在这个装置中，参与的儿童可以任意摆放并重新组合这些立方体。在这一游戏过程中，每个木块上单纯的数字变换组合使得之前领奖台的数字所表现的等级和名次含义全部都被颠覆。领奖台上的数字所代表的成就和荣耀的含义在这件装置作品中被重新定义，暗示着曾经的成功和胜利成为一片浮云和一件玩具。

第二个是《达拉维艺术之家》，2013 年至今，达拉维，孟买（Dharavi Art Room，2013—，Dharavi，Mumbai），达拉维艺术之家位于印度孟买的达拉维。达拉维是世界上最大的贫民窟，有超过 100 万人居住在这里。英国 BBC 电视台曾把达拉维区描述为"城中之城"，究其原因是因为孟买是印度的经济金融中心，其消费水准在世界范围内也是名列前茅。在这样一个生活成本极高的城市中，达拉维区的存在为从边远地区来到孟买寻求机会的人们提供了栖身之所。

《达拉维艺术之家》是一个从 2013 年开始长期持续的公共艺术计划。该艺术项目是想为社区提供一个空间，让人们在此以艺术为媒介，探索发现和表达有关看法。这个艺术项目也是以青少年为目标参与对象的，从某种角度而言，达拉维艺术之家更像一个资源中心，为艺术作品的销售提供展示空间，同时也为参与的青少年创作艺术作品提供创作空间。这个艺术项目所在的空间对有意愿开始艺术项目的不同机构和个人开放，同时艺术家也可以选择和达哈维区的居民譬如中小学学生合作，针对社区中的弱势群体如女性和青少年提供体验艺术创作的机会。在过去的一年中，已经有当代艺术家在达哈维艺术之家为女性开设了摄影工作营以及为儿童开设了艺术工作坊。这样一个开放性的艺术项目的出现为达哈维区的居民打开了一扇通往外界的窗户，让他们能够看到和分享到诸如艺术馆、大学以及来自于孟买这座城市中其他区域的信息。

PA: What kind of public art work can provoke audience /visitor's involvement?

Eve Lesmele: Art that involves people and interests them to engage with it. Interactive and participatory art projects that form meanings by collective engagements can provoke audience.Art works that are experiential in nature that get people to think about the meaning and relevance of the artwork can provoke audience. Few examples given below:

1. Podium/Cube —a public art project for children by Reena Kallat

2012. painted wood. 48 in. x 48 in. l 122 x 122 cm (display dimensions variable)

Mumbai-based artist Reena Kallat created this interactive installation for children which was installed at the beach. She explains the concept, "The wooden sculpture comprising twenty pieces that fit together to form a cube is made of sections that simulate sport-podiums. Arranged within the stable geometry and universal order of the cube, these podiums however are set in flux with their jumbled numbering and levels, ranking, grading and hierarchy gone awry; ideas of success and accomplishment reshaped to re-configure the victory stand into a plaything."

2. Dharavi Art Room in Dharavi, Mumbai

The Dharavi Art Room is a space for the community to explore issues through the artistic medium. It is a place of expression through art targeting youth. It works as a resource centre, display space for art to be sold, and a lab for art to be created. A collaborative environment open to many different organisations and individuals interested in starting art projects. It is a space where artists can plan other projects to be carried out in different parts of the community, for instance the local municipal schools. It is also be a space where work is created to share in outside spaces such as art galleries, universities, the rest of the city, etc.

Some the projects undertaken are workshops for children, photography workshops for women in these communities, art exhibitions by contemporary artists in their localities to make art easily accessible.

01 高浅正在采访沃恩·赛迪

艺术的内涵根植于艺术品之中
——研究员沃恩·赛迪访谈

Explore the Content from the Public Art Works: An Interview with Vaughn Sadie

研究员：沃恩·赛迪
采访：高浅

Researcher: Vaughn Sadie
Interviewer: Gao Qian

作为一名杰出的独立艺术家和德班理工大学公共管理技术专业的博士生，来自约翰内斯堡的艺术家沃恩·赛迪自 2002 年起，就一直活跃在国际公共艺术的舞台上。他参加过多达 50 场的个人和群展，取得了斐然的成绩。同时，他也是一名杰出的策展人，成功策划了多次个人与团体展，例如于 2005 年在德班艺术空间策划的维加讲座展览之《幻影》，2010 年在约翰内斯堡袋子工厂举办的《表演路灯》和同年在德班美术馆举行的展览《冲突情境》等。由于出色的工作经历和个人成就，2011 年，他在伊斯坦布尔被授予 iDANS 奋进奖。次年，又当选《邮卫报》南非人 200 强并入围 MTN 新当代决赛。

As an outstanding independent artist and the Doctorate of Technology in Public Management of Durban Institute of Technology, Vaughn Sadie, who comes from Johannesburg, has always been active on the international arena of public art since 2002. He participated more than 50 solo shows, collaboration and group exhibitions, and achieved great success. Meanwhile, he is also an excellent curator who curated many times of exhibitions. For example, the "Phantasmagoria" of Vega Lectures Exhibition which held in Art Space Durban in 2005, the "Performing the streetlamp", in The Bag factory of Johannesburg in 2010 and the "Conflicting Context" in The Durban Art Gallery of Durban at the same year. Due to his remarkable work experience and excellent accomplishment, he awarded 2010 iDANS Critical Endeavour Award in Istanbul, also was elected to the *Mail and Guardian* Top 200 South Africans and enter the Finalist in MTN New Contemporaries of the following year.

感谢您的支持和配合，这个采访主要围绕您的观点谈谈您对公共艺术的理解，能否请您介绍一下您在公共艺术领域中的主要工作和职责？

沃恩·赛迪：和大多数人的工作状态不同，我的工作横跨公共艺术领域的各个方面。最近主要致力于管理、设计和促进一项名为《卡斯摩城市》的公共艺术项目。这是一个与约翰内斯堡经济住宅开发相结合的项目。我代表南非视觉艺术频道，作为这个项目的主要负责人和服务商在那里工作。这个项目的目的是为了探讨在特定社区中居住的人们所承担的多种积极的社会角色类型。由于他们对社区活动参与形式的不同，可以使整个项目的定位和规模也随之发生改变。

We are going to have a briefly interview about your opinions of public art, and thanks for your support and collaboration, could you please talk about what did you exactly do in the public art area?

Vaughn Sadie: In comparison to most people, I work across different areas within public art at any given moment; I am currently managing, designing, and facilitating public art program in Cosmo City, a mixed integrated economic housing development program, in Johannesburg. I've been working there on behalf of the Visual Arts Network of South Africa, as the lead researcher and facilitator of the project. The project explores multiple strategies of participation to understand the active role residents in the specific community can play in shaping content and determining the scale, size and impact of the project though being included in multiple stages.

您的这种工作状态持续了多久？

沃恩·赛迪：我在卡斯摩城工作了差不多两年，之后的两年应该还会在那里。从2004年开始，我就作为艺术家工作并同时从事教育工作。这些年间，我与很多艺术家合作为某些场所设计定制作品，同时也进行一些公共艺术项目。和我一起工作过的人包括Jay Pather、Sello Pesa和Neil Coppen。我个人主要运用人工造明技术创建装置艺术，目的是为了阐释光是有形的，并可以与公共空间内的其他物质发生关系。这是一项体验式的作品。

How long have you been working?

Vaughn Sadie: I have been working in Cosmo City for almost two years, and possibly be there for another two years. I have been a practicing artist and educator since 2004 and throughout this period I have been developing site specific performance and public art projects with several collaborators, Jay Pather, Sello Pesa and Neil Coppen. In my individual practice I work with artificial light to create installation, which look at the role that light plays in shaping and determining of experience and relationship to public space.

所有的这一切都是您亲力亲为的么？

沃恩·赛迪：是的。项目中的很多环节都由我负责。无论是概念定义或者是基础建设的进展、设计、实施以及辅佐和应用层面。同时作为一名艺术家，我还兼顾艺术家应该做的工作。

So you actual psychically doing it?

Vaughn Sadie: Yes, I work on multiple aspects of the projects, in development of concept and infrastructure, at the design and implementation, and facilitation and application level. Then I still work as an artist in my work.

您是第一次以研究员的身份来上海吗？能否就参加这次学术会议谈谈您的感想？

沃恩·赛迪：这是我第一次来上海，但感觉非常亲切。其实，作为一位实践者，我认为比起关注理论，更加重要的是立足于观察社会，因为我对这一领域的经验大多来源于我的工作。

Is this the first time you have come to Shanghai as a researcher? Could you talk about your impression for attending this kind of academic meeting?

Vaughn Sadie: Yes, it is my first time being here, I am actually feeling a little bit intimated. As a practioner, my understanding is less theoretical and based on insights, and most of my experiences were gained from my work in the field.

那您了解中国公共艺术的情况么？

沃恩·赛迪：我并不了解中国的公共艺术状态，但是我希望现在开始我能够了解。我觉得中国公共艺术策略应该是与全球其他城市的公共艺术相同的，比如关于都市建设和地方重塑这些层面。从某种程度上说，这也可以被视作推动文化旅游的方式，并促成知名艺术家参与大型项目。但我的想法也许并不是很准确。

Do you know anything about Chinese public art?

Vaughn Sadie：No, my hope is to start developing an understanding of it. I can would only image that in some instances, that it follows similar to strategies to other global cities, where it is about urban renewal and place making. It in some way would also be used as a tool to drive cultural tourism, commission big name artist on large-scale projects. I may be wrong.

您的国家有没有颁布过公共艺术相关的政策？

沃恩·赛迪：约翰内斯堡一些地方有公共艺术政策，开普敦最近也刚开始实施。我们国家的艺术文化部围绕社会架构和国家建设，专门为公共艺术设定了一个清晰的框架。这对国家未来进一步推进"后1994彩虹国度"[1]的政治理想具有重大意义。

---

[1] 寓意不同种族的人们可以在南非和平生活，因此有"彩虹国度"之称。1994年，纳尔逊·曼德拉建立多党联合执政的民主政权，结束了种族隔离政策，这是南非历史的里程碑。

Does your country published any policy for public art?

Vaughn Sadie: The city of Johannesburg has a public art policy in place, with the city of Cape Town only recently bring theirs into effect. Our National Art and Culture department has a framework in place for public art, with the clear set of deliverable around social cohesion and national building, which are really a means to further the post 1994 myth of the "Rainbow Nation".

您如何看待艺术家、政府、社区及市民之间的关系？

沃恩·赛迪：这是一个相对复杂的问题，答案并非人们所希望的那样清晰。主要还是取决于项目的性质和所处的社会环境。通常来说，艺术家、政府、社区和市民这些都不是可以独立存在的。我最近正在进行的一个项目就需要很紧密和出色的社区网络，需要同时具有多个公民论坛和非盈利组织，以及一些社区人员在项目形成的过程中发挥至关重要的作用，并可以决定他们生活环境中应该发生的事情。与此同时，政府在其中的作用可以说微乎其微。在一个理想的国家中，你所希望看到的是一种井然有序的协调和安排，这种关系可能之前就已经形成，也可能是后天形成，但绝对不应该出现需要经常去协商与考虑权力和责任的情形。

What do you think the relationship between the artist, government, community and public citizen?

Vaughn Sadie: It is complicated question, it's not as clear as one would hope. It depends on the nature and the context of the project; often these grouping are not singular. The project that I am currently working on, has a strong and active community network, with several civic forums, NPO's and member of community playing a vital role in shaping and deciding what happens in their context, with government playing little or no role. In an ideal state you would want a hierarchical arrangement where these relationships are pre-determine or over-determine, but rather that a situation where agency and accountability are always be negotiated and reflected upon.

您前面提到了都市建设和地方重塑，能不能谈一谈您是怎么看待政府作为委托方，给予艺术家巨额预算以帮助和促进都市建设和地区重塑工作的行为？

沃恩·赛迪：我认为这不是个大问题，只要过程是透明的，且艺术家享有自主权就行。不过我由衷地认为如果政府打算作为委托方认真参与到公共艺术活动中的话，那么它至少需要认定和扩大公共艺术项目的范围和类型。政府有责任扶持自己参与的既富多样化又备受争议的新领域。

02 《发挥你的作用》 南非帕卡米萨

03 《火车上的布道》 南非约翰内斯堡

04 《安西娅.莫伊与格雷厄姆斯敦城》 南非格雷厄姆斯敦

You stated urban renewal and place making, what do you think about the government as commissioner to give the huge budget to artist to improve the urban renewal and place making works?

Vaughn Sadie: I don't see a great problem here as long as the process is transparent and the artist are given autonomy. I do think that if government is going to play a vital role as commissioner, it needs to expand the range and type of the public art projects it supports. It has a responsibility to grow and support a field that both diverse and rigours in its practice.

那您是如何看待政府作为委托人或者赞助人支持某一公共艺术项目这种情况的？

沃恩·赛迪：我认为委托人是一个很重要的角色，但是在角色选择上又经常会出现问题。因为某些委托人的所属形态和价值被添加进了公共领域这个概念。很多的委托应该被质疑，就如同政府总是在试图介入和塑造公共艺术的概念制定。我们在这个问题上的思虑并非周全。事实上，在南非这是一个非常有趣的话题。而我认为这个话题会持续变得更加直观和更加具备问题性，因为我们可以通过越来越多的案例和事实看到艺术已经被作为一个工具，运用到推动政府工作的情况中。

What do you think of the role of government as the commissioner or sponsor to support a project?

Vaughn Sadie: I think it is an important role but it is often problematic as certain ideal and values are inserted into the public realm and are over determined by type of commission. Larger commission of monuments should be questioned, as should the role that government plays in shaping our understanding of what public art should be. I don't think we give this enough consideration. This has become an interesting conversation in South Africa, and I think it will become more visible and be problematized, as we see more cases and possibility where art is being instrumentalized and used as tool for pushing the government agenda.

您为此次研究员会议提名了一些公共艺术项目，请问您选择案例的标准是什么？

沃恩·赛迪：我推荐的公共艺术项目总得来说都是一些小型的、临时性的、规模适中的，并且在项目进行的过程中与公众的行为可以有直接联系的作品。我认为案例中每个人与项目交流的影响都是可以预估的，但是同样的，这些行为也会大大增加每个项目的支出。一个公共项目中，艺术家想要突出的主体通常都是会被置换或产生变化的。因此需要仔细

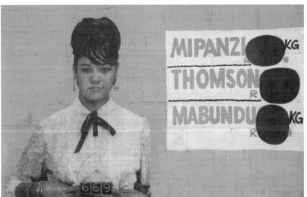

05 / 06 / 07 《燃烧的博物馆》 南非开普敦

地考虑每个项目的存在价值和意义,它们是如何在所属的空间内改变的?以及它们与其他物体之间如何相互影响的?这也给我们带来一个新的思考,即:我们应该如何定义和规划我们所理解的公共空间?

What are your criteria for selecting the cases you nominated?

Vaughn Sadie: The case studies I chose were smaller temporary projects, modest in scale and attempted to engage with a public directly through various forms of participation. In each case study the impact could be measured within interpersonal exchanges and added greatly to outcome of each project .Foregrounding the performative and taking into, account a public that is always changing, each of the projects carefully considers their presence and how they alter and affect the interaction of others in their given space, bringing into question how we formulate our understanding of public space.

您认为应该如何定义优秀的公共艺术作品?

沃恩·萨迪:我想大家已充分意识到,艺术的内涵根植于艺术品之中,它们构建艺术与空间、艺术与人之间的对话,建立和产生关系,得出艺术价值。并不是所有的人一开始就在研究环境和美学的表面呈现方式,并在空间中插入对象以强加意义。优秀的艺术作品是以刺激对话的形式创造艺术价值与艺术意义的。

What do you think defines of excellent public artwork?

Vaughn Sadie: I think something that is fully aware of that the context that it is situated in and how it responds and generates art in dialog with the space and people that constitute it. This is the opposite to some public art projects, not all of them, that do initially research at a surface level to understanding of the environment and the aesthetic register to understand how to insert an object in to space to impose meaning rather than stimulate dialog to create meaning.

您认为我们应该关注公共艺术的哪些方面?

沃恩·赛迪:这是一个很值得讨论的问题,我会思考我们是否真的需要这件作品,还是我们只是希望想要做这个项目?我认为我们需要好好思考已经存在的方案,并判定其是否适宜被放置在特定语境中。我们应该了解一个公共艺术项目在产生的过程中带动人们参与的重要性和差异性。我的工作环境是一个会随时发生变化的社会空间,这个空间中个体与空间的关系决定了个人和集体关系对这个地区的影响,而地区与项目的实施又影响到人与周围环境的关系。两者相互作用，而这种影响必须予以考虑，因为最终呈现出的结果是截然不同的。

What do you think that we need to pay attention to in public art?

Vaughn Sadie: This is a big discussion, and I would ask do we really need it, or do we just want it. I think we need to give greater consideration to the different strategies that exist and their appropriateness to a specific context. My senses of public art, is that we need to understand the importance and difference of community engagement and participation on the various process of a project. I work in an environment, where an existing and changing set of social and spatial relationships define individual and collective relationships to place, projects impact the people and these relationships. This impact has to be taken into account, to bring about an interesting outcome.

沃恩·赛迪：我认为这不是个大问题，只要过程是透明的，且艺术家享有自主权就行。不过我由衷地认为如果政府打算作为委托方认真参与到公共艺术活动中的话，那么它至少需要认定和扩大公共艺术项目的范围和类型。政府有责任扶持自己参与的既富多样化又备受争议的新领域。

Vaughn Sadie: I don't see a great problem here as long as the process is transparent and the artist are given autonomy. I do think that if government is going to play a vital role as commissioner, it needs to expand the range and type of the public art projects it supports. It has a responsibility to grow and support a field that both diverse and rigours in its practice.

01 张羽洁正在采访里奥·谭

公共艺术作为一种社会治疗
——里奥·谭访谈访谈

Public Art as a Social Therapy:
An Interview with Leon Tan

研究员：里奥·谭
采访：张羽洁

Researcher: Leon Tan
Interviewer: Zhang Jane

里奥·谭（Leon Tan）博士是一位文化艺术评论家、策展人、教育家和心理治疗师。他生活和工作在新西兰奥克兰，现任新西兰国立理工大学（Unitec Institute of Technology）设计和当代艺术专业高级讲师。里奥在奥克兰大学获得艺术史博士学位，他的博士研究是基于对心理学、美学以及新媒体和社会媒体政策方面的理论分析。他既是艺术和精神健康机构的独立咨询顾问，同时在艺术史、艺术与设计评论、新媒体与心理学领域从事本科和研究生的教学工作，负责指导艺术和设计专业的硕士研究生项目。不仅如此，他还在许多国际刊物上发表作品、出版书籍，并参加过如ISAE，Digital Arts Week Zurich，the 2012 Taipei Biennale and Malmo Konsthal等媒体艺术展。在此次采访中，他将和我们谈谈心理治疗和公共艺术的关系。

Leon Tan is an art and culture critic, curator, educator and psychotherapist. He lives and works in Auckland, New Zealand, where he is a senior lecturer in the Department of Design and Contemporary Arts at Unitec Institute of Technology. Leon holds a PhD in Art History from the University of Auckland. His doctoral research was based on the analysis and theorization of the psychology, aesthetics and politics of new media and social media.Dr.Tan's professional background includes work as an independent consultant to arts and mental health organizations, undergraduate and postgraduate teaching in art history, art and design criticism, psychology and new media, supervision and examination of postgraduate projects in art and design, publications in international journals and books, and presentations/exhibitions at venues such as ISEA, Digital Arts Week Zurich, the 2012 Taipei Biennale and Malmö Konsthall.In the interview, Leon will talk about the relationship between psychotherapy and Public art.

您能否谈谈您对公共艺术的理解？什么是公共艺术？

里奥·谭：我认为公共艺术是位于公共空间中的艺术或文化的表达。它的形式不受限制，可以是任何种类的艺术作品，包括雕塑、社会参与型的艺术、社区艺术等，是在博物馆或画廊以外的公共空间中展出的艺术作品。但是，我认为我们还必须认识到很重要的一点，即"公共"一词还可以包含很多不同的事物。从某种程度上来说，网络也可以被认为是公共领域，或者甚至说是一种公共设施。许多在网络上的艺术作品也可以被认为是公共艺术。同样，用无线电广播作为媒介的作品也可以被认为是公共艺术，例如与Franco Bifo Berardi合作的意大利自治论者《广播爱丽丝》项目就是一个案例。我想很重要的是能够对公共艺术的概念有一个更为宽泛的理解。

Could you please talk about your understanding of public art? What is public art in your opinion?

Leon Tan: Public art is any kind of art or cultural expression that takes place in public space. Traditionally we can think about sculpture, socially engaged projects, community art projects that occur outside museums or galleries. But, I think it's also important to realize that the term "public" can mean a lot of different things. In a way, the parts of the Internet can be considered as a public sphere, or even a public utility. A lot of things happening online could also be considered public art.

Works that take place in the air, in radio waves, for example, similarly can be considered public art too (an example might be the Italian Autonomist 'Radio Alice' project, with which Franco Bifo Berardi is associated). I think its important to have a very broad conception of what public art means.

作为公共艺术协会研究员，你的研究兴趣点是什么？

里奥·谭：公共艺术作品种类繁多，就"公共艺术"这一宽泛的概念而言，并没有必要都以改变社会为其创作的目标，因为并不是所有的艺术家都对此感兴趣。但这恰恰是我的兴趣点之一。我对于艺术用其独特的方式改变世界特别感兴趣。为什么呢？一方面是因为我自身的专业背景，另一方面是因为我们所生活的环境始终充满着各种未能解决的问题。例如气候变化、社会暴力等问题，似乎并没有真正的解决途径。但有趣的是，突破性的见解和解决方案常常源自艺术家。不幸的是，在某些圈子里，对于用艺术来寻求改变世界的想法常常被认为是过时的。一些艺术评论家，例如 Claire Bishop 用一种轻视的态度来描绘社区艺术，认为社区艺术缺乏批判性。而我完全不同意这一观点，相反，被称为社区艺术的作品常常是非常具有批判性的。我不知道你是否注意到人们有的时候用一种鄙视的眼光看待"艺术可以作为社会福利"的这一观点。我不理解这是为什么。在我的研究中，我试图挑战这种思想。

As an IPA researchers, what are your research interests in public art?

Leon Tan: There are all kinds of public art. Public art doesn't need to have the objective to change society. Not all artists are interested in that. But that happens to be one of my interests; I am very interested in art that changes the world in some way. Why? Partly because of my own professional background (I trained in psychology and psychotherapy, as well as art history) and partly because we are surrounded by so many problems for which there seem to be no solutions. For example, the problem of climate change, no one has really come up with a clear solution. The problem of violence, there seems to be no real solution to that. Interestingly, breakthrough insights and solutions, where they occur, frequently come from artists. Unfortunately, in certain circles, art that seeks to change the world has been considered quite unfashionable. Some critics, for example Claire Bishop, have even written disparagingly of community art suggesting that it lacks criticality. I really disagree with this perspective. I think what is called "community art" can often be extremely critical. I don't know if you have noticed that people sometimes look down on the idea of art contributing to social good? I don't really understand why. In my research I try to challenge that kind of thinking.

据了解,您除了是一位艺术评论家之外,还是一位心理治疗师。您能否谈谈心理治疗与公共艺术之间的联系?心理治疗如何被运用到公共艺术中?

里奥·谭:我认为公共艺术之所以能够产生影响力,在很大程度上是由于公共艺术能够改变和塑造人们思考问题的方法。特别是当人们在思考自身与他人在公共空间中的关系时所采用的方式。我认为两个人之间的私下交流与在公共空间中公开地讨论是有很大不同的。因为当我们在公共空间中讨论时,其他人也能听到我们的谈话,甚至对我们所交谈的内容提出挑战。公共空间给予我们一种机会,让我们更好地体验不同的社会角色,从而构建更好的未来。公共艺术可以被认为是一种社会治疗,但并不是必须的。

As I know you are not only an art and culture critic, but you are also a psychotherapist. Could you talk about the link between psychotherapy with pubic art? How the psychotherapy could be used in Public art?

Leon Tan: I think public art's impact largely comes from its ability of alter or shape the way people think. Particularly the way people think about themselves and each other in public space. I think talking to a person one on one in a private setting and discussing something in public are very different situations. Because when we have a conversation in public, other people can hear what we say, and even challenge it. Public space offers opportunities for us to collectively experiment with different social roles, thereby potentially shaping preferable futures. I think you can relate public art and psychotherapy. Public art can be, but is not necessarily, a form of social therapy.

如果公共艺术具有社会治疗的作用,那么您能否从心理学的角度分析一下,公共艺术是如何改变人的?是怎样的一种心理机制在起作用?

里奥·谭:如果观察那些存在精神或心理问题的人,往往会发现他们存在一些社会行为的障碍。心理治疗实际上是试图改变我们生活世界中某一方面或多个方面的思维习惯、感觉习惯或行为习惯。我们所面临的大多数问题,许多心理上的经历,常常都与社会有关,与社会中遇到的人或事有密切的关系。许多人遇到的问题常常与人际关系有关,与人们如何看待他人有关，与他人如何看待自己有关。我认为公共艺术提供了一种机会，来扰乱我们的习惯性思维、感觉和体验生活。例如，当人们染上毒瘾时，只有通过改变吸毒这一习惯，才能使事情发生转机。这一点不仅对毒品如此，对于社会暴力也是如此。暴力同样也是一种习惯，公共艺术具有打破人们习惯性思维的潜质。我们要改变自身，就必须要经历一些新的体验，例如一种新的观看事物的角度，一种新的行为，一种新的与世界、与我们所身处的环境联系在一起的方式。例如，有一些公共艺术项目是关于全球气候变化或环境污染的。这些作品向我们展现了一个完全不同的视角，让我们看到我们自身的各种行为、我们的消费文化、我们对待这个世界的方式，以及与气候改变之间的关系。

If the public art can be a form of social therapy, could you talk about how the public art change peoples? What kinds of psychological mechanism happen?

Leon Tan: If you look in general at peaople who have mental or psychological problems, these often manifest in difficult social behaviors or patterns. Psychotherapy is really about attempting to change those habits of thinking, feeling or behaving in relation to one or more aspects of the world in which we live. All of our problems, and much of our mental experiences, are usually related to society, related to someone or something that is happening in society. Lots of problems that people have are with their relationships, or with how they perceive other people, or how other people perceive them. I think public art provides the opportunity for disrupting our habits of thinking, feeling and experiencing life. For example, when people are addicted to drugs, things can only change if the habit is broken. This applies not only to drugs. It's also the case for violence in society. Violence is also a habit. Public art has the potential to disrupt such habits. That's how we change ourselves, by experiencing something new, such as a new way of looking at things, a new way of behavior and a new way of relating to the world, relating to the environment. For example, some public art projects that deal with issue of climate change or environmental pollution. They work by showing us different ways to relate to the world, to relate to consumption, to reflect on the kinds of activity that created those problems in the first place.

我们刚才谈到了公共艺术具有改变人的力量，可否谈谈艺术家、公众、政府他们三者在这场改变的过程中扮演了怎样的角色？

里奥·谭：我认为改变是一个非常复杂而又缓慢的过程。我们中的许多人会认为改变没有发生，但事实上，事物总是在发生改变。问题的关键是变化的频率。当变化的频率非常缓慢时，我们很容易会认为没有发生过改变，因为这种改变常常需要很长的时间。人们能够感受到的视角常常是很有限的，5年、10年，或者是人的一生。而大规模的变化常常需要比人的一生更长的时间。我想强调的是，改变始终在发生，而我们有可能使变化的频率加快或减慢。公共艺术有潜力成为一种激发改变的力量，激发人们的讨论，特别是在政府、公众、艺术家三者之间。我特别想谈论的是艺术家角色的问题。我认为艺术家是具有较强优势的角色。这是因为，在我们现有的各种不同的专业中，艺术这一专业所具有的想象力和技能要远远大于其他专业，例如在法律、机械、工程这些专业中，可以发挥想像力的空间是比较有限的。而艺术家可以想象整个世界，可以想象他们所喜爱的任何事物。也正是由于艺术家所拥有的想象力，带给了公众、政府同时也包括艺术家自身，能够体验一些与众不同的经历。我想艺术家在公共艺术中的角色是想象并创造一种新的可能性。当然，政府和一般公众也可以发挥这种创造力。但艺术作为一种专业，其特点就在于具有巨大的想象力，发现各种可能性，并因此而不被固有的习惯、规定所束缚。公共艺术可以向我们的社会呈现一种挑战，而只有在公众、政府、艺术家三者共同来应对这一挑战时，一种有意义的改变才能发生，并贯穿于三者之间。

The artists, the public and the government are the three groups of the structure of public art. We just talked about that public art has the power to transform people, could you talk about the role of these three groups play in the process of change?

Leon Tan: I think the issue of change is complex one. Many of us like to think that things are not changing. But in fact, things are always changing. The only thing is the rate of change. So when the rate of change is very slow, it's easy to think that nothing is changing because it takes so long. That's only from very limited perspective such as 5 years, 10 years, or even one lifetime. The large-scale change can actually take more than one lifetime to have its impact. What I want to say is change is always happening. It's possible for the rate of change to be accelerated or slow down. Public art has the potential to create provocations, to provoke people to have discussions, to have debates particularly on these three sectors, the government, the public and artists. And I want to mention the special role of artists. I think artists have a very privileged role. Because if you think of all the professions we have, the artists' profession seems to have the biggest scope for imagination, much more than the accountant, lawyer, mechanic or engineer. In those fields the potential for imagination is quite limited. Artists are different. Artists have the whole world. They can imagine whatever they like. The power of imagination that artists bring allows for these different sectors including artists themselves to experience something new. I think artists have an important role of imagining new possibilities. Of course, those ideas can also come from government or the general public, but I think art as a discipline or profession is where you find the greatest amount of imagination, the greatest amount of possibility, relatively unconstrained by existing structures, existing habits. Public art can present society with provocations. Only when these different sectors meet that provocation will some kind of meaningful change happen across all three parties.

您是否创作过一些公共艺术作品用到您心理学方面的专长？

里奥·谭：2012年我受到台北双年展的委托，与瑞典的一位电影制作人弗兰妮·侯柏格 (Virlani Hallberg) 合作创作作品。我们去了中国台湾地区，共同完成了一部40分钟纪录片《倒退三角广场》(Receding Triangular Square)。作品最终的形式是一个单通道视频装置。在这部件作品中，我们试图调研殖民对人们的思想带来的影响以及人们从这种影响中得以解脱的可能性。我们知道中国台湾地区曾被不同的国家殖民，包括日本和美国，这使得那里的发展受到一定程度的影响。

一方面，就心理健康的概念而言有很强的美国式倾向，这是由日本人引进的。而另一方面，还有一些非美国式的、非西方式的关于精神、心理健康方面的思考。所以我们试图探索这些不同的可能性。虽然我们的作品最终是在博物馆展出，其展示的方式不怎么公共，但作品的整个过程非常公共。我们去了不同的社区，去采访了不同的人，例如中国台湾地区的原住民、传统的治疗师和精神科医生。通过这一过程，我们和不同的群体互动，我想我们可以说，一种心理分析正在这一过程中产生了。因为心理治疗和心理分析是通过对话来完成的，而对话正是我们在这个项目中所使用的工具。在这一过程中，我们通过与这些人交流，向他们提出挑战，思考为什么中国台湾地区会有目前的状况。事实上，我们发现许多人对此都很感兴趣。而我们的这个项目为人们提供了一种机会，促使人们能够更为清晰地思考，究竟什么是他们想改变的，而他们又应当对中国台湾地区过去的历史做出哪些不同的回应。不少人认为，这一过程对于他们思考未来的人生道路有重要的意义和价值。其中一位精神科医生在采访中说到，他受此启发，试图创造一种新的心理健康模式。他开设了一家完全不同的心理健康诊所，作为官方健康系统之外的又一选择。而这一想法是在与我们分享和交流的过程中产生的。自2012年以后，这件作品在世界各地展出。而每次展出，我们都会试图创造一个公共艺术的组成部分，即必须包含一次公共的散步，我们与不同国家居民一同散步，并共同探讨特定地区的历史文化。这也成为了作品与当地联系的一种方式。在中国台湾地区之后，我们在菲律宾、旧金山、斯德哥尔摩、新德里的马尔默等地展出，每次展览都会包含一个公共散步环节，而这正是通过公共艺术来经历心理治疗的方式。我可以这么做，是因为我受过心理治疗的训练，我和电影制作人合作。通过合作，这一公共作品很令人兴奋。原因很简单，因为它向人们提出挑战，促使人们使用一种新的方式思考和构建社会，传递一种经历，同时对社会问题进行批判并促使我们思考如何改变。我的兴趣点不单单是对事物的批判，同时还希望能够找到解决的方案。这也是为什么我们在公共的对话中向人们提出挑战。我想这可能是一个将心理治疗和公共艺术相结合的案例。只要将心理治疗中的一些步骤和原则，例如人是如何改变的原则，巧妙地运用在公共艺术的创造中，就可以使得公共艺术的干预更为有效。这其实并不苦难，困难的是人们甚至从来没有想过要这么做。心理学是一个接口，公众通过它和所有公共艺术作品产生回应。所以，如果你能了解人的心理是如何工作的一些原则，那么，你就能够更好地设计出一种经历，使得作品更具有针对性。

Have you done any project used the knowledge of psychotherapy in the public art?

Leon Tan: The project I did in 2012 was commissioned by Taipei biennial to produce a work with a Swedish filmmaker Virlani Hallberg-Rupini. We went to Taiwan, China. We produced a documentary called "Receding Triangular Square" which is realized as a single channel installation. The project investigated the impact colonization on people's minds and the potential or lack of potential that might result from that. Taiwan, China has been colonized/occupied by several different countries including Japan and America. Because of that, Taiwan, China has developed in slightly different ways.

On one hand, there is a very strong focus on the American conception of mental health, which was imported by the Japanese. On the other hand, there are all the other non-American, non-western ways of thinking about mind and mental health. So we wanted to explore these different possibilities. Even though the result was not a public work, it was in the museum, the process itself was very public. We went out to different communities, to interview different people like aboriginal Taiwanese, traditional healers and psychiatrists. Through these processes of interacting and engaging with different parties, I think it's possible to talk about a kind of psychological analysis taking place. Psychotherapy or Psychoanalysis fundamentally takes place through conversation. Conversation is the tool we used in this project. During the process, we have conversations with these people and we challenge them to think about why the situation in Taiwan, China is the way it is. Actually, we found lots of people are very interested to do this. This project is the opportunity for them to think more clearly about what is they want to change, and how they might respond differently to the historical circumstances of Taiwan, China. A number of them actually said that it was very valuable for them in terms of shaping what they are going to do in the future. One of the psychiatrists in the interview said that he has initiated a new kind of model of mental health. He set up a very different kind of clinic, as an alternative to the government system, based on ideas he shared with us. Since 2012, we have exhibited the work around world. Every time we exhibit, we also try to have a public art component. That component essentially involves public walks. We took walks with people in different countries and have conversations about the history of that place. After Taiwan, China we exhibited in the Philippines, San Francisco, Stockholm, Malmo, New Delhi. Each of the these exhibitions was accompanied by a public walks program. That's really a way of experimenting with the psychotherapeutic dimension of public art. I can do that because I have the training as a therapist and collaborate with a filmmaker. Through our collaboration, the public work is actually quite exciting simply because of challenging people to think of alternative ways to shape society, to deliver certain kinds of experiences, and also to critique the problems in society with a view towards how we can change them. I am not only interested in criticizing problems. I am also interested in coming up with some solutions. That's why we challenge people in the pubic conversation. I think that's one example of how psychotherapy might be integrated with public art. Just a small step of taking certain psychological or psychotherapeutic principles such as the principle of how people change, and applying that to a public art work, can actually produce much more engaging or successful public art interventions. It's not difficult. The difficulty is that people may not even think to try that. Psychology is the interface by means of which all audiences respond to public work. So it makes sense that if you can understand the principles of how the mind works, you should then be able to craft experiences that are more targeted.

从您的视角度来看，您是否认为公共艺术家应该学习一些心理治疗方面的技巧，这或许能够助于他们更好地创作？

里奥·谭：如果他们想这么做的话，我想是的。那是因为，在传统的博物馆和美术馆体系中，艺术家、艺术作品和公众之间是存在着一个中间媒介的。这一媒介包括艺术品商人、艺术评论家、策展人。在公共空间中，虽然也有策展人和艺术批评家，但是他们的媒介作用要相对减弱。比如，在公园或者街道的公共艺术作品中，很难有策展人或艺术评论家站在观众和艺术作品之间，观众将直接地感受作品而不再经过艺术评论家的诠释。这就意味着艺术家不再能简单地依赖策展人或艺术评论家作为他们与公众交流的媒介，他们必须依赖自身的能力。如果他们不具有这些技能，他们的作品就可能不怎么容易被理解，不怎么容易能与公众互动，或许也是一些作品失败的原因。最后有一点我觉得很重要，我不认为公共艺术应当完全以其社会影响力做为唯一的评估标准，同时也应当基于美学、历史和政治环境进行评估。

In your perspective, do you think artist should learn some psychotherapy skill, which might help them to create public art works?

Leon Tan: I think so, if they want to. That's because in more conventional settings like the museum and gallery, the interface between the artist, art works and the public is more mediated. It mediated by the dealer, curator and the art critic. In the public, of course you still have a lot of that, but its less mediation. For example, if a public art project takes place in the public park or in the street, you don't have curator or critic standing between you as an artist and audience. That mean the artist cannot simply rely on a curator or critic to do that aspect of interfacing with the public. They have to do this themselves, or not. If they don't have those skills, then of course their work might not be so accessible or engaging. That's perhaps why some of public art works fail. One important last point, I don't think that public art should be evaluated solely on social impact. It should also be evaluated on the basis of aesthetic considerations, as well as geo-historical and political context.

01 张羽洁正在采访茱茜·乔克拉

公共艺术与城市身份
——研究员茱茜·乔克拉访谈

Public Art as an Urban Identity:
An Interview with Giusy Checola

研究员：茱茜·乔克拉
采访：张羽洁

Interviewer: Giusy Checola
Interviewer: Zhang Jane

茱茜·乔克拉是意大利研究者和策展人，工作于意大利及海外。她是法国巴黎第八大学的博士研究生，研究方向为艺术的美学、科学和技术，她还是意大利贝加莫大学人文文化研究部的博士研究生，主要关注点是艺术的在地性、区域化过程及公共空间的时空纬度在与文化地理学、地理哲学对话中构成的关系。目前，她与 Angelo Bianco 联合指导意大利实验性的《马泰拉 Maverick 校园》项目，该项目由南方传承当代艺术基金会推广，是典型的促进多学科研究的基础设施原型。她还是意大利南部和近地中海地区首个专门存放艺术和公共领域档案的档案馆创始人和领导者，是 2015 年意大利 UNIDEE 大学创意 Pistoletto 基金会项目的导师。其工作内容包括国际驻地项目、以艺术项目为引导的艺术家实践，以及对公共领域、城市和乡村空间、集体维度内的文化和社会需求进行相关的理论研究，其策展项目大多基于国际艺术机构间的合作与伙伴关系，通过艺术与策展的交换项目与北欧、地中海、中东的艺术家和机构建立交往。

Giusy Checola is an Italian researcher and curator who works in Italy and abroad. She's PhD researcher at University Paris 8 Vincennes Saint-Denis (France) in Aesthetics, Sciences and Technologies of Arts and University of Bergamo (Italy) in Humanistic and Intercultural Studies, where she mainly focuses on the relationship between artistic place making, territorialization processes and spatial-temporal dimension of the public space and sphere, in dialogue with cultural geography and geofilosophy. Currently she's co-director with Angelo Bianco of the experimental Maverick Campus of Matera (Italy), a prototype of infrastructure for pluridisciplinary research promoted by SoutHeritage Foundation for Contemporary Art; she's founder of Archiviazioni, a platform for research and first archive specialized on art and public domain in Southern Italy (since 2010); she's tutor of the program 2015 of UNIDEE - University of Ideas at Pistoletto Foundation (Biella, Italy). She has worked for international residency programs and art projects produced in the urban and rural spaces related to specific cultural and social requests within the collective dimension. The projects she has curated have been often produced in partnership and collaboration with international networks of art institutions and residencies and programs of artistic and curatorial exchange between North- European, Mediterranean and Middle Eastern organizations.

能否请您谈谈您对公共艺术的理解？在您的观点里，什么是公共艺术？

茱茜·乔克拉：我认为公共艺术在今天可以有不止一个定义。在某些语境中，公共艺术是一种通过美和希望复兴日常生活空间的方式，正如我们这次在上海的会议中讨论到的。在另一些情况下，公共艺术为重新定义个人与空间、环境、景观以及不同形式的社会关系创造了一种新的可能性。我所理解的公共艺术是一个开放的领域。在这一领域中，艺术家可以创造一种形式和方法，使人们对公共问题和日常的生活有不同的思考。这种形式和方法可以是一种体验、一种语言，一个关于特定地域的讲座和艺术家调研的方法，例如通过与不同领域的人们交流关于城市和乡村空间的规划和研究等。

就人们普遍接受的"造地"公共艺术和以社区为主导的艺术家项目，我认为它们同样可以被理解为一种空间的临时性方案，它允许艺术作品超越公共空间的物理形态，成为临时性的公共艺术作品（即作品如果是临时性的可以短暂地存在，如果是永久性的则可以用"临时悬置"的方式呈现）。自从公共空间可以被理解为物质的（物理空间）和非物质的（网络空间），公共艺术的影响也不再仅仅局限在艺术领域。在一个长期的过程中公共艺术可以塑造集体行为，影响公共空间的规划，创造社区或是创造自身。

在某种意义上，公共艺术的功能和其最终形式是不能被事先定义的。因为它会随着地方的政策、历史性的事件、周围的环境、社会文化因素而改变自身的发展潜质。"公共"一词是指向某一个特定时期的。

Could you please talk about your understanding of Public art? What is public art in your opinion?

Giusy Checola: I think that public art nowadays could have more then one definition. In some contexts it could be the way to regenerate the daily life spaces through beauty and hopefulness, as underlined during the network meeting in Shanghai; in others it could be the possibility to redefine our relationship as social individuals with the place, the environment and the landscapes and to create different forms of social relations. Actually I understand public art as an open field in which artists could create forms and ways of thinking differently about public issues and common life, in which it is possible experiment the medium, the language, the lecture of the place and the methods for artistic research, for example through the interaction with other disciplines involved in the studying and the planning of urban and rural spaces.

In its acceptation of place-making and community-led artists' projects, I understand it also as spatial-temporal formula, that allows the artwork to go over the physicality of the public space and the temporality of the public artworks (that could be ephemeral if they're temporary or "temporally suspended" if they're permanent), since public space can be nowadays understood both as material (physical) and immaterial (net), and since its "effects" could not only affect the art field but the collective behavior and, in cased of long term processes, it could also affect the way in which public spaces are planned and communities are created or create themselves.

In these terms the function and so the destiny of these kind of public artworks can't be predefined, since it could change according with its transforming potential, with the local political and historical events, and with the environmental, social and cultural factors that define that "publicness" in that specific time.

公共艺术的社会功能是什么？

茱茜·乔克拉：我认为除了作为创造复兴的一种形式和对公共空间的美化之外（这一点 Lewis Biggs 在介绍在地性在区域改造中的作用的相关出版物中有过解释），公共艺术的社会功能包括两个主要的方面：对艺术家而言，它为其提供了一种面对公共空间的挑战机会，艺术家必须面对不可预知的未来，针对公共领域中正在发生的事件进行工作，这些工作不仅仅是材料层面的。第二，创造一种新的形式或版本的现实，这将促使人们重新思考"人类的居住区"（城市区域、社区、农村空间和其他形式的人类居住空间）与作为生物体的领土，作为"文化产品"之间的关系。我们的生活质量已被认为与我们所生存的环境密不可分。

在这些方面，我目前关于公共艺术社会功能的思考与坐落于布里斯托（Bristol）的机构"情境"（Situations）于 2013 年提出的"公共艺术新规则"中的规则 3 有所不同。规则 3 指出，公共艺术是一种"创造无规划空间"的方式。

What's the function of public art in the society?

Giusy Checola: I think that beside the creation of forms of regeneration and beautification of the public space (in the terms explained by Lewis Biggs in its introduction about the place-making in Local Remodeling publication), I see the function(s) of public art in the society as related to two main possibilities: for the artists, it could be the opportunity to measure themselves with the challenges that public spaces offers, by dealing with its unpredictability and with the opportunities given by working in and on the live matters of public sphere, which is not only a material dimension; second, to create new forms and visions of the reality that could allow to rethink the relationship between the "human settlements" (urban areas, communities, rural spaces and other kind of forms of human living spaces) and the territory as living organism, as "cultural product", since the quality of our life can't be considered as separated from that of the environment that hosts us.

In these terms I currently think the functions of public art in the society as related in different ways to what the Bristol based institution Situations named as a "new rule of public art" on 2013, the rule n.3, that indicates public art as the way to "create space for the

02 《墙(档案)4号》 工作坊和档案展 Pietro Gaglianò 与档案馆项目的合作项目
意大利莱切市 艺术与建筑实验室 2011年

03 《谁画的线》 在地装置艺术 Devrim Kadirbeyoğlu
意大利莱切市 艺术与建筑实验室档案馆项目 2011年

unplanned".

据了解，您的研究曾聚焦在公共艺术的"造地"性，聚焦在创造一种文化条件以促进艺术在公共领域的实践。那么究竟怎样的文化条件能够促进艺术在公共领域的实践呢？

茱茜·乔克拉：我认为这主要取决于以下因素：源自公共艺术被要求产生效应的所在区域的背景特征，源自于我们自身对有效性的理解，以及因此带来的我们对公共艺术可能发挥出的潜质的期待和目标的设定。

As I know, your research work used to focus on the placemaking, on the creation of cultural conditions for the improvement of the art's action in the public and common domains. So, what kind of cultural condition could improve the art's action in the public?

Giusy Checola: I think that it depends mainly from the following factors: from the characteristics of the contexts in which public art is required to be "effective", from what we mean for "effectiveness" and thus from our expectations about public art potentiality and objectives.

您能否谈谈您所做过的一些公共艺术项目和取得的成绩？

茱茜·乔克拉：2010年在意大利南部的阿普利亚（Apulia）区域，我启动并指导了称为"生成性档案"的项目。一份档案和一个实验室空间用于艺术和公共领域的研究和计划。第一年，它位于莱切（Lecce），2011年以后，它成为一个移动档案。最近，我们打算在邻近区域建立一个新的物理空间。

当我抵达莱切的时候，这个地方吸引我的是其历史与记忆、地理位置与环境特征之间的存在关系。这种关系使这一区域在一特定时期里，从国家层面被确定为合法的旅行者的"新乐园"。这为以这一特定空间为基础的自我展现和陈述形式得到了创造和繁殖的契机。带有区域环境和人类学特征的象征性画面，通过这种创造地域以及其地理特征的方式在过去的几年中不断出现。这一过程既是一种工具，同时也带来了大量日渐增长的游客。现在每年，特别是在夏天，来自国内和国外的游客数量惊人，也因此为当地的居民带来了可观的经济收益。

但我的问题是，人们的生活质量和这一空间日后的岁月将会怎样？对我而言，改善生活质量意味着为公民推进城市复兴而不是为旅游者；为文化研究和产出打开其流失的空间；创造讨论公共问题的场合；产生地域文化的丰富性和复杂性知识，以此来激发更多转变性的公民行为，减少代表性的公民行为，从而带来更为美好的未来。

因此，首先我们试图重新激活一个位于昔日一家烟草工厂内的未能被使用的空间，一个私人建筑工作室的储存空间被改造为公共艺术档案和实验室，对观众开放用于咨询，举办研讨会和会议。国内和国际的艺术家被邀请参加短期的驻地、工作坊、区域性的调研和展览。这些咨询、工作坊和会议是向观众开放的。同时还有一个空间用于推广由孤立的乡村空间而可能带来的机会，利用温和的天气和周围美丽的风景，从春天到秋天，这一空间允许我们、艺术家以及周围的人们在此交替工作以"复兴"这一区域。

这一空间主要由当地的大学生和艺术学院的学生来创造。这是他们第一次有机会拥有一个真实空间的处置权。他们通过对国内和国际在公共区域相关项目进行调研，与我们共同探讨他们的实践，参与该项目中被邀请艺术家的一些活动，以拓展他们的项目发展思路。他们还有机会浏览专门针对为国内外本科生和研究生创造机会和驻地项目的那一部分档案内容。

然后，我们开始试图建立地方和国家文化机构，与当地的生产者之间的联系。就像 Campolisio Sarruni，一位生物农场主，成为我们的合作伙伴（和旅游业一样，农业依然是这一地区最主要的经济支柱）。创造国内和国际艺术机构的网络，以收集档案馆的材料为起始点，也因此经常带来合作项目和合作生产。

最后，这个档案馆有另外两个主要的目标：在该区域以外的地区推广当地的当代艺术，以及在该区域内推广与这一地域不同的知识，从而使区域转而获得一种更为宽广的叙事模式。这些都将以特定地域的旅游以及与艺术家、社区和当地的合作者所分享的瞬间为基础。

Could you please talk about the projects or achievements you have done?

Giusy Checola: On 2010 in Apulia region (Southern Italy) I have initiated and directed what my colleagues and me have called a "generative archive", an archive and laboratorial space devoted to research and planning on art and public sphere. For one year it has been located in Lecce, after 2011 become a mobile archive and currently we're going to establish a new physical space in a neighbour region.

When I was arrived in Lecce, what interested me about that place was the results of the relationship existing between historical and popular memory on the one side, and the geographical position and the environmental characteristics on the other side. This relationship, in that specific time in which that area was going to be definitely legitimated on national level as "new paradise" for tourists, gave birth to the creation and proliferation of forms of narration and self-representation based on specific spatial, environmental and anthropological symbolical images that in a way "created" the place and its geography in the last years. This process has been both a tool and the cause of a very fast increasing of massive tourism that nowadays every year bring there an impressive number of people from the rest of the country and abroad, that especially in summer time generates financial benefits for inhabitants.

But my question has been this one: what about the quality of life and the spaces in the rest of the year? Improving quality of life for me means to promote urban regeneration for citizens over than for tourists; to open missing spaces for cultural research and production; to create occasions for discussion about public issues; and to produce knowledge about the cultural richness of the place and its complexity, in order to stimulate a more transforming and less representative civic behavior that could lead to a better future.

So, first of all we have tried to reactivate a not-used space located in a former tobacco factory, that of a former storage of a private architectural studio has been transformed in a public art archive and laboratory, opened to the audience for consultation, workshops and meetings, in which national and international artists have been invited to participate to short term residencies, workshops, territorial investigations and exhibitions. That has been also a space for promoting the possibility offered by an isolated rural space, by taking advantage to the mild weather and the beautiful surrounding landscape, that from spring to autumn allowed us, the artists and the people attending it to alternate working moments with "regenerative" ones.

The space was mainly created for local students of the University and the Academy of Fine Arts, that for the first time could have at their disposal a place to make research on national and international projects related to public domain, and to discuss with us about their practice and their projects and develop ideas by participating at the activities with the artists invited to be part of the program. They could also have access to a section of the archive devoted to the national and international graduate and post-graduate opportunities and residency programs.

Then we started creating connections with local and national cultural associations and with local producers, like Campolisio Sarruni, a biological farm that became our partner (since, together with tourism, agriculture is still the main economy of the region). The creation of a network with national and international artistic organizations, started with the collection of the archive's materials, often brought to the cooperation and the co-production of projects.

Finally, the archive had other two main aims: that of promoting outside the region the knowledge of local contemporary art. And that of promoting inside the region a different knowledge of the place, alternative to the widespread narrative, based on site-specific tours and sharing moments with artists, communities and local partners.

您能否谈谈艺术家、策展人、政府以及公众之间的关系，他们在公共艺术中扮演什么角色？

茱茜·乔克拉：他们之间的关系在公共艺术中扮演着很重要的角色。一个能够解释他们之间的这种关系对公共艺术社会文化角色的重要性的案例是我向 IAPA 递交的案例之一，斯特丹·森兰的作品《Atelier / TRANS305 》，起始于 2007 年。在这一案例中，艺术家成功地在某一特定时间采取行动，将艺术家、政府和公众之间的社会政治冲突转化成了一种前所未有的可能性。他利用了法国当时的"大巴黎"城市计划造成的紧张局势，该计划试图将法国中央政府，塞纳河畔伊夫里郊区（Banlieu of Ivry-sur-Seine）的地方政府以及郊区的居民都包含在规划之内，引起了社会争议。"大巴黎"项目由前总统尼古拉斯·萨科齐（Nicolas Sarkozy）发起，包括巴黎中心管理局管辖下的外围区域。森兰说服了塞纳河畔伊夫里地区当选的议员，将这一大规模的城市改变作为创造一种创新的、艺术的动态社会的契机，通过构想 HQAC(Haute Qualité Artistique et Culturelle：高超的艺术和文化质量），创造一种新的城市特征。HQAC 是一个品牌，使得艺术家能够成功地融入到与城市规划相关的开发商所签署的新建筑和新建绿地区域的建设合同中。这一品牌带来的是在空间决策的过程中，与那些经过讨论的和已被建立起来的在外围地区组织的项目活动进行对应，一个临时性的建筑模块由艺术家创造出来，其中涉及到艺术家、研究人员、城市规划师、拆除公司和居民。在这种方式下，艺术家能够创造一个以社区为主导的模式，以重新定义城市空间。

另一个可以解释艺术家、策展人、中介人与政府建立较好关系的案例，我想是以艺术方式聚集艺术性思考的"Identità al centro"（市中心身份）项目。该项目于 2010 年在意大利蒙特瓦尔基市开展。在这个案例中，艺术作品所关注的是一个复杂的存在于地方政治家和不同社会文化团体之间的关系策略。其目的在于创造一种不可预知的居民和官员之间的交互。其背景是关于个人安全问题上的社会分裂和冲突，起始于城市的历史中心与其周边区域的分离。艺术家和市民制作了三份集体地图，并呈现在主广场的路面上。这也是地方政府在重新定义 2011 年的城市战略规划时所参考的部分基础。最后，在这个项目中，艺术和文化成为了地方政策中的重要因素。为此，蒙特瓦尔基市的公共部门获得了

04 《艺术的思考方式》 意大利蒙特瓦尔基 2010 年
市中心的身份项目 参与性公共艺术项目

05 《Atelier TRANS305 》 斯特丹森兰 法国塞纳河畔伊夫里 2007~2012 年
建筑模型 原型和参与性公共艺术项目

3000万欧元的公共基金用于社区的发展。

Could you talk about the relationship between artist, curator, government and public? What's the role they play in the public art?

Giusy Checola: This relation plays a very important role in the public art, according with the "statements" and objectives of all these actors. An example of how this relationship could lead to an important social and cultural role of public art could be one of the projects I recommended for IAPA, which is Stefan Shankland's "Atelier / TRANS305 ", iniciated on 2007. In that case the artist managed to realize the project because he acted in a specific time in which the relationship between artist, government and public was characterized by a social-political conflict, that he managed to transform in an unprecedent possibility. Basically he took advantage from the tensions that at the time in France were going on between the French central government, the local authorities of the banlieu (periphery) of Ivry-sur-Seine, and the banlieu's inhabitants, as consequence of the urbanization plans included in the so called Le Grand Paris (The Greater Paris). The project Greater Paris has been launched by the former president Nicolas Sarkozy and provided the inclusion of the peripherical areas under the Paris-central administration. Shankland convinced the elected members of Ivry-sur-Seine to consider this large-scale urban changes as opportunity to create innovative artistic and social dynamics that would characterize the new city by conceiving the HQAC, Haute Qualité Artistique et Culturelle (High Artistic and Cultural Quality). HQAC is a brand that the artist managed to make integrate in the contracts signed by the urban plan's developers for the construction of the new buildings and the new green areas. The qualities brought by the brand in the space making process corresponded to those discussed and established within the activities of the project organized in the Atelier, a temporary architectural module created by the artist, in which were involved artists, researchers, urban planners, demolishing companies and the inhabitants. Thus the artist created a community-lead model for the redefinition of the urban spaces.

Another example of results of a good relationship between artists, curators, mediators, government and public (participants and audience), I think is the project Identità al centro (Identity at the centre) by art collective Artway of Thinking, produced on 2010 in Montevarchi (Italy). In that case the artwork has been focused more on a complex relational "strategy" between local politicians and different social-cultural groups, aimed at create an unprecedent interaction between the residents and the authorities, in a condition of social divisions and conflicts about individual security, which started with the separation of the historical centre of the city with its periphery. The artists and the citizens produced three collective maps, realized on the big pavement of the main square, on which the local authorities based part of the redefinition of the Strategy Urban Plan 2011. Last but not least, with this project, art and culture became a key factor into local politics' strategies, and for that the Public Administration of Montevarchi City obtained from public funds 30 millions euros from for the development of the community ideas.

您为"国际公共艺术奖"推荐了许多公共艺术项目,哪一个是你最喜欢的?为什么?

茱茜·乔克拉:事实上并没有我最喜欢的项目,我推荐它们是因为我认为它们可以被认为是好的公共艺术项目,每一个都有不同的原因,与它们所处的社会环境相关联。但是我想说,我喜欢的作品是艺术家能够在概念上促进交互,在审美纬度上体现态度转变的作品。

You have recommended some public art projects to IAPA? Which one is your favorite? Why

Giusy Checola: Actually there is no a favorite one, since I recommended them because I thought that they could be considered good public art projects, each one for different reasons in relation with the contexts in which they took place. But I can say that I like the ones by artists who are able to make interact the conceptual quality and the aesthetic dimension with the transforming attitude.

我听说您近来在南部意大利建立了一个公共学校。你能否谈谈这所学校以及您为何决定这么做?

茱茜·乔克拉:这是我同艺术家以及意大利马特拉 SoutHeritage 基金会的艺术指导安吉洛·比安科(Angelo Bianco)共同建立的。我负责 Maverick 校区,一系列的实验室和网上研究、培训和艺术跨界计划平台,将在三月份于马特拉拉开帷幕。这将涉及不同的村庄和巴西利卡塔区域(意大利南部)的城市——乡村空间。这是一个试点项目,目的在于在接下来的几年里建立一些现如今正在消逝的文化基础设施。该项目与意大利以及国外的艺术机构和大学合作。

该项目被巴西利卡塔区域(归属于城市视觉项目)与欧洲社区(欧洲领土视觉的 FESR 计划)推广,同时还归属于欧洲文化资本 2019 年马特拉计划中。第一年,2015—2016 年,我们与地方和国际上的创意中心、大学以及艺术学院建立了合作关系,由他们向我们推荐实验室参与者的可选名单(主要是博士研究生和艺术家)。Maverick 校区的第一个国际合作试点项目的合作伙伴是位于卢森堡的 Mudan 现代艺术博物馆 (Musée d'Art Moderne Grand-Duc Jean, Luxembourg) 和位于巴黎的"容量———你看到的就是你听到的"基金会（VOLUME—What you see is what you hear）。在城市视觉项目中，区域公共管理机构试图振兴地方社区的五栋被废弃的建筑，这些建筑在此之前一直被认为是对公共基金的浪费。在这一项目中，它们也会成为这一项目中推进创新活动，激发创造性能量的中心。

Maverick校区的想法源自于艺术研究和产出缺乏空间，以及艺术家和研究者在区域和邻里关系中缺少机会。这一点在今天依然以大量的教育和职业移民为特征。一方面，这个项目可以成为具体的建议之一，通过建立本国的和全球的艺术产品，研究和组织之间的联系，以打破该区域在物理上和文化上的隔阂。另一方面，该项目是对与日俱增的计划的一种回应。为其提供空间、实验室、不同的方法和机会，以获得一种体验和理论分析。该项目是近期来公共基金对教育和研究机构削减开支的结果。

I heard currently you are setting up in a public school of Southern Italy. Could you talk about it and why you decide to do it?

Giusy Checola: Together with Angelo Bianco, artist and artistic director of the SoutHeritage Foundation (Matera, Italy) I will direct the Maverick Campus, a series of laboratories and on line platform for the research, training and artistic interdisciplinary planning, that will opened in March in Matera and will involve different villages and urban-rural spaces of Basilicata Region (Southern Italy). It's a pilot project that aims to be established in the following years as cultural infrastructure that currently is missing, in partnership with art institutions and universities in Italy and abroad.

The project is promoted by Basilicata Region (within a project called Urban Visions) and European Community (FESR programs for European territorial cooperation), within the program of Matera 2019, European Capital of Culture. For the first year 2015—2016 we have established local and international partnerships with Centres for Creativity, Universities and Academies of Fine Arts based in the region, were the participants at the laboratories will be selected (mainly PhD candidates and artists); with Mudam-Museum of Modern Art (Musée d'Art Moderne Grand-Duc Jean, Luxembourg) and VOLUME Foundation (VOLUME-What you see is what you hear"Paris). With the project Urban Visions the regional public administration, aimed at revitalize and bring back to the local communities five buildings which it has restored, that before this intervention were perceived as symbols of wasting of public funds, with this project they are going to become propulsive centers for innovative activities and creative energies.

The Maverick Campus's idea was born because of the lack of spaces that could bring together artistic research and production, and lack of opportunities for artists and researchers in the region and neighbour ones, that still nowadays are characterized by massive educational and professional emigration. On the one hand, this project would be one of the concrete proposal concerning the breaking of the physical and cultural isolation of the region, by connecting it to national and global art production, research and organizations; on the other hand, the project would be a reply to the request for the increasing of programs that will offer spaces, laboratories, different methodologies and opportunities for making experience and thematic analysis, as consequence of the current cuts of public funds in educational field, that will trend to reduce the educational and research institutes.

您目前对公共艺术领域的研究兴趣点在哪里？对于未来您有什么计划？

茱茜·乔克拉：我现在是巴黎八大的博士研究生，目前我博士的研究课题是将公共艺术实践与文化地理和司法因素相结合，其目的在于创造一种不同区域空间的共识。从地方的角度来看，我打算研究是怎样的一种关系制造了社会想象的意义及其变化的可能性。从全球的角度来看，我试图思考地域自身的文化产出模式和形式是如何建立的。关于未来，我将遵循我研究公共艺术的轨迹，专业的机会将会向我打开。

What's your current research interest in the public art area? What's your future plan?

Giusy Checola: Currently I'm PhD candidate at University Paris 8 Vincennes Saint-Denis. At the moment for my PhD I'm working on the ways in which public art practices, geocultural and juridical factors are combined in order to produce the common sense within different territorialized spaces. In local terms I'm going to study how this relationship produces social imaginary meanings and the possibility of their alterity; in global terms I'm going to consider how it establishes forms and models of knowledge production of the place itself. About the future, I'll follow the routes that my research on public art and the professional opportunities will open.

案例研究
CASE STUDY

大洋洲

南亚 / 中西亚

中南美洲

欧洲

东亚 / 东南亚

北美

非洲

基于子区域的公共艺术研究分区方案

作者：冯莉（上海美术学院在读博士） 2014年11月

2014年11月，第二届国际公共艺术研究员会议讨论了针对国际公共艺术奖而做的区域划分方案。该方案本着清晰、可视化表达的设计目标，结合前两届公共艺术奖案例分区中发现的一些问题，就自然地理学、政治地理学、文化地理学以及人口地理学角度四大因素做了深入研究，最终得出七大区域划分方案地图及区域国家目录。

一、存在的问题

2013年第一届国际公共艺术奖提名作品（案例）共142件，按照欧洲地区、亚洲地区、北美洲地区、南美洲地区、大洋洲地区以及非洲地区分为6类，评审委员从每个地区案例中选出一个作品作为地区获奖作品。但6个地区案例分布不均衡，欧洲（42个案例）、亚洲（37个案例）及北美洲地区（25个）案例明显多于大洋洲（13个）、非洲（10个）等地区。

2014年第二届国际公共艺术奖提名作品（案例）共计125件，考虑到地区特点共分为7大类，分别是：非洲、东亚、欧亚大陆、中南美洲，大洋洲及东南亚、北美洲、中东及中亚。这一次案例在7个区域中分布比较均衡，但区域名称以及具体国家的归属并无明确规定。比如：区域名称中使用了中东和中亚的称呼，但具体包括哪些国家并未清晰说明。

此外，在2014年第二届国际公共艺术奖评审会议中，有评委将大洋洲的案例和东南亚的案例放在一起评选提出异议，认为这些案例属于完全不同的两种文化体系（文化圈）。类似的情况也出现在墨西哥与北美地区的案例中。

案例分布不均。区域内国家无明确统一规定。文化体系的差异。区域名称的不规范，这些实际操作中暴露出的问题，促使我开始做进一步的研究并尝试做出界定。

二、针对问题的研究

1. 确定区域与国家的对应关系

寻找一种国际公认的区域划分方案，使得国家与区域的对应关系清晰明确，保证所有的国家都被包括在区域划分中。联合国统计署（UNSD）

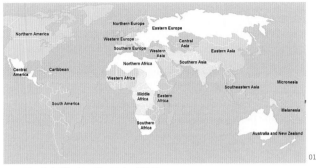

01 联合国统计署地理子区域（22个）划分图

提供的区域（Region）—子区域（Sub-Region）—国家列表（Country list)提供了比较权威的界定[①]。它的优势在于：

（1）方便查询：它将六大洲（除南极洲外）以国家为最小单位分为22个子区域（Sub-region），并提供了每个子区域包括的国家列表，并统一编码（M49）。

（2）灵活性：可以根据实际需要适当调整个别国家在子区域中的划分。例如：子区域南亚(Southern Asia)中包括了伊朗、印度等9个国家，可以根据需要将伊朗从南亚划分到西亚，以保证未来区域划分的需求。

34	Southern Asia
4	Afghanistan
50	Bangladesh
64	Bhutan
356	India
364	Iran （Islamic Republic of）
462	Maldives
524	Nepal
586	Pakistan
144	Sri Lanka

02 子区域南亚包括的9个国家

2. 文化圈分类

文化圈是文化人类学的概念，分类方法众多，较为常见的有6种和11种分类方法。结合公共艺术案例的特点，确定出国际公共艺术协会的9大文化圈分类法，即：

（1）非洲文化区：包括东、西、南、中部非洲子区域中的国家。

（2）伊斯兰文化区：北非、中亚、西亚子区域中的国家。

（3）东亚文化区：东亚子区域中的国家。

（4）东南亚文化区：东南亚子区域中国家。

（5）西方工业化文化区：北美、澳大利亚、新西兰、西欧、南欧、北欧子区域中国家。

（6）大洋洲文化区：大洋洲三大群岛中的国家。

（7）拉美文化区：中美、加勒比地区、南美子区域中国家。

（8）斯拉夫文化区：东欧子区域中国家。

（9）南亚文化区：南亚子区域中国家。

3. 世界人口分布

根据 2008 年的统计数据，为方便统计合并若干子区域，最终得出区域人口分布统计如下图：

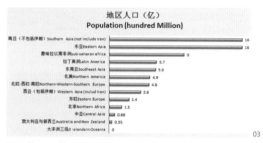

03 区域、子区域人口分布情况（2008 年）

4. 前两届 267 个案例的区域分布研究

对 2013 年、2014 年国际公共艺术奖共 267 个案例，做子区域分布统计（为方便统计合并若干子区域，如撒哈拉以南非洲即由子区域西非、东非、中非、南非合并而成）如下图：

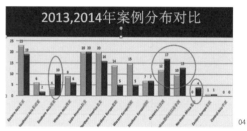

04 2013 年、2014 年案例子区域数目分布

图中红圈所示：南亚、大洋洲、北非的案例有明显增加，而北欧、西欧、北美、东南亚、东亚的案例有所减少。这一方面显示了该地区公共艺术的活跃程度，另一方面也可能说明对该地区案例研究有待增强，从一个侧面反映了公共艺术协会研究员案例研究的覆盖情况。如中亚案例为 0，是否需要增设对该地区案例的研究员？

三、结论

综合考虑以上因素，得出分区方案的划分策略为：地理区域为主，文化区域为辅，关注案例增加区域。最终划定以子区域为基础的国际公共艺术七大区域：

（1）Africa 非洲。

（2）East and Southeast Asia 东亚和东南亚。

（3）Oceania 大洋洲。

（4）Latin America 拉丁美洲。

（5）Europe 欧洲。

（6）Northern America 北美。

（7）South Asia and Central and West Asia 南亚和中西亚。

05 七大区域划分方案

06 七大区域划分方案地图

---

① 联合国统计署地理子区域划分：UNSD Geographical sub-regions
The list of countries or areas contains the names of countries or areas in alphabetical order, their three-digit numerical codes used for statistical processing purposes by the Statistics Division of the United Nations Secretariat, and their three-digit alphabetical codes assigned by the International Organization for Standardization (ISO).1 In general, this list of countries or areas includes those countries or areas for which statistical data are compiled by the Statistics Division of the United Nations Secretariat.
https://unstats.un.org/unsd/methodology/m49/

案例研究
大洋洲

CASE STUDY
Oceania

Pigs in the Yard	院子里的猪
Landshaft	Landshaft
Untitled	无题
Evening Echo	夜晚的回声
Whole House Reuse	整个房屋重复利用
Hei Korowai Mo Ruatapuwahine	Hei Korowai Mo Ruatapuwahine
Mo'ui Tukuhausia	无家可归的生活
Gap Filler	空隙填补者
Port Workers	港口工人
Luxcity	灯光城
If You were to Work Here	如果你将在这里工作
Level Playing Field	公平的竞技场

院子里的猪
Pigs In The Yard

艺术家：凯里索拉伊特·犹赫拉
地点：新西兰奥克兰
形式和材料：行为艺术
时间：2011 年
推荐人：布鲁斯·菲利普斯

Artist：Kalisolaite 'Uhila
Location：Auckland，New Zealand
Media/Type：Behavior Art
Date：2011
Researcher：Bruce Phillips

凯里索拉伊特·犹赫拉的《院子里的猪》是发生在公共场地内的两项独立的行为表演艺术。第一个在南奥克兰的曼格里艺术中心的庭院里，第二个在奥克兰市中心的奥特安城市广场。在第一个作品里，犹赫拉通过允许一群猪自由跑，让艺术家和观众被限于在围栏后面 8 小时，颠倒了人类和动物的关系。在作品以后的发展中，他和一只小猪一起分享了一个集装箱船长达一周。在这两种情况下，"犹赫拉被迫建立与猪的关系，因为在公众眼中它们除了吃就是睡在一起"。犹赫拉总结了所有杀死或者吃掉其同伴猪的行动，不但描述了他对动物发自内心的尊重，也表达了他的要求维持生命的需求。

就像约瑟夫·博伊斯 1974 年的作品《我喜欢美国，美国喜欢我》一样，犹赫拉的《院子里的猪》试图从字面上并象征性地度过了殖民造成的伤口。然而，与博伊斯相反的是：犹赫拉的作品通过并置波利尼西亚和欧洲社会结构之间的不协调，使得这些关注点达到了另一层的复杂性。在许多波利尼西亚社会中，猪被保留作为神兽象征财富、声望，作为礼仪食品，由于这种地位，他们可以自由漫游地球。在直接的对比中，大多数西方社会认为猪为不洁的象征，通常养殖在不利条件下。为了使这个波利尼西亚和西方的分歧更加突出，犹赫拉命名猪为"殖民者"。这种命名混淆了两种不同文化的记号，生动地还原了跨太平洋岛国殖民的历史——这种分歧在现在的社会功能紊乱的各个方面还明显存在。

除了博伊斯，犹赫拉还引用了乔治·奥威尔的小说"动物庄园"中的名叫拿破仑的猪作为直接影响。通过引用"动物庄园"的政治讽刺，犹赫拉增加了人类创造社会结构的强制性以及革命性的剧变，还有空间占有的频繁性的进一步思考。在奥威尔的小说中，猪征服了人类，在犹赫拉的作品中，是开拓殖民的人类消除了殖民者的猪。这个转折点意味着殖民后国家的挑战与不安，如新西兰以适应和处理过去的创伤的现实意义——更不用提对土著人的那不平等的规范。

Pigs in the Yard (2011) by Kalisolaite 'Uhila is a long duration performance work that has been performed in two separate iterations taking place in public plazas. The first in the courtyard of the Mangere Art Centre, in South Auckland, and the second in Aotea square, downtown Auckland. In the first iteration 'Uhila reversed the relationship of humans and animals by allowing a group of pigs to run free while the artist and the audience were confined behind fences over an 8 hour period. In the later development of the work he shared a shipping container with a single piglet for an entire week. In both instances, 'Uhila was forced to establish a relationship with the pigs as they ate and slept together in full public view. Despite this close relationship 'Uhila concluded each performance by ultimately killing and eating his pig companions and thereby demonstrating both his heartfelt respect for the animals but also his requirement to sustain life.

Much like Joseph Beuys' seminal 1974 work I Like America, America Likes Me, 'Uhila's Pigs in the Yard attempted to literally and symbolically live through the wounds created by colonisation. However, in contrast to Beuys', 'Uhila's work takes these concerns to another level of complexity by juxtaposing key incongruities between Polynesian and European societal structures. Within many Polynesian societies, pigs are reserved as a sacred animal linked to wealth, prestige, used as a ceremonial food and due to this status they are allowed to roam the earth freely. In direct contrast, most Western societies consider pigs as a symbol for uncleanliness and are often farmed in adverse conditions. To complicate this Polynesian vs Western dichotomy even further 'Uhila named the pig 'Colonist'. This naming muddles two different cultural signifiers and bluntly addresses the history of colonisation across the Pacific Island Nations–the ramifications of which are still very present today in all manner of societal dysfunctions.

In addition to Beuys', 'Uhila also cites the pig Napoleon in George Orwell's novel Animal Farm as a direct influence. By referencing the political satire of Animal Farm, 'Uhila adds a further reflection on the compulsion of humans to create social structures and how this often involves revolutionary upheaval and the occupation of space. While it is the pig that conquers the humans in Orwell's fiction, in 'Uhila's performance it is the colonised human that laterally dispenses with Colonist the pig. This turning of the table suggests both challenge and disquiet for post-colonial nations such as New Zealand to accommodate and deal with very real implications of past trauma– not to mention the ongoing hegemonic paradigms that perpetuate inequalities for indigenous citizens.

[解读]

"殖民者"的猪

犹赫拉的《院子里的猪》曾经在两个公共场所公开表演，在第一个场所，赫拉通过允许一群猪自由跑，而艺术家和观众被限于在围栏后面8小时，颠倒了人类和动物的关系。在第二个场所，他进行了为期一周的——和一只小猪一起分享了一个集装箱船。在这两种情况下，犹赫拉被迫建立与猪的关系，因为在公众眼中它们除了吃就是睡在一起。在许多波利尼西亚社会中，猪被保留作为神兽象征财富、声望，作为礼仪食品。而大多数西方社会认为猪为不洁的象征，通常将猪养殖在不利的条件下。以此来突出波利尼西亚和西方的分歧，他将猪命名为"殖民者"，并引用乔治·奥威尔小说"动物庄园"的政治讽刺，在奥威尔的小说中，猪征服了人类，在犹赫拉的作品中，是开拓殖民的人类消除了殖民者的猪，进一步意味着殖民后国家的不安。（祁雪峰）

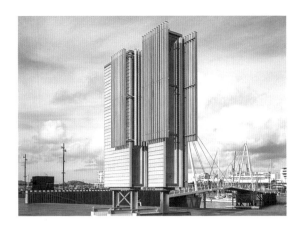

Landshaft

艺术家：德里克·彻里	Artist：Derrick Cherrie
地点：新西兰奥克兰	Location：Auckland，New Zealand
形式和材料：雕塑	Media/Type：Sculpture
时间：2012年	Date：2012
推荐人：布鲁斯·菲利普斯	Researcher：Bruce Phillips

艺术家德里克彻里的作品《Landshaft》是一个重约7吨、2层楼高的建筑雕塑，作品作为一个永久建筑物，安装在奥克兰的两个不同地点。每处安装约花费3个月，整个工程花费了两年时间。艺术家将雕塑设计成拼装结构，使之可以重新安放在任何新的空间环境中。Landshaft被安装在两个截然不同的地方：Ti Rakau公园——一个半商业和住宅郊区，以及位于奥克兰最近开发的市中心海滨——Karanga广场。

雕塑采用传统建筑结构的形式来呈现材料的品质，但以一种指示不确定空间和身体关系的方式展示。插入其内部的成对香烟，仿佛该结构本身正在试图通过自身的皮肤吸收物质。当观众从周围游走观看时，雕塑外形并不像一个确定的固体建筑，其狭窄的空间亦不适合人居住。但是整个雕塑构成足以吸引路人的好奇，同时将异常元素带入到了周围的建筑环境。《Landshaft》企图通过构建模糊的建筑形式，破坏建筑环境材料和建筑环境体验的一致性，整个作品为了唤起人们对其中蕴含的复杂社会政治与经济意识形态的质疑。

这一种方式恰好切中奥克兰城市环境的要害。在八个地区议会合并成一个市政府之后，奥克兰启动了一个内容充实的永久计划，该计划预计将在20年内使人口增加一倍，同时加大中央商务区，建造各大公路和公共交通基础设施，以此解决无家可归者及受首次购房者无法进入市场的问题。然而，住房短缺和房地产泡沫，又膨胀了奥克兰市中心的物价。

彻里的《Landshaft》很偶然地使用了不适宜居住的游牧式结构，它表明了建筑和城市变革的思想。围绕Ti Rakau公园和Karanga广场周围地区被指定为未来的高层住宅及商业发展区，以面对未来城市扩建的需要。在这个意义上，作品包含了显著的城市巨变的概念和后果，超越了物理城市而存在。

除了这种和奥克兰的特定关系，《Landshaft》还参照了建筑和雕塑的历史，探索身体与建筑环境之间的关系。《Landshaft》分享了诸多艺术家，如布鲁斯·瑙曼、理查德·塞拉、丹·格雷厄姆、安德烈泽塔尔和艾特利尔·范·李肖特创造的准建筑形式。特别是瑙曼的许多走廊作品与此很相似。《Landshaft》仿佛瑙曼的走廊通道一样通过限制身体动作，在心理上操纵我们的建筑空间，进而被认识存在。然而，与瑙曼的作品相反，彻里的走廊并不面向公众开放，它是一个无人居住的密封大厦。因此，该作品的经验是主观的投射，而不是现象。

从架构上，这个作品参考了北欧现代主义者阿尔瓦奥拓的设计。事实上，雕塑采用国际现代主义，从地板到天花板的玻璃和桦木衬里的内部，通过精密的直线形式形成犹如陶瓷散热片的外观。然而，该作品作为一个生活功能体却并不适合居住，作者借此来抵抗现代主义建筑的乌托邦意识形态。

所有这些因素综合在一起，使得《Landshaft》变得具有一种不可思议的性格并悄悄地嘲弄着城市。它是一个物理存在，符合二元论形式的物理存在却又抗拒定义，希望被使用又拒绝被进入，活在当下却又不存在。

Landshaft (2012) by artist Derrick Cherrie is a large two story and 7 tonne structure that was installed in two different sites over a two year period spending approximately 3 months in each location within Auckland, New Zealand. While the work could exist as a permanently installed piece it was designed by the artist as a kitset construction allowing it to be temporarily re-situated in new spatial contexts. Landshaft was installed in two vastly different locations Ti Rakau Park, a semi-commercial and residential suburb, and Karanga Plaza in Auckland's recently developed downtown waterfront.

The sculpture draws on the material qualities and forms of conventional architecture but in a way that directs a range of uncertain spatial and bodily relationships. Inside, are pairs of cigarettes perplexingly inserted into the interior cladding as if the structure itself is attempting to imbibe substances from within its own skin. As the viewer moves around the sculpture its form appears more indefinite as a solid building. Too narrow to allow comfortable human habitation, the sculpture is constructed to entice passerby curiosity but also to inject an element of abnormality into the surrounding built environment. By constructing a hybridised and ambiguous architectural form Cherrie attempts to disrupt the material and experiential coherence of the built environment so that the complex socio-political and economic ideologies of their making might be questioned.

This approach was particularly poignant at this point in time for Auckland's urban enviroment. Following the rapid merger of eight regional councils into one city government, Auckland was embarking on a period of substantial and irrevocable change with the population expected to double within 20 years, plans to intensify the CBD, and with major roading and public transport infrastructure to be built. These changes were influenced by major issues of homelessness and the inability of first home buyers to enter the market. These problems have been attributed by many to a housing shortage and real-estate bubble that has inflated the value of properties in central Auckland.

Cherrie's Landshaft incidentally engaged with this given context by being an uninhabitable nomadic structure that suggests the ideology of architectural and urban change but simultaneously resists this possibility. This reference was all the more emphasised by the chosen locations that the work came to be installed. The areas that surround Ti Rakau Park and Karanga Plaza are earmarked for future high-rise residential and commercial developments to help future proof the cities need of expansion. In this sense the work engaged with notions of significant urban upheaval and the ramifications that such developments inevitably has upon the bodily presence within cities.

Aside from this specific relationship to Auckland, Landshaft also references significant movements in architectural and sculptural history that have explored the relationship between the body and built environment. Landshaft shares the quasi-architectural forms created by artists such as Bruce Nauman, Richard Serra, Dan Graham, Andrea Zittel and Atelier Van Lieshout. In particular, Nauman's many corridor works share a strong similarity to this work. Landshaft, like Nauman's corridors are passages that limit bodily movement and as such psychologically manipulate our preconceptions of architectural space and in turn the awareness of being. However, in contrast to Nauman's work Cherrie's corridor is not accessible to the public to enter rather it is an uninhabited hermetically sealed edifice. Therefore, the experience of the sculpture is one of subjective projection rather than the phenomenological.

Architecturally the work references the design of Scandinavian modernists such as Alvar Alto. Indeed, the sculpture shares the elegance of warm domestic modernism through the combination of a sophisticated rectilinear form with a cedar fin façade, floor to ceiling glass and a birch lined interior. Yet since it is uninhabitable the work essentially resists the utopian ideology of modernist architecture to be a functional machine for living.

All these factors combined makes Landshaft a suspect character which quietly taunts the city. It is a physical presence that exists in simultaneous dualisms by conforming to regulations but also resisting definition, by inviting use but also denying entry, and by being present but also absent.

[解读]

叛逆者

鳞次栉比的建筑充斥着整个城市,但是由于各式各样的原因,诸多建筑或搁置,或废弃。原本为了宜居的建筑物,却与人们形成了深深的隔阂,而在当下,此类建筑更是犹如流水线生产般接踵而至,彻里的《Landshaft》以建筑的外形、狭窄的空间,嘲弄现代建筑的乌托邦意识,其中更是蕴含着建筑与城市的变革,然而这种变革却割裂了空间与身体的关系,原本应该居住的生活功能体却拒绝着人们的进入,使原本既定的概念充满矛盾。《Landshaft》不断拓展着建筑定义的边界,同时却又破坏着既定的法则,预示着城市的变革却又嘲弄着城市的存在。艺术批评家、策展人罗伯特·斯托在描述瑙曼的影响时这样说:"如果瑙曼的作品不能让公众迷惑,公众也就无法真正理解他们",彻里的《Landshaft》亦是这样一个充满叛逆和迷惑的矛盾体。(闫丽祥)

无题
Untitled

艺术家：卢克·威尔斯·汤普森
地点：新西兰奥克兰
形式和材料：装置艺术
时间：2012 年
推荐人：布鲁斯·菲利普斯

Artist：Luke Willis Thompson
Location：Auckland，New Zealand
Media/Type：Installation
Date：2012
Researcher：Bruce Phillips

卢克·威尔斯·汤普森的《无题》是一个以消除一宗谋杀案的公共符号，而成为新西兰奥克兰种族刺激辩论温床的大胆的艺术行为。从而需要与业主进行困难的交流，获得与皮海玛·卡梅隆死亡有关的车库门，车库门属于布鲁斯·埃默里，埃默里是一位经常去教堂且以家庭为中心的白人。像许多谋杀案一样，这起案件涉嫌种族歧视。作家香农特敖简洁编写这个有争议的故事，揭穿了埃默里的后续轻度刑罚的结果：

在 2008 年 1 月 26 日凌晨，皮海玛·卡梅隆被杀于一场对抗中，当时他和他的侄子发现在标记南奥克兰居民布鲁斯埃默里的车库门前……据报道，埃默里最初逮捕他们，他持刀追赶他们大约 300 米，埃默里随后被判谋杀青年，对不合理的判决，埃默里脸上露出愤怒。用刀杀人的凶杀案，一般判决通常是五年半到六年，但埃默里只服 11 个月的刑期后获释。

卢克·威利斯汤普森与业主交流，想获得这些车库门同时保留和删除公祭的网站。汤普森还通过与房主谈判、快速装修，使建筑和街道看起来

137

和之前不再一样,从而消除该事件的所有视觉关联。这种消除掩蔽行为也作为艺术家的一部分颠覆思维来创作,从毁灭中保存生命最后剩余的痕迹。

不过,保存的这种行为是复杂的,因为卡梅隆被刺伤后,埃默里随即清理掉涂鸦。由于这个原因,埃默里的清洁痕迹在车库门左侧的喷漆轮廓的摩擦处很明显。因此,汤普森收集的这个门也就保留了被害人的痕迹,正是那双杀人的手。杀手的标志和生活中实物的标志是息息相关的;创伤的时间标记被同时擦除也被从现场移走而保留;得失的复杂性被注定分离为两方面。这种心理上的难题后来是由许多相关的新闻标题和报告进一步补充的。涂鸦标记的车库门成为了很多新闻报道突出的点,再一次说明卡梅隆被认定为"恶搞"而不是暴力行为的悲惨受害者,这是荒谬至极的。

在随后的 2013 年,车库门在奥克兰美术馆(新西兰最著名的公共艺术博物馆)收藏。因此,汤普森试图通过它转移到博物馆,使得这扇门获得永久的公共纪念,达到纪念和安抚这条街的创伤的目的——这种永久的沉思比有偏见的头条新闻更引人注目。

Untitled (2012) by Luke Willis Thompson is a work that consisted of a bold action to remove the public symbol of a murder that became a hotbed of racially fuelled debate in Auckland, New Zealand. This removal required a logistically difficult exchange with a property owner to obtain the garage doors associated with the death of Pihema Cameron, a young Maori man, at the hands of Bruce Emery a white, church going, family orientated home owner. Like many murders in which there is suspected influence of racial bias, the reported motivations of the incident were skewed via the popular news media outlets. Writer Shannon Te Ao succinctly frames up this controversial story and Emery's subsequent light sentencing:

"[in the early hours of] 26 January 2008, Pihema Cameron was killed during a confrontation, which occurred when he and his cousin were found tagging the garage doors of the property of South Auckland resident Bruce Emery ... It was reported that Emery initially apprehended the pair, he chased them approximately 300 metres down the road armed with a knife. Emery was later convicted of manslaughter over the death of the teenager. More anger surfaced in response to the apparent disproportionate sentencing of Emery after his conviction. Emery was sentenced to four years and three months imprisonment for manslaughter. The starting point for sentencing of a homicide with a knife is usually five and a half to six years but ... Emery was released after serving only 11 months of his sentence."

Luke Willis Thompson's exchange with the property owner to obtain these garage doors simultaneously preserves and erases a site of public memorial. Thompson also negotiated with the home owner to fast track building renovations so that the building would no longer look the same from the street and thereby erase all visual association to the incident. This act of erasure is also intended as an act of subversion on part of the artist to save, from inevitable destruction, the last remaining trace of a life.

However, this act of preservation is complicated because after stabbing Cameron, Emery then proceeded to clean off the graffiti. Due to this, Emery's cleaning marks are also evident on the garage doors through the abrasions left in the outline of the spray paint. Therefore, Thompson's act of collecting the doors preserves the trace of the victim but also of the hand that killed. The mark of the killer is inextricably bound in the mark of the life lost; the time marker of the trauma is simultaneously erased and conserved as it is removed from the site; the complication between what is being saved and lost is bound in the impossibility of separating the two. This psychological conundrum is further added to by the many news headlines and reports that have surrounded the incident. The graffiti-marked garage doors became the salient point for many news reports, through which Cameron was identified as the "tagger" rather than the tragic victim of a violent act.

Subsequently in 2013, the garage doors were purchased and accessioned into the collection of the Auckland Art Gallery (New Zealand's most prominent public art museum). Therefore, by attaining the doors for permanent public remembrance Thompson attempted to heal the trauma and remembrance from the street by shifting it to the context of the to the context of the museum—where slow thinking and contemplation are prioritised over attention-grabbing and biased headlines.

[解读]

有创伤的车库门

卢克·威尔斯汤普森的这组名为《无题》的装置作品表现了一个有关种族歧视的敏感性题材。创作背景是源于一个名叫皮海玛卡梅隆的叔侄俩和南奥克兰居民布鲁斯埃默里打杀的事件，埃默里最后被判谋杀青年，然而通常用刀杀人的凶杀案判决是五年半到六年，但埃默里只服 11 个月的刑期后获释。在所有人看来这是极其不公平的，艺术家卢克·威利斯汤普森找到业主并与之交流，来获得这些车库门作为保留收藏。在行凶后，埃默里对门进行涂鸦清理，所以在车库门左侧的喷漆轮廓摩擦处很明显，汤普森收集的门保留了被害人的痕迹。该作品已被购买和登记在奥克兰美术馆，通过转移它到博物馆获得这扇门以永久的公共纪念，让人们时刻记得埃默里的罪恶是永远洗不清的，以此来纪念受害者的死亡，有力地抨击了种族歧视这个社会问题。（祁雪峰）

整个房屋重复利用
Whole House Reuse

艺术家：新西兰多位艺术家
地点：新西兰克赖斯特彻奇
形式和材料：空间艺术
时间：2013—2015 年
推荐人：凯利·卡迈克尔

Artist：New Zealand Artists
Location：Christchurch, New Zealand
Media/Type：Space Art
Date：2013—2015
Researcher：Kelly Carmichael

在 2011 年发生的一系列毁灭性的地震后，坎塔布连人发现他们以前居住的城市已经不复存在了。在那里缺少了建筑物、地标和心爱的人。然而自从新西兰南部岛屿的城市克赖斯特彻奇地震以来，发生了一个创造性的觉醒。许多小的创新住所、人们对住所关系的疑问，以及令人兴奋的计划好像一夜之间突然出现了。这个城市特有的构造感受到了积极的变迁状态。《整个房屋重复利用》计划就是这样一个初步行动。

地震过后，克赖斯特彻奇的一个 10 000 多所房屋的"红色区域"被确定为被拆迁区域。如此大数量的房屋拆迁任务不仅对城市中已损坏的基础设施来说很困难，而且对那些庇护所、往日的记忆以及自己的家在突然间消失的人们来说也非常困难。在新布莱顿郊区的 19 号海军上将路就是一个这样的家，而且变成了一个为了流通社区，为来自整个国家的艺术家和手艺人的社会和环境初步行动提供场所。在 2013 年 8 月和 9 月中共 7 天的时间，一个职业救助船员拆除了全部单层红色区域的家，与一个志愿者团队一起，除了混凝土地基房屋，剩余的全部房屋材料被用手拆除、打捞、用文件证明和储存，填满了 6 个汽车修理厂，产生了一个 65 页长的目录，号召新西兰的艺术家和有创意的实践者，让他们把残骸转变为有目的并且漂亮的作品。这些作品的展览计划于 2015 年在坎特伯雷博物馆举行。

事实上，《整个房屋重复利用》计划突出了拆除产生的废物，并且鼓励了围绕着解决和创新材料未来使用的创造性难题。作为一个社会的初步行动，这个项目为一个灾难后环境中紧需的社会和文化需求提供了一个安慰。在震后的克赖斯特彻奇，这个项目是一个把资源从受创伤的地点移开，并且将它们作为一个复兴创作动力提供出来的社会事业。它要求我们重新定义这些场所以及我们和它们的关系，整项计划不是来自破坏而是由新创意所定义的复员的住宅空间。

《整个房屋重复利用》计划提供了一个有创意并且有疗效的出路——给予材料第二次生命，为了从战争中创造有意义且美丽的事情，为了给陷于混乱的城市和人口提供一个比喻意义上的春天提供了机会。当艺术家利用可循环材料，将工作的缺陷和瑕疵变成了材料的声音，这样的不完美证明了场所，促进了对材料怎样把对一个场所的感觉带入到一个新形式中的理解。震前的生活、家庭和土地使用期限的回音开启了所创作的作品和事物。《整个房屋重复利用》项目开创了材料、制造者以及新物品所有者之间的对话。

富有创意的实践者把拆除的废料通过《整个房屋重复利用》计划变成了艺术品、家具、珠宝和其他物品。这些新作品不单单是它们所采取的形式，而且也表明了它们之前的生活——19 号海军上将路的故事和社会历史，被层层的字画、墙纸以及在这个家庭里所创造的回忆密封的时间。这些物品纪念了无穷尽的家庭以及流失的土地。它们代表历史、遗产，更为重要的是，作为希望和复兴的物品。

After a series of devastating earthquakes in 2011, many Cantabrians found that the city they lived in was no longer there. In its place were absences—of buildings, landmarks and loved ones. However since the earthquakes the city of Christchurch, in the South Island of New Zealand, has had a creative reawakening. Many small, innovative organisations and exciting projects that question place and people's relationship to place have cropped up seemingly over night. The very fabric of the city feels transitional and in a state of positive flux. Whole House Reuse is one such initiative.

After the major earthquake, a "Red Zone" of more than 10,000 houses in Christchurch were identified for demolition. The task of demolishing such a large number of buildings was not simply difficult for the city's already damaged infrastructure, but also for the people whose shelters, memories and homes were suddenly gone. 19 Admirals Way in the suburb of New Brighton was one such home, and became the site for a social and environmental initiative that mobilised the community and involved artists and craftspeople from across the country. Over a period of 7 days in August and September 2013 a professional salvage crew fully deconstructed the single storey red-zoned home. With a team of volunteers, the entire material of the home aside from the concrete foundation was dismantled by hand, salvaged, documented and stored; filling six garages. A catalogue of materials—65 pages long—was produced and a call went out to artists and creative practitioners throughout New Zealand, asking them to transform the debris into purposeful, beautiful works. An exhibition of these works is planned for mid 2015 at the Canterbury Museum.

Practically, Whole House Reuse highlighted the waste generated by demolitions and encouraged creative problem solving and innovation around future uses for materials. As a social initiative, however, the project offered a salve for urgent social and cultural needs of a post-disaster environment. In post-earthquake Christchurch, the project was a social enterprise that removed resources from a site of trauma and offered them as an impetus for a creative process of renewal. It asked that we redefine such places and our relationship to them, proposing deactivated residential space as not defined by destruction, but instead by new creation.

Whole House Reuse offers a creative and therapeutic outlet—a second life for materials and the opportunity to create something meaningful and beautiful from disaster, a metaphorical spring for a devastated city and population. The flaws and the blemishes artists must work around when using recycled material became the voice of the material. Such imperfections speak of place and promote an understanding of how materials might carry a sense of place into their new form. An echo of pre-earthquake lives, homes and plots of land lives on in the art works and items created. Whole House Reuse initiates a conversation between material, maker and owner of the new item.

Creative practitioners have turned demolition waste from the Whole House Reuse project into art works, furniture, jewellery and other objects. These new pieces are not simply the form they take, but speak of their former life—the stories and social history of 19 Admirals Way, the time encapsulated by layers of paint and wallpaper and the memories created within the home. The items celebrate the countless homes, family possessions and land lost. They operate as items of history, of heritage and, most importantly, of hope and renewal.

[解读]

废墟上的重生

每一次灾难都伴随着一场毁灭,我们无可避免,但是毁灭之后是沉沦苦海,唉声叹气,惶惶不可终日,还是破茧成蝶,重新积攒力量,进行鱼跃龙门的蜕变,选择权却在我们手中。2011 年一系列毁灭性的地震过后,新西兰克赖斯特彻奇陷入一片废墟,但在不久以后,这座城市发生了一个创造性的觉醒。这就是众多新西兰艺术家联手打造的《整个房屋重复利用》,艺术家们将拆除产生的废物,制作成艺术品、家具、珠宝等物品,这些物品形式上的改变蕴含着希望和重生,它们不光是带着记忆的历史遗产,更是富有新生命的特征。《整个房屋重复利用》将资源从受创伤的地点重新扶起,转变成复兴的动力,为灾难后的社会和文化提供了一个安慰和希望。(闫丽祥)

Hei Korowai Mo Ruatapuwahine

艺术家：乔治·努库
地点：新西兰霍克斯湾，
　　　蕨山，毛利人会堂
形式和材料：雕塑
时间：2014年
推荐人：凯利·卡迈克尔

Artist：George Nuku
Location：Omahu Marae Fernhill,
　　　Hawke's Bay, New Zealand
Media/Type：Sculpture
Date：2014
Researcher：Kelly Carmichael

乔治·努库是一个德高望重的雕刻家，他遵循毛利文化的传统，在石头、骨头、数目和贝壳上雕刻。但是令人惊讶的是，他也在现代的材料，如聚苯乙烯和塑胶（也称为有机玻璃）上雕刻。2014年，努库在新西兰乡下毛利人会堂的毛利棚屋（公用的餐厅和娱乐大厅）中完成了一件著名的雕刻作品。

毛利人会堂是毛利部落的中心，位于毛利乡下。会堂由建筑和开放的空间组成，并且承载了浓重的文化含义。毛利会堂有着重要的宗教和社会职能——传承毛利文化。人们通过使用毛利语言，重述地方的历史，使这里得传统和风俗得以维持并传授给年轻一代。

这里还是重要纪念典礼的场地，比如纪念死者、欢迎客人、部族间会见，或庆祝喜庆。

毛利人会堂天生就是公共、公开、为社区服务的地方。传统毛利会堂的建筑都大量装饰有雕刻、编织和彩绘板，成为保持这些艺术形式的活的博物馆。在当代社会,他们生活的城市中心仍保留有那些乡村的毛利会堂。

乔治·努库最新、最大的项目，是在祖屋 Omahu 会堂的聚苯乙烯天花板上雕刻而成。雕刻在毛利文化中享有很高的美誉，并且努库显示了当代人对于古老艺术形式的态度，展示了 21 世界的新型材料和毛利装饰品的融合。

努库把这看作是传承生机勃勃的、生活的、动态的毛利文化的一部分。他评论道，"传统是创新的持续过程……我想创作出我祖先以前没见过的东西……我不能仅仅只照搬他们创作的……他们也不想我是这样做的。我们需要认识到，每一个素材，每一件事情都有自己的神灵"。

而努库蚀刻的有机玻璃作品，如鬼魅雕塑般出现。聚苯乙烯雕刻品运用强烈的雕刻形式，表达了复杂的政治和文化思想。在 wharekai Ruatapuwahine 的雕花天花板上，意喻着努库部落祖先 Ruatapuwahine 的外衣。

天花板作为一件建筑外衣，被理解为保护和培养她的子女或后代和会堂的客人。许多传统的雕刻图案运用其中，这个作品展现了古代传统图案如何运用在当代作品中。

在社区和当地其他会堂的共同参与建造下，《Hei Korowai Mo Ruatapuwahine》拥有自然世界、传统毛利神话统一的逻辑。顺着 Ruatapuwahine 斗篷下来的细节雕刻，展现了栖息在毛利会堂周围的河流、水路沿岸的生灵，以及部落的重要食物来源。

这些嵌板被认为是表达动物的威信和生命精神，同时作为保护他们环境的呼吁。天花板上雕刻的三角形元素是 ngā tini marae，所有毛利会堂和展览馆都有这样的元素存在，并会一直存在下去。

努库所采用的图案"kaokao"，来自于上身的姿势，在毛利战舞（祖先的舞蹈或宣称实力和威力）中隐含着身强体壮的含义，以及步调一致、共同合作、获得运动的动力。

从最大的嵌板上可以拆下 1000 多个小三角形，它们描述了 Ruatapuwahine 零散分布在世界各地的后代，就像艺术作品中的三角形一样。努库选择在天花板上采用几何设计，以传统的编织板修饰 wharekai 的墙以及螺旋图案（入口上方的门楣上雕刻）。室内悬挂的灯饰雕刻成毛利星座，暗含毛利人对宇宙的理解。浮动的立方体，蕴含着对惯例的挑战，是艺术家藐视重力的媒介。

Sculptor George Nuku is a highly respected carver in the tradition of his Maori culture. He works in stone, bone, wood and shell but also —and strikingly— in contemporary materials such as polystyrene and perspex (also known as Plexiglass). In 2014 Nuku completed a carved work at his marae in rural New Zealand in the wharekai Ruatapuwahine (communal dining and recreation hall). Marae are central to Maori communities, especially in rural areas. They consist of both buildings and open space and carry great cultural meaning. Marae serve important religious and social purpose—a place where Maori culture can be celebrated, the language spoken, histories retold and where customs can be upheld and taught to a younger generation. They are the site of important ceremonies and of the occasions that mark our lives—such as farewelling the dead and supporting the grieving, welcoming visitors and meeting intertribal obligations, or celebrating happy occasions. Marae are by nature public and communal places, serving the community. Traditionally, marae buildings are richly adorned in with carving, weaving and painted panels and remain living museums of these art forms. In contemporary society their centrality to life remains, some existing as urban marae in the heart of city centres.

In George Nuku's biggest project to date, a carved ceiling at his ancestral home Omahu Marae has been created in polystyrene. Carving is highly prized in Maori culture and here Nuku presents a contemporary face to an ancient art form, offering a 21st century reinterpretation of materials and marae decoration. Nuku sees this as part of his role as an artist in the continuum of a vibrant, living and dynamic Maori culture, commenting "Tradition is about the ongoing process of being innovative⋯I want to make things my ancestors haven't seen before⋯ I can't just photocopy what they did... I don't think they want us to do that. We need to realise there is a divinity to every material, to every thing" . While Nuku's shaped and etched Perspex works can appear as ghostly sculptural offerings, the polystyrene toiwhakairo (carvings) express complex political and cultural ideas in strong sculptural forms. The carved ceiling in the wharekai Ruatapuwahine is a metaphor for the cloak of Ruatapuwahine, an ancestor of Nuku's tribe. Acting as a cloak, the ceiling is understood to protect and nurture her tamariki (children or descendants) and the marae's manuhiri (visitors). Many traditional carving patterns are used, the art work offering ancient koru, pūhoro, kaokao, niho taniwha and mangopare patterns in a contemporary execution.

Created with the participation of the community and other local marae, Hei Korowai Mo Ruatapuwahinehas a unifying logic based on the natural world and traditional Maori myths and understanding of the world. Detailed panels running down Ruatapuwahine's cloak are carved representations of living creatures that inhabit the rivers and waterways surrounding the marae, important food sources for the tribe. These panels are considered to express the prestige and mauri (life spirit) of the animals while acting as a call to preserve their environment. Carved triangular elements in the ceiling are ngā tini marae, all the marae and habitations that existed and continue to exist in this place. One pattern employed by Nuku is known as "kaokao". This derives from an upper torso posture and is a metaphor for the marae's people to be strong as in the haka (an ancestral dance or challenge proclaiming strength and prowess), and to act and work collectively in unison to attain movement and momentum. Over 1,000 small triangles were cut from the larger panels, depicting Ruatapuwahine's descendants who are scattered across the world, like the triangles over the art work. Nuku's choice of geometric designs in the ceiling speak to the traditional woven panels lining the wharekai's walls and the spiral patterns to the pare (lintel carving above the entrance). Lights inside suspended carved cubes form Māori star constellations and suggest a Māori understanding of the cosmos. The cubes appear to float, defying gravity as their medium defies convention.

[解读]

雕塑民族神灵

"民族的才是世界的",正是每个民族独特的文化存在,才构成了丰富多彩的文化世界,每个民族的文化,都蕴含着本民族特有的精神,而能将之表现出来的艺术品,才能真正的流传不朽。乔治·努库作为一位著名的雕刻家,始终坚持用自己的刻刀表现毛利的精神。他遵循传统又不墨守成规,将毛利传统的文化精神,以现代的艺术手法表现出来,如鬼魅般存在,却又饱含生机和活力。他位于毛利人会堂的雕塑《Hei Korowai Mo Ruatapuwahine》,由附有动物威信和生命精神的嵌板,源自毛利战舞上身姿势的图案,形如毛利星座的灯饰组成,每一部分蕴含着毛利文化勃勃的生机,既有对传统文化的继承,又有对惯例的挑战,艺术家天马行空的思想甚至打破重力的存在,雕塑出附有民族韵味的神灵姿态。(闫丽祥)

无家可归的生活
Mo'ui Tukuhausia

艺术家:卡里索雷特·尤拉　　　Artist:Kalisdaite 'Uhila
地点:新西兰奥克兰　　　　　　Location:Auckland, New Zealand
形式和材料:行为艺术　　　　　Media/Type:Behavior Art
时间:2012 年　　　　　　　　 Date:2012
推荐人:凯利·卡迈克尔　　　　Researcher:Kelly Carmichael

卡里索雷特·尤拉出生于汤加,成长在新西兰,他通过行为艺术、装置和社会行为,试图探索太平洋岛屿和其人民的刻板印象。尤拉成长在世界上最大的波利尼西亚人城市——奥克兰。

他的作品关注的是通过太平洋岛民,把新西兰的奥特阿罗理想化为一个充满机会的城市,实现按照季节,干净地就业劳作。"尤拉的表演,是基于艺术品的耐力,他的作品企图破坏这些乌托邦式的价值观,并强调亚太移民在低工资,手工劳动阶层的经济作用"。

《无家可归的生活》首次创作是在 2012 年,可以看到卡里索雷特·尤拉是一个无家可归的人,在奥克兰的一个公共资助的艺术画廊,生活粗糙,吃当地游客捐赠的食品,睡在一个纸板和破旧的帐篷里。《无家可归的生活》的表现无不顾及个人安危,并要求艺术家不断地做出决定,在自然和社会给出的位置中存活。

在 2014 年,当被提名沃尔特斯奖的时候,该作品重新面世。沃尔特斯奖，是颁给那些在过去两年间，在新西兰当代的艺术和展览中脱颖而出的作品。如同第一次展出，卡里索雷特·尤拉没有地方住，偶尔进入画廊睡在里面，而其他的候选艺术家如同度假般的干净的美术馆里展出。

《无家可归的生活》引发了社区自然性、陈旧规则以及归属感问题的思考。该作品为艺术家——一群当地的粗糙沉睡者，创建了一个新的社区，并要求观众行走在伦理学的钢索上，徘徊在 尤拉所表现的身份和"一个新西兰最著名艺术奖、当红的年轻艺术家"的头衔之间。"尤拉以前经常躲猫猫，当你能找到他时，他身着笨重的衣服，脸被帽子遮住。"不同于脑海中那个雄心勃勃的当代艺术家，此时的尤拉成为一个典型的无家可归的人——隐藏，毫无个性，并试图保持不被发现以保安全。

《无家可归的生活》展览最初设定的标题是"你的意思是什么？我们吗？"这个展览把国际上所有精选的作品聚集在一起，而这些艺术家的作品检验观众的心理偏见，把他的标题和一个"独行侠和他的美洲土著人的搭档拖通"的老笑话联系在一起，在被激烈的阿帕奇战士包围时，独行侠托恩托问道："我们现在做什么？"《无家可归的生活》的两个化身都是公众、警方和全国新闻媒体审查的主题。

在第一次创作和两次修订.之间的两年时间内，尤其是波利尼西亚男性的无家可归，已经成为奥特兰以及新西兰其他城市迫切需要解决的问题。该作品表露出暗流涌动的社会和种族的紧张关系，并成为无家可归者，和许多对新西兰右倾政府经济政策感到脆弱的新西兰人中的一个催化剂。这项工作是扎根于新西兰及南太平洋范围内，但却生动的谈到全球社会底层的漏洞。《无家可归的生活》是最后四个有望夺得新西兰最著名、最杰出艺术大奖的项目之一。虽然没有夺得第一，艺术家在经过一番道德的深思熟虑，也最终决定参加黑领带奖晚宴。对比早先展览进行过程中的事件，此事件脱颖而出。他和无家可归的同伴被警察从画廊外面用力推向前，当时里面也有类似的晚宴正在进行。《无家可归的生活》是一个充满活力的，挑衅性和开放式结尾的艺术作品，它提供给典型的画廊观众一个窗口：在社会条件和法律形式，以及警察的干预下，城市是如何被控制的。

Kalisolaite 'Uhila's practice straddles performance, installation and social intervention, seeking to explore stereotyped images of the Pacific Islands and their people. Born in Tonga and raised in New Zealand, 'Uhila is part of a generation who grew up in the largest Polynesian city in the world—Auckland. His work is concerned with the idealisation of Aotearoa New Zealand by Pacific Islanders as a land of opportunity, a concept sharply at odds with the reality of minimum-wage factory, cleaning and seasonal labour jobs most obtain. 'Uhila's performances are endurance based artworks that seek to undermine these utopian values and highlight Pacific migrants' role in supporting the low-wage, manual-labour strata of the economy.

First created in 2012, Mo'ui tukuhausia ('a life set aside') saw Kalisolaite' Uhila living rough as a homeless person in the grounds of a publically funded art gallery in Auckland, eating food donated by local visitors, sleeping under a cardboard shelter and old tent. The performance of Mo'ui tukuhausiawas not without the considerable personal risk and required the artist to constantly make decisions to survive in the physical and social sphere of a given location. In 2014 the work was restaged when nominated for the Walters Prize, an award given for an outstanding work of contemporary NZ art produced and exhibited during the past two years. As with the first performance, Kalisolaite 'Uhila lived homeless, occasionally entering and sleeping in the gallery, while the other nominees exhibited exclusively within its white walls.

Mo'ui tukuhausia provoked questions as to the nature of community and troubled stereotypes of community and belonging. The work created a new community for the artist—a group of local rough sleepers (the people who quite literally watched his back) and demanded that audiences walk an ethical tightrope between 'Uhila's performance identity and his other 'hat' as a hot young artist in NZ's richest art prize. 'Uhila was often tricky to find and, when you could locate him, was an unusual figure in bulky clothes with his face continually hidden under a hoodie. Flipping the idea of ambitious contemporary artist on its head, 'Uhila became an archetypical homeless person—concealed, devoid of individual character and attempting to remain invisible, and therefore safe.

Mo'ui tukuhausia was originally created for an exhibition titled What do you mean, we?. The exhibition brought together an international selection of artists whose work examined the psychology of prejudice and took its title from an old joke about The Lone Ranger and his Native American sidekick, Tonto. Surrounded by fierce Apache warriors, The Lone Ranger asks Tonto, "What do we do now?", to which Tonto replies "What do you mean, 'we'?" Both incarnations of Mo'ui tukuhausiawere the subject of intense scrutiny by the public, police, and national news media. In the two years between first performing the work and restaging it, homelessness amongst Polynesian men in particular, has emerged as a pressing issue for Auckland and other urban centres in New Zealand. The art work gave voice to simmering social and racial tensions and became a catalyst for debate about not only the homeless, but also the vulnerability of many New Zealanders to the right-leaning government's economic policies. The work is strongly located within a NZ and South Pacific context, but speaks vividly of the vulnerability of a global social underclass.

That Mo'ui tukuhausia was one of four shortlisted projects for New Zealand's richest and most prominent art prize did not pass uncommented on, and after serious consideration of the ethics involved, the artist did decide to attend the black tie prize giving dinner. This event stood out in contrast to an earlier occasion in the duration of the performance when he and fellow homeless had been pushed on from outside the gallery by police when a similar dinner had taken place inside. A dynamic, provocative and open ended art work, engaging with Mo'ui tukuhausia offered typical gallery-going audiences a rare window into how urban space is controlled, by both social conditioning and forms of legal and political enforcement.

[解读]

来自底层的呐喊

一面是新西兰当红的年轻艺术家,一面是流浪街头的无家可归者,人们很难将这两者联系在一个人身上,而艺术家卡里索雷特·尤拉就是以这种方式告诉人们,来自于底层人群的呐喊。新西兰表面平静的社会下,却暗流涌动。流浪于街头的无家可归者,社会底层的移民者,受种族偏见歧视的人群,他们慢慢被社会边缘化,然后隔离,整个社会鼓吹着乌托邦似的理想,却无视挣扎在社会底层的人群。艺术家以无家可归流浪者的身份,穿梭于城市内,跻身于著名的艺术奖项中,艺术家双重的身份使观众徘徊于内心的偏见之间,既尊敬又歧视,同一个人只是由于头衔、装束的不同,所受的待遇也有你天壤之隔。这只是一个社会矛盾的催化剂,全球社会底层的漏洞始终存在,打开良心的门窗,就可以听到来自社会底层的呐喊。(闫丽祥)

空隙填补者
Gap Filler

艺术家：赖安·雷诺兹（信托公司主席）和科拉莉·温（艺术总监）
地点：新西兰克莱斯特彻奇和利特顿
形式和材料：不同媒介的各种不同的装置
时间：2010
推荐人：彼得·尚德

Artist : Ryan Reynolds (Chair of Trust) & Coralie Winn (Artistic Director)
Location : Christchurch and Lyttleton, New Zealand
Media/Type : Various installaitons of different media
Date : 2010
Researcher: Peter Shand

《空隙填补者》是一个城市重建倡议项目，以信托为模型进行构建，包含7位全职或兼职员工。他们与艺术家、创意专业人士以及社区群体一起工作，以到达在克莱斯特彻奇和利特顿的空置地盘上启动发展该项目的目的。在2010年9月和2011年2月，克莱斯特彻奇发生了严重地震，两次地震破坏了该城市的大量建筑，并极大地改变了该城市住宅区及周围地区的样貌，而《空隙填补者》项目也因此形成。给克莱斯特彻奇和当地居民带来积极变化的，与其说是对单一项目或多个项目的提名，不如说是《空隙填补者》在实现长期公共艺术和给予创意调解上做出的努力。作为一个整体，《空隙填补者》的活动都致力于对公共雕像、壁画的价值修复，以及改善与群众互动交流的表现。

这个目的组合（古典公众艺术在实施和开发涉及社区的项目中的价值）在克莱斯特彻奇和利特顿的重建时期具有特殊的价值。同样重要的是，《空隙填补者》标志着一种重新流行的方法论，而这种方法也是《空隙填补者》承担其工作的方法。

《空隙填补者》具有高度的灵活性与适应性。它给克立斯特彻奇小镇及镇内居民带来大量的项目，且项目涵盖范围广。以下有两个项目说明作为展示。第一个项目是在2014年启动的朗尼哈金森的《我喜欢你的形式》，该项目的目标是制作一个大型的暂时性雕塑。这个雕塑的原型是坐落在克立斯特彻奇镇中心的本土渔网塑像，而镇中心这块公共用地又是空隙填补者在克立斯特彻奇活动的核心。这块公共用地形状蜿蜒，提高了其自身的使用率，也令市民回想起传统的打渔活动和一个神话水精灵的形状，这个水精灵叫尼瓦，负责保护人民和土地，却是个十分差劲的守护者。第二个项目是《舞垫》，这个项目从2012年起就在许多地点得到实施，在这些项目实施地上，每个参与者只要支付两美元就可以将自己的便携音乐播放设备接到这个定制的舞池，播放音乐，营造出舞蹈的氛围。这个既放松又巧妙有趣的项目将荒芜的地方重新改造成一个可以举行有趣的舞蹈活动和社交聚会的地方。

社群的参与和感受是活动的核心。项目活动具有极高的转换率和非重复性，这保证了活动持久的活力，广泛的参与度，和实施的可能性。《空隙填补者》虽然是命题性的项目，但这些命题都是可以在世界上真实实践的。尊重并推广特定领域的艺术和艺术实践的创新和想象使得项目保持新颖。项目的智谋也在基于持续性和适应性而进行的工作中得到体现。作为艺术与市民之间的连接线，《空隙填补者》项目中不会有限定个人参与者这样失策的行为，但会要求集体行动与合作，一次激发创新的潜能。

《空隙填补者》战略项目的特殊成就在于这个项目的每个公共艺术发展推广活动中都有市民的参与。从这个角度看来，《空隙填补者》填补的不仅仅是公共用地上因地震而出现的裂痕，而且每个活动以及所有活动作为一个整体来说，都给生活在这个充满着戏剧性变化的城市中的市民，填补对这个城市在历史上、知识上以及心理上的理解的缺口。另外，社群的参与也填补了项目所属权的空白，并让市民认识到他们对自己所属社群的价值。

无论是从全城范围来看，还是从救灾角度来看，这些完整统一的结果都是公共艺术领域上巨大的成就。

Gap Filler is an urban regeneration initiative. It is structured as a Trust with seven full- and part-time employees who work with artists, other creative professionals and community groups in order to develop and realize projects in vacant sites across Christchurch and Lyttleton. It was formed in the aftermath of the Christchurch earthquakes of September 2010 and February 2011 that resulted in considerable damage to buildings and caused significant changes to the inhabited landscape of the city and surrounding region. Rather than nominate an individual project or projects, it is the work of Gap Filler in realizing an on-going series of public art and creative interventions that has created a positive change for the city and its citizens. When taken as a whole, Gap Filler's activities show a sustained engagement both in the healing value of public sculpture, mural and interactive performance. This as well as realizing a strategy that situates public art at the heart of urban experience and the positive impact it has on the lived environment.

It is this combination of purposes (the classic public art value of installing or creating projects that involve communities) that is of particular value to the cities of Christchurch and Lyttleton at a time of rebuilding. Importantly, too, Gap Filler marks a revitalized methodology by which it undertakes that work. It works across different sectors, realizing projects of civic scale and significance but where the impetus for that regeneration comes from communities, creative and volunteers rather than through more normative channels of government or corporate institutions.

Gap Filler is nimble and highly adaptive. It has given back to the citizens and Christchurch a significant number and range of projects. Two are illustrated here to give an indication of its work. The first is Lonnie Hutchinson's I Like Your Form, 2014, a large scale temporary sculpture based on the shapes of indigenous fishing nets situated in "The Commons" in central Christchurch, itself a piece of urban land that is a centrepiece of Gap Filler's activities in the city. It activates the public space with its serpentine shape, which recalls not only the traditional activities of fishing but also the shape of a taniwha or mythical protective water-creature that is the terrible guardian of people and land. The second is The Dance-o-Mat, a project installed in various locations since 2012 and where for $2 participants can plug-in their portable music devices to create a dance environment on the piece's custom-made dancefloor. It reimagines desolate spaces as places of social coming-together coupled with the joyous activity of dance in a project that is simultaneously therapeutic, witty and fun.

Gap Filler's philosophy is underpinned by six core value statements that shape both their overarching activity and establish the conceptual parameters of the projects realized. Community engagement: is at the heart of the activity, through both participation and experience. Experimentation: projects are high turnover, non-repetitive which creates an on-going sense of vibrancy, involvement and possibility. Leadership: Gap Filler is propositional but propositions that are made real in the world. Creativity: underpinned by innovation and imagination that promotes and honours the particular domain of art and artistic practices. Resourcefulness: the work is predicated on sustainability and adaptability. Collaboration: Gap Filler is a connector, it understands that in the regeneration of a city it is a false move to valorise individual players but, instead, requires collective action and co-operation to unlock creative potential.

The particular achievement of the Gap Filler strategy is that it involves citizens in the very activity of developing and helping to make actual public art projects. In this sense the "gaps" filled are not only those of the public space. The projects individually and collectively fill historical, intellectual and psychological gaps in how citizens understand and inhabit an urban region so wrought by dramatic change. Moreover, community involvement "fills the gap" of a sense of ownership of projects and a sense of their value and appropriateness for those very communities.

To achieve these integrated outcomes on a whole-city scale and in the context of disaster relief is a significant achievement in the field of public art.

[解读]

痛苦之上舞蹈

地震可以摧毁建筑，可以让街道布满裂痕，可以让一座城市成为废墟，可是不能阻止人们依然可以在其之上祈福和舞蹈。《空隙填补者》项目在被地震暂时摧毁的空地上，试图组织一个让人们寄居的精神家园，地震改变了这座城市的建筑和格局，但是《空隙填补者》可以让这座城市重新燃起信仰和希望，以等待无数崭新的建筑和设施渐渐兴起。暂时性的雕塑渔网一让人们联想到当地神话水精灵尼瓦的形状，他可以保护人民和土地，他的现身唤起人们的原始信仰和对这片土地的情感，也给人们的创伤一种精神的慰抚。《舞垫》让人们可以在灾难之后依然翩翩起舞，作废墟之上的舞者，这种公开的活动也会让更多的人受到彼此的感染。虽然有些灾难我们无法左右，但是精神上的胜利可以让阳光与快乐来得更快一些。（李田）

港口工人
Port Workers

艺术家:菲奥娜·杰克
地点:新西兰奥克兰
形式和材料:摄影作品

时间:2014 年
推荐人:彼得·尚德

Artist : Fiona Jack
Location : Auckland, New Zealand
Media/Type : Multiple photographic prints installed
Date : 2014
Researcher: Peter Shand

菲奥娜·杰克的项目《港口工人》是一个极其简单却十分细腻的公众政治艺术创作。在政治艺术这个方面,这个项目与一些著名的现代艺术作品(如安迪·沃霍尔的《13 位最高通缉犯》,巴可·狄米崔杰影响深远的肖像画)有相似之处。虽然如此,出于艺术家要与各种各样摄影对象进行道德接触的需求,这个项目另外引进了一系列更加复杂的目标。《港口工人》的影响跟公众场所的关系和政治化区域跟先前艺术作品的关系相似,而杰克的目的和方法的蕴涵要求经过思考后做出选择来展示港口工人的肖像。

这个项目的提出回应了奥克兰海港城市一再拖长的劳资纠纷。在这场劳资纠纷中,形成工会的工人们和他们的在奥克兰港口的雇主们无法解决更改集体劳工协议中产生的分歧。在这次事件中,联合成工会的工人经过投票,决定用罢工来支持他们的声明。

这场罢工运动包含了另外两个含义。首先，1951年在相同场所爆发的罢工运动导致了新西兰二战后劳工历史中最漫长，双方意见分歧最大的一次纠纷。其次，媒体对2012年纠纷的回应中，尽管陈明了在仔细审查后得到的严重性，却没有清楚说明最急迫的事件，更不用说罢工工人试图吸引公众注意到的事件背后的本质。

杰克对工会行动的贡献在于她与160位罢工工人协商后，给他们拍摄肖像时做出的努力。这些肖像都以宣传贴图的形式印制出来，有些贴图由注册的私人公司贴到公共区域作为海报使用，另外一些则由通过许可的活动或游击活动传播到城市的各个角落。

这个公共艺术的目的是让公众重新意识到工人之间的亲密感和工人自身的个性，而这些工人才是在谈判破裂以及后续罢工行动中最受影响的。这个公共艺术项目深刻地给公众提醒了那些涉及政治活动的人的人性，特别是被大家认为是新自由主义经济意识终结者的那些人。这些简单匿名的肖像不仅给肖像本人找到服务机构（通过和艺术家的协商以及自身在公众地区的出现），而且是展现艺术团结的强有力的政治举动。

《港口工人》没有选择使用不露面的罢工者和工会成员，而选择肖像照片是为了给港口工人重建个人尊严，同时也是给他们在合同谈判中为取得理想成果的集体努力致敬。从这个方面来说，这是一个具有重建性和连接性的项目，因为它从拙劣且过度炒作的媒体处理中把当代劳动规则的脸面救了出来。这个项目也让广大群众记起他们和被卷入这次政治行动的同胞们共有的基本志向抱负。

《港口工人》在公众区域得到引入，并且在工人十分熟悉的条款下，这个项目为工人赢得一个独特的代理机构，这个机构赞扬了工人们通过撤回劳动力赢得合同更改的行动。在这一方面来说，《港口工人》慎重的使用了公众可见性中的一些属性而不是服从景观化的空洞举动。直到最后，这个项目依旧牢牢守住了在人文主义中的尊严和庄严的道德观。

凭借它的政治连贯性和服务并致涉及此项目的敬港口工人的方法，这个引起群众共鸣的《港口工人》应该获得公共艺术奖。

Fiona Jack's project Port Workers is a deceptively simple, carefully nuanced work of political artmaking in the public space. In this respect it has echoes of well-known modern and contemporary artworks such as (Andy Warhol's 13 Most Wanted or Braco Dimitrijec's monumental portraits). Nevertheless, the project recruited to a suite of potentially more complex objectives because of the demands of the artist's ethical engagement with the diverse subjects of the project's photographic component. While Port Workers impact has similar connectivity to the public space as a politicized space as those antecedents, the implications of Jack's intent and methodology require considered alternatives to a re-presentation of the workers' portraits.

The project arose in response to a protracted industrial dispute at the Auckland seaport. In that dispute, unionized workers and their employers, Ports of Auckland, had failed successfully to resolve a dispute about changes to those workers' collective employment agreement. This against a complex backdrop of neo-liberalist economic policies that impacted not only on changes to New Zealand labour regulation but also to such issues as the privatization of public infrastructure, such as vital transport or trade networks. In this instance, unionized labour voted to strike in support of their claims.

That action contained two additional implications. First, a strike at the same facility in 1951 resulted in one of the most protracted and divisive incidents in post-World War II labour history in New Zealand. Second, the response of the media to the 2012 dispute, though intense in its scrutiny, was not necessarily clear in it's articulation of the most immediate issues, let alone the underlying principles that the striking workers sought to draw attention to.

Jack's contribution to the union action was to negotiate with 160 of the striking workers to make a photographic portrait of each. These photographs were then printed as bill-sticks and, both through a private company registered to put up posters in public spaces and some both sanctioned and guerrilla activities, distributed across the city.

The purpose of this was to bring back into public consciousness the intimacy and personality of those most impacted by the breakdown in negotiations and in the subsequent strike action—the workers themselves. In this, the project represented a profound reminder of the humanity of those involved in any political action, most particularly those that are perceived as on the sharp end of neo-liberal economic ideology. Not only did the simple, anonymous portraits achieve agency for those workers portrayed (through the negotiation with the artist and their subsequent presence in the public space), they also served as a powerfully politicized act of artistic solidarity.

Instead of a faceless mass of "strikers" or "unionists", the project restored individual dignity to the port workers whilst also paying homage to their decision to work collectively to seek to achieve the desired results from their contract negotiation. In this respect it was a restorative and bridging project because it retrieved from clumsy or sensationalized media treatment the human face of contemporary labour regulation. It also served as a reminder or spur to the wider public of the shared basic aspirations shared with their fellow citizens caught-up in this particular political action.

Introduced to the public space and on terms with which they were fully cognizant, the project retrieved for the port workers a distinct agency that complimented their adoption of action through with the withdrawal of labour. In this respect Port Workers critically utilized some attributes of public visibility without succumbing to hollow gestures of spectacularisation. To this end, it remained firmly grounded in an ethos of dignity and gravitas that accentuated its humanist agenda.

It is because of the sense of political coherence and the means by which it both served and honoured the eponymous port workers involved in the project that this spare, resonant work should be considered for a public art award.

[解读]

威武不能屈

自资产阶级兴起到如今，工人运动就没有停止过，人们企图建立一个消除阶级差异的世界，可是理想主义只存在于理想之中，我们还需要漫长的岁月去斗争、去建设、去协调。《港口工人》是一个公众政治艺术创作项目，艺术也能够影响和参与政治。艺术家菲奥娜·杰克以一种类似于波普艺术中的展示方式将不同的工人的肖像组织在一起，他们是一个群体，也是一个个独特的个体，艺术家给予了他们肖像正面的拍摄效果，让我们想起一些名人的肖像，我们可以通过艺术家的视角看到他们的形象，健康的、温和的、明朗的。这种展示和斗争方式给了他们以平等的尊严，而并没有夸大他们的苦难以博取同情，可以看到艺术家对他们面对这次劳资纠纷的态度所表现出的赞赏，首先，这组摄影作品本身的态度就是要将这些工人放在一个平等的位置上，他们值得被尊重、被重视。（李田）

灯光城
Luxcity

艺术家：杰西卡·哈利迪和乌韦·赖格尔	Artist：Jessica Halliday and Uwe Reiger
地点：新西兰克赖斯特彻奇	Location：Christchurch，New Zealand
时间：2012年	Date：2012
推荐人：彼得·尚德	Researcher: Peter Shand

《灯光城》是2012年在新西兰克立斯特彻奇举办的传统建筑节的主题活动。它把在5个三级建筑的350名学生的教学机会与包含16个规模庞大的临时性公共艺术作品的新西兰设计计划结合了起来。《灯光城》活动的构想与实现都是以支持、鼓舞克立斯特彻奇市民为目标的。

克立斯特彻奇和利特顿两个城市在2010年9月及2011年2月受到地震的重创。克立斯特彻奇的中心城区已经荒废，在2月份的地震后许多建筑（包括历史建筑和20世纪晚期建筑）已经不适宜居住。市中心的许多街区已经重新标上"红色区域"的标志，这就意味着市民不仅不能进入那些倒塌的建筑，而且不能使用市中心的民用区域。在受到灾难的打击和人员的伤亡后，市民又接着被政府从他们的社区中心里疏离出去。这样的处境加剧了市民在灾后的离家感和失落感。因此平和地去参加那些对他们有巨大影响的活动，对于克立斯特彻奇的市民来说是一项更大的挑战。

概念上，《灯光城》项目是围绕着光和光亮所代表的希望来构想的。实际上，这项在多数"红色区域"上的创新型干预点亮这个城市的大多数街区，而这些街区从拆除和清理工作开展以来已经被封锁了18个月了。而在《灯光城》项目实施的时候，这些工作依旧在进行着。象征意义上，在城市清理工作过渡到城市重建工作时，这个项目将会对城市未来的发展起着指引作用。项目内的活动在艺术环境下，在城市标准活动内（例如聚餐、饮酒、跳舞、交流）是相互影响的，并且能够活化社交活动。

一方面，这个项目提供了深刻的乐观感和责任感，同时带有实质程度上的激发性。另一方面，这个项目也表示出项目参与人对克立斯特彻奇市民的慷慨支持，也有利于鼓励市民积极参与到城市新未来的建设中。

《灯光城》活动是以为期一晚的灯饰节的形式呈现出来，试图缓解开启"红色区域"形成的床上，并几年那些在这些区域以及更广阔区域内遇难的人。这个活动将一个隐藏的温暖的纪念仪式和在这个城市幸存下来的人的目的和可能性结合起来。此次参加活动的人数是活动组织方估计的五倍还多，从而也可以看出人们是如何迎接这一份礼物的。此外，观众的年龄跨度非常大，其中大部分观众都是以家庭为单位参与。在这方面，这个活动影响了整体社区支持的来源的转变，从来自家庭和朋友的支持转变到集体共有活动的真实经历提供的力量与支持。这次的活动在点亮灾难地区的同时，也使完成清理建筑瓦砾工作的地区重获生机，变得适宜居住。

另外一个值得注意的是，这些项目都是由专业人才、学院、学生共同承担的。这样的团队提供了一个极好的教学机会来进行课程相关的实践，而这些实践积极地根植于道德的社交活动。组织方意识到与其用灯光城项目来推行单一庞大的组织，不如用它来践行能够好找群众个人想象和集体想象的渐变的暂时性的创意命题——做出知识重建、现象学重建和心理重建的努力来回应市民的城市重建工作，同时加强其重要性，引出新的可能性。

《灯光城》项目包含了社会道德的交织、创新的试验、慷慨的奉献、严谨的知识体系、情感的概念化。

Luxcity was the central event of the 2012 Festival of Transitional Architecture (FESTA) held in Christchurch, New Zealand. It combined significant pedagogical opportunities for 350 students in five tertiary architecture and design programmes in New Zealand with a cohesive series of 16 temporary public artworks of large scale. Luxcity was conceived and realized as an action of support, resilience and provocation to the citizens of Christchurch.

In September 2010 and February 2011 the cities of Christchurch and Lyttleton were devastated by two major earthquakes. Much of the Christchurch central city district became derelict, with a majority of its buildings (historic and late twentieth century) rendered uninhabitable in the aftermath of the February catastrophe. Much of the city centre was re-designated as a "red zone", meaning that people were denied access to much of the civic space of the central city not simply buildings condemned on safety grounds. Already traumatised by the disaster and the attendant loss of life, citizens were consequently alienated from the civic centre of their own community. That condition exacerbated a sense of dislocation and loss that occurred in the aftermath of the earthquakes. It also became an increasing challenge for the citizens of Christchurch to come to be at peace with the events that had affected them so significantly.

Conceptually, Luxcity was conceived around the central notion of hopefulness that accompanies the central theme of light or lightedness. Literally, the creative interventions in most of the "red zone" illuminated areas of the city from which citizen had been barred for over eighteen months as demolition and clean-up commenced. Much of that work was continuing when Luxcity was realised. Metaphorically, the project acted as a beacon or light for what the city might become as it moved from clean-up to re-build. Many of the projects were interactive and invited social activation, both within an artistic context and the normative activities of any urban centre (food, drink, dancing, communication).

It was in this sense that it offered a profound sense of optimism and possibility but with a substantial degree of respectful provocation. It was, in this sense, an act of generosity and support offered by those involved in the project to the citizens of Christchurch, an act with the potential to be healing whilst also encouraging active engagement with the city's possible new futures.

Presented as a one-night festival of light, Luxcity sought to ameliorate the trauma of opening-up the "red zone" and honouring those who had lost their lives there and in the wider region. It combined an underlying memorial activity with the warmth and a sense of possibility and purpose for those who remained in the city. As a reflection of how this gift was received, attendance at the event was more than five times organisers' estimates. Moreover, the audience was multigenerational and included a large number of family groups. It affected a shift, in this respect, from the support of family and friends to a real experience of a communal and collective act of strength and support for the community as a whole. This at an event that both lit-up the area of devastation whilst simultaneously enlivening and humanizing the spaces newly cleared of buildings and rubble.

It is also important to note that the projects undertaken were partnerships between practitioners, academics and students. This component afforded a critical pedagogical opportunity to undertake course-related practice that was firmly and positively rooted in ethical social engagement. Rather than imposed monoliths, then, Luxcity realized a suite of transitional and temporary creative propositions that called upon citizens' individual and collective imaginations—inviting acts of intellectual, phenomenological and psychological re-building that echoed the civic activity of reconstruction but also helped to advance criticality and to raise new possibilities.

It is this interweaving of social ethics, youthful experimentation, generosity of offer and rigorous intellectual.

[解读]

希望之光

光让人们想到黎明，想到太阳，想到希望，这片光华带给这座受到重创的城市以美好，也让因为倒塌而黑暗封闭许久的市中心重新充满活力。这场灾难让克立斯特彻奇几乎被摧毁，人们被迫离家，而这座城市曾经美丽古老的建筑也摇摇欲坠，面对大家共同的家园，许多人团结起来，一起去修复他、点亮他。危险的"红色区域"终于重新被灯火环绕，人们在看到这些放射的，旋转的灯光时，会想到克立斯特彻奇曾经的样子，还有对遇难者的默哀和祝福。艺术家以这种形式将人们从沉痛中拯救出来，市民们一起观赏这些美丽的光，彼此慰藉，这场灯光展示其实对市民来说是一份礼物，他们结伴而来，对未来的这座城市的恢复充满期待。艺术可以在黑暗和苦难里创造美，可以是城市创伤的修复者，同时也可是对受难者心灵的拯救。（李田）

如果你将在这里工作
If You were to Work Here

艺术家：彼得·罗宾森
地点：新西兰 克赖斯特彻奇
和利特尔顿
形式和材料：毡棒和金属棒
时间：2013 年
推荐人：彼得·尚德

Artist：Peter Robinson
Location：Christchurch and
　　　　　Lyttleton，New Zealand
Media/Type：Felt and metal rods
Date：2013
Researcher: Peter Shand

《假如你将在这里工作》这个项目是皮德·罗滨逊在 2013 年奥克兰三年展的约请项目，而这一年奥克兰三年展的策展人是侯瀚如，展题是"假如你将在这里工作……"。这是一个复杂的项目，它关系到奥克兰这个城市的两大公共机构，分别是这次三年展的主办方——奥克兰美术馆，以及奥克兰战争纪念博物馆，此博物馆是这个国家有历史意义的主要机构之一，还有着世界上最多的毛利遗产珍品收藏（这些都是毛利有形的文化遗产）。在罗滨逊项目发展的一系列过程中，他不仅想要把这两个机构联系起来，还希望在这个项目能够进行的地理空间中把各种复杂的感情及具有创意的行动交织在一起。

这个项目要制作出数百根的长棍及礼拜堂里放置的那种木棍。制作这些的人是大学生和高中生，这个项目给他们提供了一个重要的教学实践机会，让他们为由一个出色的艺术家提出的项目作出努力，同时也将他们带到了这位艺术家的工作室中进行实践。尽管如此，这个项目还是有一层模糊的含义，这个含义是此项目概念化的要素，尤其是要怎么把参与者之间的联系和参与意义具体表现出来。在大量生产鲜艳毛毡的过程中存在着一种微妙的暗示。这个暗示来自于约瑟夫·博伊斯，他认为毛毡是一个传递理解与支持的一个关键媒介。

长棍上的颜色能让人愉快地想起人体生物化学中的四种体液表现形式，以及想起在人体健康方面，这四种体液所起的作用。在参与项目的第二阶段时，罗滨逊遇到了一些在奥克兰战争纪念博物馆的工作人员，他们正要概述出罗滨逊要制作那么长棍的意图。在那里，那些职员可以选择一根最能代表他们对自己工作场所感受和心情的有颜色的木棍。然后他们受邀在博物馆内选择任意一处（包括公共展示区）防止他们所选择的木棍。而这个举动能让员工以一种新的方式参与机构的工作。而参加了这一项活动员工表示，这个举动让他们在长时间放置木棍的无聊过程中常常感到有趣和触动。通过意料之外的颜色鲜明的毛毡杆，整栋建筑都让访客感到惊喜，让馆内珍贵的收藏充满了活力，同时，也让访客在参观的过程中将历史珍宝与当代艺术结合起来欣赏。

这个项目里更进一步的参与方式是与这些长棍如何运到博物馆的方法是紧密相连的。在奥克兰三年展那天，那些长棍要根据毛利礼拜堂和其他教区里的织面板的梯形摆放方式（通向大国的阶梯）来摆放，而且是要在公共门廊那里摊开放好这种摆放方式暗示着人类从众神那里得到的知识。

那时大约有200多个人受邀要步行3 000米到奥克兰战争纪念博物馆，而且还要手持一到两根他们自己选择的长棍。这次的步行被选定为一次游行，是这个项目里非常重要的一步。这次游行就类似于毛利传统中的大规模公众运动，为了反映这些活动（不管是抗议类型的还是庆祝类型的），罗滨逊故意地让人们举着成百上千的各色鲜艳长棍以巡行展示的方式穿过城市的大街小巷，以此来营造一种视觉效果。这项活动马上就引起了宗教仪式、市民庆祝活动和政府行动的响应。与传统游行相一致的是，它是一种仪式，它触动人们的感情，给予他们温暖，仪式中所带有的特质是与这项长棍转移活动有关的，同时也与这两个前后储存长棍的机构有关。

在新西兰，这次游行所带来的社会政治影响特别强烈，尤其是因为它明确反映了毛利的政治意向。最著名的游行发生在1975年，那时距离毛利人与新西兰君主正式确立殖民关系已经过了135年了，毛利人走过了约有整个新西兰北岛周长的路程渠道当时的中央政府，要求土地赔偿。罗滨逊不仅想起了那件事情，还给公众制造了一次类似的游行，让他们干预且参与到政府一些越发模棱两可的政治事务上。例如，参与者是抱着一种欢乐的庆祝心情参加活动的，但又夹杂着一丝挑衅的态度。这种心情刻画了公共行为，当然包括公共艺术在内，具有变化的潜力，能够在理解和任意妄为之间切换。在这个方面，它是一个虽然有趣，但也有力的一个提示。它提醒了我们公共艺术的重要性在于唤醒市民责任以及加大市民的参与度。而这也是"如果你将生活在这里……"这个概念的关键因素。

这是一个内容丰富的公共艺术计划，它将一系列不同的、也是免费的体验和影响交织在一起。由于这些原因，这个计划理应被纳入公共艺术奖的考虑范围中。

If You were to Work Here…was Peter Robinson's commissioned project for the 2013 Auckland Triennial If you were to live here…curated by Hou Hanru. It was a complex project that linked the two major public institutions of the city, the Auckland Art Gallery that is driving institution of the Triennial and the Auckland War Memorial Museum, one of the country's leading historic institutions and with the world's richest collection of taonga Maori (objects of Maori tangible cultural heritage). In an on-going suite of actions, Robinson sought not only to link the two sites but to weave a sophisticated sense of emotional and creative action across the interceding geographical space.

The project involved the making of many hundred tokotoko or oratory sticks. Included in their making were tertiary and high school students, which gave them a key pedagogical opportunity to contribute to a leading artist's project but also to be introduced to the studio practice of such an artist. This diffuse implication was, nevertheless, a key component of the conceptualisation of the work, especially for how it embodied the sense of connection and meaningfulness of participation. In the brightly coloured felt of manufacture was a subtle suggestion of Joseph Beuys' identification of felt as a key medium of shared understanding and support.

The colours chosen for the tokotoko were playfully reminiscent of the humours and the role of this conception of human biochemistry in the achievement of good health and well-being. In the second moment of participant engagement, Robinson met with members of the staff of the Museum to outline his intention for that part of the project. Here, those staff were able to choose a coloured stick that best represented their feeling or mood about their place of work. They were then invited to select a space in the museum (including in areas of public display) into which the chosen stick or sticks were then installed. In this gesture, staff were able to mark their involvement with the institution in new ways—and the narratives articulated by participants during the lengthy installing process were frequently very interesting and moving. The unexpectedness of coloured rods of felt also served to surprise and enliven the experience of the building and its important collection for visitors and connected for them experiences of historical taonga and contemporary art.

A further participatory component of the work was attached to the arrival of the tokotoko at the Museum. On the opening day of the Triennial, they were laid out in the public atrium of the Art Gallery according to the poutama (stairway to heaven) pattern of Maori woven panels that feature in meeting houses and other spaces of congregation and learning. The pattern's implications are concerned with the human receipt of knowledge from divine sources.

Some 200 members of the public were then invited to select one or two tokotoko and to take them on the three-kilometre walk to the Museum. This walk, the pivotal moment in the work, was named as a hikoi. Hikoi are akin to mass movements of public action in Maori tradition and may variously be regarded as marches, protests or parades. In reflection of those activities (whether protest or celebratory) Robinson knowingly anticipated the visual impact of hundreds of coloured sticks being paraded through the city streets—an activity that at once echoes religious observance, civic celebration and political action. In terms aligned to the tradition of hikoi, it marks a ritualized performance that activates and warms the public space—qualities that are pertinent to this transitional activity but also to the two institutional sites that housed the tokotoko before and after.

In New Zealand, the socio-political resonances of hikoi are particularly strong, especially with their reflection of specific activity of Maori political intent. The most famous hikoi was undertaken in 1975, when Maori marched the length of the North Island of New Zealand to present to central government claims for land restitution some 135 years after the formal declaration of a colonial relationship between Maori and the Crown. Robinson not only recalled that and like events he also created the opportunity for public intervention and engagement that operated in more ambiguous terms. The mood of participants, for example, was openly celebratory and somewhat defiant, a mood that captured an understanding and wilful adoption of the transformative potential of public action including, significantly, public art. In this regard it was a powerful if playful reminder of the importance of public art to notions of civic responsibility and engagement—a critical element of what it means "if you were to live here ..."

This, then, was a replete public art project; one that interwove a suite of different though complimentary experiences and implications. It is, for these reasons, a project that ought to be considered for a public art award.

[解读]

毛利的象征

自英国库克船长踏上这片土地,到后来新西兰成为英国的自治殖民地,原住民毛利人与移民者已相处几百年,但是民族关系对于新西兰来说一直是一个十分复杂的问题。新西兰的国徽左侧是手持新西兰过期的欧洲美女形象,右侧是手持棍状武器的毛利人,代表着这个国家的一种民族融合。如今毛利人占新西兰总人口的近15%,相对于欧洲移民来说,仍不是主流民族,艺术家皮德·罗滨逊运用毛利人的特有的彩色棍子,给这个传统的毛利人物品带来了当代的想象和意义。当现在的人们拿着毛利人的棍子,行走在街上的时候,感受到的是一种传统的集会,一种毛利人争取主权的示威,还是一种娱乐的欢庆,也许每个参与者连同观看者的感受都是不一样的,这便是艺术家想要我们体验到的,我们不仅会关注到毛利人的传统,也会因自我的参与而又新的发现。(李田)

公平的竞技场
Level Playing Field

艺术家：大卫·柯若思
地点：新西兰 克赖斯特彻奇
形式和材料：互动装置
时间：2013 年
推荐人：布鲁斯·菲利普斯

Artist：David Cross
Location：Christchurch，New Zealand
Media/Type：Interactive Device
Date：2013
Researcher：Bruce Phillips

《公平的竞技场》是大卫·柯若思设计的一个临时的公共艺术作品，其目的是将人与被地震蹂躏的新西兰基督城结合在一起。在 2011 年 2 月，基督城遭受了一系列地震，其中最大的测得 7 级，185 人死亡，把城市夷为平地。从二战德累斯顿遭受燃烧弹后，基督城再次被认为是破坏最严重的城市。

对此，柯若思设想了一个大规模的参与式装置，包括特别为当地坎塔不连人团队演奏活动设计的充气结构安装。五颜六色的长方形充气装置占据着基督城市中心，拆除高层建筑的整个足迹，与城市震后环境中的瓦砾以及飓风围栏的灰色背景形成鲜明对比。然而，尽管这个引人入胜的装置是以公众的社会参与为目的，但这个参与有很大影响，使得人民重新参与受灾城市的重建。

《公平的竞技场》包含了 48 队，每队 5 人，争夺 6 周参与者的资格。这项运动的创造是针对不稳定地面进行挑战。规则要求参与者以长丘

169

堆的形式在不稳定的波浪起伏的表面奔跑。团队轮流以一个继电器电路方式奔跑在正在运行的充气圈上，而对方球队，藏在结构内部，以拉动绳索扰乱表面稳定性使得上面奔跑的人失去资格。在有限时间内，完成更多奔跑回路的团队就是胜利者。

这非常直接的社区参与的形式，其设想试图改变关于心理相关的——不确定的地震会变成真正的生活灾难，通过容易理解的所参与的游戏，给那些悲惨的不愉快的社会困境的中注入了治疗性活力。

这个装置暂时激活城市的心脏，但仍显著缺少活动和社区的存在。鉴于基督城体育文化的疯狂，这个活动不仅吸引了240多个参与者，而且还有大量的围观观众。《公平的竞技场》用一种识别概念的方法来证明：临时性的公共艺术会吸引有意义的社区参与，并且具有持久的令人难忘的结果。

Level Playing Field (2013) was a temporary public work by artist David Cross which was designed to socially engage with the people and place of earthquake ravaged Christchurch New Zealand. In February 2011, Christchurch was hit by a number of earthquakes the largest of which measured a magnitude 7 killing 185 people and levelling the city to the ground. Christchurch has since been considered as the worst damaged city since the firebombing of Dresden in WWII.

In response, Cross envisioned a large-scale participatory installation consisting of an inflatable structure designed specifically for an invented team sport that was played by teams of local Cantabrians. The colourful rectangular inflatable occupied the entire footprint of a since demolished high-rise building in downtown Christchurch and stood in stark contrast to the grey backdrop of concrete rubble and hurricane fences that typified the post quake environment of the city. Yet despite this engaging formal presence it was the social engagement of the work that truly made the significant impact and reengagement of people with the disaster stricken city.

Level Playing Field involved 48 teams of 5 people who responded to an open call for participants who competed for 6 weeks. The creation of the sport was in direct response to the idea of unstable ground. The rules required participants to run over an erratically undulating playing surface in the form of a long mound. Teams took turns running in a relay circuit over the inflatable while the opposing team, hidden inside structure, pulled upon strapping to manipulate the stability of the surface with the goal of unseating those traversing over top. The more circuits each team completed, within the limited time frame, became the victors.

This very direct form of community engagement was envisioned in attempt to change the psychological association of unstable ground being something truly life shattering and difficult to overcome to being momentarily reconceptualised as something that locals could control. The easily understood parameters and energetic experience of the game injected therapeutic activity and humour to a dire and unpleasant social predicament.

This work also temporarily activated the heart of the city which was still significantly absent of activity and community presence. Given the sports crazed culture in Christchurch, the work drew in not only 240 participants but also scores of spectators. Level Playing Field is a testament to how a discerning conceptual approach to temporary public art can result in meaningful community engagement with lasting memorable outcomes.

[解读]
游戏背后的意义

2011 年 2 月，新西兰基督城遭受了一系列地震，其中最大的测得 7 级，185 人死亡，把城市夷为平地。《公平的竞技场》是大卫柯若思设计的一个临时的公共艺术作品，以五颜六色的长方形充气装置占据着基督城市的中心。鼓励公众参与到这个装置游戏中来，规则要求参与者在不稳定充气装置表面奔跑，对方则藏在结构内部，以拉动绳索降低表面稳定性以此来让对手摔倒从而失去比赛规则。一方面，这个游戏说明像地震这样的灾难防不胜防，另一方面，希望人们在参与游戏的过程中，忘记那些悲惨的不愉快的因素，给他们注入了治疗性活力。这个互动装置公共艺术提供了有意义的社区参与，给予灾难中的人们一种温暖的安抚。（祁雪峰）

案例研究
南亚 中西亚

CASE STUDY
South Asia, Central and West Asia

From Gulf to Gulf to Gulf	从海湾到海湾到海湾
House Of Comfort Art Project	舒适的家艺术项目
Berbeda dan Merdeka 100% (Different and Free 100%)	Berbeda 丹默迪卡 100%（不同的折扣和完全免费）
The School Never Asked	无人问津的学校
Chess Park	象棋公园
Cigondewah Cultural Centre	Cigondewah 文化中心
Public Art Mourning & Resilience	公共艺术、哀悼和恢复
Gebran Tueni Memorial	吉卜兰·图韦尼纪念馆
[En]Counters	计数器
In Context : Public · Ar · Ecology	背景：公共·艺术·生态
"Jaya He" (Glory to India)	"贾亚赫"（印度的荣耀）
Meanwhile I Elsewhere	与此同时在别处
Phoenix Market City Art Project	菲尼克斯市

从海湾到海湾到海湾
From Gulf to Gulf to Gulf

艺术家：阿肖克·苏库马兰、
　桑杰·邦格阿尔、谢纳·阿南德、
　泽尼亚·安巴帕迪瓦拉
地点：阿联酋沙迦

形式和材料：调查研究，
　广播视频投影
时间：2014 年
推荐人：纳赫拉·阿尔·塔瓦

Artist：Ashok Sukumaran，Sanjay
　Bhangar，Shaina Anand，Zinnia
　Ambapardiwala
Location：Al Shuwaiyheen,
　Arts Area, Sharjah, UAE

Media/Type：Research, Radio and
　Video projection
Date：2014
Researcher：Tabbaa

自 2009 年起，CAMP 工作室开始对沙迦的海上身份产生浓厚的兴趣，它的船只和水手时不时地在这个具有历史意义的港口出发和返回，如今它已经是阿联酋的主要工业终端。他们发现这些船员大多出生于吉吉拉特邦，伊朗南部和巴基斯坦。出于做贸易的本性，他们习惯性地在阿联酋、伊朗、印度和索马里海岸之间穿梭，在不同的海港之间传递着他们文化的相同性和差异性。通过会见这些水手并追踪他们的贸易航线以及船只所载物品，CAMP 为大家呈现出这些船只和水手航行中的亲密关系和全球话语。

《从海湾到海湾到海湾》是一部收集采访镜头，电话视频，歌曲以及于古吉拉特邦修建船只的单独航行之旅。通过这部电影，人们会了解来自一个国家的一群人构建的工程是怎样载着带有恐惧与欢欣情绪的水手，去经历多种语言、文化、法律体质、风暴、海盗占领水域以及制裁等多种挑战的。

这部电影是以一艘叫做"Sabir Piya"（有耐心的）船的旅行为线索的，这艘船制造于 1999 年古吉拉特邦，并于迪拜安装了引擎。它的第一次航行是在 2002 年运送糖，然后负责从科威特向索马里运送捐赠物，接着被包船向迪拜运送山羊。2009 年，"Sabir Piya"在沙迦的一个水湾发生了火灾并下沉，最后只留下一些残骸和木材，但是它随后便在古吉拉特邦进行了重建，变得更大、更结实。

关于"Sabir Piya"的信息是通过官方港口记录以及它水手的一些故事收集到的。CAMP 与水手之间的对话、友谊以及信息交流引导了一种掩盖在社会下方的文化，明显是隔离的状态。出现了很多的有趣事情，比如水手和商人可以说多国语言的能力，这常常被认为是取得许多社会中最有竞争力职业阶梯的途径。这样的语言交流时刻显示出他们用北印度语言或马尔都语来交流的才能，其实这两种语言有相同的口语形式，尽管这并不是港口所在地的母语但是它们通常用作港口语言。

这部电影的首次放映是在沙迦 Biennial 11 活动中，并被投影在一个公共墙壁上。它的成功之处在于，这部电影的初始日常观众包括一定数量的水手，他们对这个电影所显示的情节，表现出很深的情感依恋，这时他们可以与周围观众如馆长，家庭以及其他的社会元素融合在一起分享公共空间，这在以往是不曾发生的。这可以为更深了解国际水域区的文化交流提供集体性的论坛。而这个电影已经连续放映了三个月，并还会被邀请参与更多的合作活动，这一事实即可反映它的成功。

From 2009, studio CAMP started to take a deep interest in the maritime identity of Sharjah, through its boats and sailors constantly arriving and leaving the historic port town, today a major industrial terminal in the UAE. They discovered that these sailors hailed from Gujarat, Southern Iran and Pakistan mostly. Through the nature of their trade, they would habitually forge the shores of the UAE, Iran, India and Somalia, carrying the effects of their stark cultural similarities and differences between each port. Through meeting these sailors, tracing their trade routes and ship contents, CAMP unfolds the intimacy and global discourse of every journey these ships and sailors embark on.

"From Gulf to Gulf to Gulf" is a film that collects footage of interviews, phone videos, songs, and the journeys of a single ship built in Gujarat. Through this film, one discovers how a single construction built by a dozen people in one country can pass through multiple languages, cultures, legal frameworks, winds and storms, pirate waters, sanctions, carrying the fears, joys and emotions of the sailors aboard them.

This film follows the journey of one boat called "Sabir Piya" (the Patient One) which was built in Gujarat in 1999 with an engine installed in Dubai. It made its first trip in 2002 carrying sugar, and was then responsible for delivering donations to Somalia from Kuwait, followed by a charter to carry goats to Dubai. Other cargo entailed a single ostrich, the odd limousine, coal, all highlighting the distinctly unique and infinitely varied purposes, and limitless identities, of the vessel and its sailors onboard. In 2009, "Sabir Piya" caught fire on the Sharjah creek, sunk, surviving as debris and timber which was subsequently rebuilt in Gujarat, bigger and stronger.

Information about "Sabir Piya's" was collected through official port records and the stories of its sailors. The dialogue, friendship and exchanges between CAMP and the sailors draws on a culture that is hidden from society, and distinctly separated. Interesting facts emerge, such as the sailors' and traders' ability to speak multiple languages, a talent that is often cited as a route to the top of many of society's most competitive career ladders. One such linguistic moment highlights their ability to converse in Hindi or Urdu, which shares the same spoken form, and is the commonly used port language despite not being the native language of where the port is located.

The film was first released during Sharjah Biennial 11 and shown as a projection on public walls. In its success, its daily audience habitually included a number of sailors, who displayed a deep emotional attachment to what they watched, as they mingled with visitors to the biennial, curators, families and all elements of society that they would usually have never shared such a space with. It provided a collective forum for a deeper understanding of cultural exchange on international waters. Its success can be seen in the fact that it was screened for three months, and inviting further collaboration to happen.

[解读]

新时代的 "辛巴达"

有不同地域和文化激烈撞击和交流的地方往往会充满了神奇的故事和惊人的传奇。CAMP 工作室用电影记录下他们不为人知的生活和故事，他们的奇遇，他们的文化，他们的歌曲还有他们的快乐和恐惧。《一千零一夜》里辛巴达七次激荡人心的航海经历，九死一生，不灭的大概是他的活力和决心，而这个时代的 "辛巴达" 们在船上形成了一种交融式的氛围，因为他们本身来自于不同的地方，也要往返于语言文化都异常迥异的地域。他们的所见所闻所感所想，都与那些终生的绝大多数时间都生活在一个地方的人们大为不同，这艘古吉拉特邦船 "Sabir Piya" 毁于火灾之后，又变得更加结实强大，而这些 "Sabir Piya" 的水手们将继续在海上御风而行。这个影片不仅是呈现给观者不一样的世界和文化，更是水手彼此的一种交流，这个影片让他们得以心心相惜的连接点。（李田）

舒适的家艺术项目
House Of Comfort Art Project

艺术家：阿尔玛·昆托
地点：菲律宾 马尼拉
形式和材料：装置艺术
时间：2007 年
推荐人：史黛拉·普拉瑟塔

Artist : Alma Quinto
Location : Manila , the Philippines
Media/Type : Installation
Date : 2007
Researcher : Stella Prasetya

《舒适的家》是一个合作的艺术项目，该项目涉及来自菲律宾的特殊艺术家和边缘群体，他们建造一个方便安装的"梦想之家"，这个装置可折叠，模块化，具有易于安装的结构。而这个结构有一个类似与参与者艺术品的房子，房顶、墙壁、窗户和门是缝在一起的。这个房子是一种设想，通过房屋的每个元素表达了参与者的生活。

项目本身放大了阿尔玛·昆托和受虐待儿童在 CRIBS 基金会、Bantay 巴塔、女农民工、日、菲儿童、和平记者、孤儿、妓女，比科尔撤离人员和来菲律宾所有地方的其他参与者一起参与的各项活动。通过该项目，阿尔玛·昆托确认这个项目能改善社区参与和协作艺术创作的情况。阿尔玛·昆托认为这是一个重要步骤，将艺术带到菲律宾的不同区域。

2006—2007 年，日本国际交流基金会、全球妇女基金和舒适房子获得国家文化和艺术委员会（NCCA）拨款支持，与幸存的社区举办了一系列的研讨会。这些研讨会举办了22场，有一千多名参与者。通过创造性的心理治疗、创意写作和讲故事、戏剧艺术和视觉艺术课程，参与者共享赋权个人的知识和技能。

《舒适的家艺术项目》是创伤、中断的项目，该项目于2007年陈列在菲律宾的文化中心；由Flaudette May V. Datuin和来自菲律宾艺术大学（UP DAS）的团队所策划。房子由阿尔玛昆托概念化，用柔软，舒服的雕塑和由妇女、儿童、艺术家、记者、教师和受自然、认为伤害的人做的挂毯作品织成。这所房子象征着希望和恢复力的象征性的房子，聚集了所有合作者可见的梦想和故事。

《舒适的家艺术项目》已经演变成一个非政府组织，舒适艺术之家的网络。该项目将继续与全国各地的群体合作，特别妇女和儿童，他们深受社会经济现实、性别不平等和自然灾害的伤害，处于不利地位。

阿尔玛·昆托认为艺术能安慰受害者；通过视觉艺术的创作手法，它也可以成为感知受害者潜感情的有效工具。她家项目的强项在于转达众多受害者的记忆和希望，通过艺术短时间拥有一种改造他们生活的潜力。

House of Comfort is a collaborative art project that involves selected artists and marginalised communities from all over the Philippines in building a portable installation of a "dream house" based on a collapsible, modular, simple installation structure that has a house like resemblance with the artworks of participant sewn on the roof, walls, windows, and doors. Which is an envisioned of participant's life representation through each element of the house.

The project itself serves as an extension of Alma Quinto activities in working with abused children at CRIBS Foundation, Bantay Bata, women migrant workers, Japanese-Filipino children, peace journalists, orphans, prostitutes, Bicol evacuees, and other participants from communities all over the Philippines. Through the project Alma Quinto identify this as a project that strengthens community participation and collaborative art making. Alma Quinto considers this as a major step to bring art to the communities in different area of the Philippines.

Through a funding support from the National Commission for Culture and the Arts (NCCA), Japan Foundation and Global Women's Fund, the House of Comfort held series of workshops with traumatised survivor communities in 2006—2007. The series of workshops were held in twenty-two venues with more than one thousand participants taking part of the workshops. The workshop is a means of knowledge and skills sharing in empowering individuals through sessions of creative psychotherapy, creative writing and storytelling, theatre arts and visual arts.

The House of Comfort Art Project is the centrepiece of the trauma, interrupted project that was exhibited at the Cultural Center of the Philippines in 2007; curated by Flaudette May V. Datuin and a team of the University of colleagues from the University of the Philippines Department of Art Studies (UP DAS). As conceptualised by Alma Quinto, the house is made out of textile grids of soft and comforting sculptures and tapestry works by women, children, artists, journalists, teachers, and others who were affected by natural and man-made disasters. This house is a symbolic house of hope and resilience that gathers all the visual dreams narratives of the collaborators.

House of Comfort Art Project has now evolved into an NGO, the House of Comfort Art Network, which the project will continue to work with communities all over the country especially with women and children disadvantaged by socio-economic realities, gender inequalities and natural disasters.

Alma Quinto believes that art has a comforting value for victims; through creative approach of visual art it can be an effective tool in perceiving the unspoken feelings of the victims. The strong piece of her house project has conveyed multiple layers of memories and hopes of the victims that for one moment through art there is a potentiality in transforming their lives.

[解读]

安居之所

安得广厦千万间，大庇天下寒士俱欢颜，这是古人的梦想，杜甫在草庐里却想象着能为天下寒士建造一个坚固的居所。可见有一个温暖舒适的家是人类共同的愿望，有时候我们生来就拥有的东西却是很多人终生都追寻而不得的温暖，阿尔玛·昆托让女农民工，日、菲儿童，和平记者，孤儿，妓女，比科尔撤离人员都参与其中，建造了一个"梦想之家"，这个家柔软而舒适，象征着希望和恢复力，里面有着各种代表他们的故事和梦想的象征元素。艺术可以激发一些潜在的情感，能营造出一种现象，他像一种魔术，但是却比魔术真实，这样的艺术作品可以安慰受害者，慢慢地改造和挖掘他们生命中美好的潜力。可以说艺术本身就是希望的代名词，他是黑暗里的微光，是邪恶世界里的挽救者，艺术本身并不能带来实质性的东西，但是他可以点燃和唤醒沉睡的愿望。（李田）

Berbeda 丹默迪卡 100%
Berbeda dan Merdeka 100%

艺术家：Respecta 街艺术画廊
地点：印度尼西亚和新加坡
形式和材料：壁画、涂鸦、标签、
  海报贴纸和模板
时间：2011 年
推荐人：史黛拉·普拉瑟塔

Artist : Respecta Street Art Gallery
Location : Indonesia and Singapore
Media/Type : Mural, Graffiti,tagging,
  poster, sticker and stencil
Date : 2011
Researcher : Stella Prasetya

《Berbeda 丹默迪卡 100%》是一项始于 2011 年 2 月 13 日西爪哇的社会文化运动。运动由 Respecta 街艺术画廊发起，涉及一系列抗议，把街头艺术用作对一系列宗教屠杀的反应，这些街头艺术在一天之内遍布印尼和新加坡的 20 个城市。这项运动是一个开放式的邀请，它号召印尼人尊重差异和实践。运动成为应对印尼少数民族宗教暴力的反应，如阿赫默德教派伊斯兰教派成员遭袭，印度尼西亚爪哇教的纵火。全国各地已经发生了这些威胁少数民族的宗教冲突；作为一个民主国家的一部分，该运动声明，拒绝对近期发生的各种象征性和生理的暴力行为保持沉默。

这样做的目的是传播宗教宽容和人权方面的主要信息，提醒这个社会，一个多元化的价值必须像一个民主国家的基本原则那样得到尊重。这一运动对社会产生了巨大影响，尽管他们的作品很容易在次日遭到替换。除了涂鸦和壁画，海报和贴纸是扩大传播所使用的形式。此外《Berbeda

丹默迪卡100%》是街头艺术转化在印尼社会运动的第一种形式。尤其是印尼的年轻人，他们深受影响，这一点有助于消息在全国各地的蔓延。参与这种公众活动，让他们通过流行文化来积极表达自己对问题的关注。线上已经组织并了大量运动，协同街头艺术社区内使用的网站和社交媒体。

同时举行的传达和谐共处愿景的运动可能实现，因此"多元化"这个概念需要不断提醒。这种类型的街头艺术运动有一种使命，充分利用解决公共问题的公共空间；因为公共场所有潜力成为提高社会意识的工具的陈设艺术。Respecta街艺术画廊坚信可以公开发现的视觉历史有助于政治思想和社会文化意识的发展。

Berbeda dan Merdeka 100% (Different and Free 100%) is a socio cultural movement that began on February 13th 2011 in West Java. Initiated by Respecta Street Art Gallery on going movement involves series of protest with the use of street art that spread over twenty Indonesian cities and Singapore within one day as a reaction toward a series of religious killings. This movement is an open invitation that acts as a call to people in Indonesia in order to respect differences and practise tolerance. The movement emerged as a response to interfaith violence to the minorities in Indonesia, such as the attack on Ahmadiyah Islamic sect members, and fire setting on churches in Central Java, Indonesia. These religious conflict that threatened minorities has been occurred across the country; as part of a democratic nation, this movement is a state of declaration to refuse to remain silent of different act of violence both symbolic and physically that recently occurred.

The aim is to spread the key message of interfaith tolerance and human right as a reminder for the society that a diversity value has to be respected as a ground principle for a democratic nation. This movement has produced an enormous impact in society in spite of their inherently ephemeral works that easily to be replaced on the following day. Posters and stickers are the form of expanded medium that were used apart from graffiti and mural. Furthermore Berbeda dan Merdeka 100% (Different and Free 100%) is the first form of street art that transformed into a social movement in Indonesia. The impact has been developed particularly amongst the youth in Indonesia, with street art as the vital component that contributes in spreading the message throughout the country. This public engagement activity has enabled them to express their concern of the issues vigorously through popular culture. The movement has been broadly organised and coordinated online with the use of website and social media within the street art community.

The movement held simultaneously to convey the vision that living together in diversity harmoniously is possible to achieve, therefore the concept of "Diversity" is something that needs to be reminded continuously. This type of street art movement has a mission to make a good use of public spaces that serves public issues; since public spaces have a potential to be display art that can be use as a tool to raise social awareness. Respecta Street Art Gallery convinced that visual histories that can be found in public could contribute to the development of political thought and cultural consciousness of the society.

[解读]
宽容与多元

曾经战争是一种各民族相互征服和交流的常态，他们打着宗教的旗号，发动了一场又一场基于信仰的战争。如今千百年过去了，不同的信仰依然存在，并且共同存在于同一个地区或者国家，不同的宗教和信仰不再需要彼此攻击和征服，而是可以通过更为和平的方式共存和彼此尊重。Berbeda 丹默迪卡社会文化活动对一系列宗教屠杀做出反应，用街头艺术的方式来倡导这种文化的多元性，这种走近大众生活的街头艺术更能够演化为一种群体性的思考与行动，这场活动蔓延 20 多个城市，用这种更能够为大众所接受的艺术的形式表示我们并不会对宗教暴力事件表示沉默，我们有所回应，并且希望整个社会也能够对待不同的信仰时更为宽容，我们可以用不同的语言向自己心中的真神祷告，但是对他人的友善与尊重也是神希望我们能够做到的。（李田）

> This comic book contains true stories about 3 girls and what they learnt about life, love and sexuality while growing up in the Normal Technical stream in Singapore.

无人问津的学校
The School Never Asked

艺术家：费罗西亚·洛	Artist：Felicia Low
地点：新加坡	Location：Singapore
形式和材料：网络漫画	Media/Type：Web comic
时间：2014年	Date：2014
推荐人：史黛拉·普拉瑟塔	Researcher：Stella Prasetya

《无人问津的学校》是关于新加坡性教育体系而献给新加坡年轻一代的项目，该体系与现状无关；它只关注科学基础。洛先前在新加坡与年轻人有过艺术项目的合作经历，他根据他们的好奇心和资源问题，从不同的视角观察年轻人如何处理来自现实世界的挑战。因此，洛利用媒体资源开创了一个项目，来反映年轻人在学习课堂内无法学到的人生价值。

它以连环画的形式，探索了爱、性及成长的主题，这些基于对三个有不同性趋向的年轻女子的一系列采访：同性恋、双性恋和异性恋。被采访主体曾是普通工艺流的学生，年龄均为24岁。该项目的目的是接触普通工艺流的学生对他们高等经验的印象。这些经验被转变为一系列场景来帮助学生处理有关性教育的问题，而这些问题不存在于现有教育制度；比如无论性别，他们如何确认自己对另一个人的感情，如何处理外在的，比如父母和社会的观念。费罗西亚·洛试图强调书中基于关系情感方面的性教育内容；而这易于被具权威的人压制。洛为学生提供更多可利用的材料；因此这些材料免费在网上发布和传播。

作为艺术家居留项目的一部分，本项目受到配电站两年居住的资助。书中插画部分也是与 Joy Abigail Ho 进行合作，在创作过程中，她年轻时的经历也帮助她联系并理解这一话题的涵义。

这本书本身是要为新加坡的年轻人提供信息资源，使其成为年轻人决策时的支持和思考方法。这一方式强调成熟和循序渐进的办法来支持那些与遵纪守法和常规学生相反的秘密生活的学生，法律和纪律将社会禁锢了社会对性教育和性取向的观念。

之前，费罗西亚·洛通过与边缘人群，如老人、危险少年、唐顿综合征人群、变性工作者开展合作项目进行社区艺术项目。该项目过去鼓励了对话，加强了社区。对于她过去的经历，洛将该项目重点放在新加坡相关社会问题上，而与政府议程或同一性目的无关。

The School Never Asked is a project dedicated for the Singapore young generations in regards to the system of sexual education in Singapore that is no longer relevant with the current situation; that only focuses upon scientific basis. Having had experience in collaborative art project with young people in Singapore, Low has identified different perspective based on their curiosity and resource issue regarding with how young people cope with challenges of real world. Therefore Low has created a project with the use of media materials as a means of reflections for the young people to learn the value of life that cannot be found inside classroom.

In the form of comic book, it explores the themes of love, sexuality, and growing up, which based on a series of interview of three young women with three different sexual orientations; homosexual, bisexual and heterosexual. The subject interviewees were former students of the Normal Technical stream, with the same range of age, 24 years old. The project intention was to reach students from the Normal Technical stream in having an impression of their senior's experience. These experiences were transformed into a series of scenarios in order to help the students addressing some issues regarding with sexual education that was not covered by current education system; such as how they identified their feeling towards a person regardless to the gender and how to deal with external perspective such as parents and society. Felicia Low attempt to emphasize the sexual education content of the book based on the emotional side of relationship; which tend to be suppressed by the students from the authority figure. Low has made the materials to be more accessible for the students; therefore the materials were published and distributed online without charged.

As part of the Associate Artist in Residence Programme, the project is funded by the Substation through a two years residency. The illustration component of the book was also collaboration with Joy Abigail Ho which in the progress of illustrating, her youth experience has helped her both to relates and comprehend the issue conveyed.

The book itself was aimed to provide an information source for the youngster in Singapore that could support as well as becoming a means of reflection by the youngster in regards with decision making. The approach was emphasised on mature and progressive approach to support students living in confidence as opposed to live by law and regulation restriction that bound the society with limited social perspective towards sexual education and sexual orientation.

Previously, Felicia Low has been conducted community art project through collaborative project with marginal groups such as the elderly, at-risk youth, people with Down syndrome, transgender sex worker. The project was used to stimulate dialogues and empower the community. In regards with her past experience, Low focuses this project to be not in relation with governmental agenda or identitarian purposes, instead it focuses upon a pertinent social issue in Singapore.

[解读]

教育的另一种形式

我们在成年之前的教育主要是在学校完成的，而学校没有教给我们的，我们应该从哪里学习？我们应当如何思考？面对我们一生都要面对的爱、性和成长，学校大概不能教给我们什么，我们跌跌撞撞地探索，这似乎是每一个经历过青春期的人都曾思索和迷惘过的。Low 以连环画的形式，从不同的视角观察年轻人如何处理来自现实世界的挑战。Low 给学生们提供了一种新的信息和视角，这种更为多元的价值观和性取向，给了普通学生关于爱和性的更多的了解和参考，同时也给了特殊的群体打破社会禁锢的机会。我们在成长的过程中不应该只接受单一的性爱和价值观，去了解别人的想法和生活，会教会我们如何更加平和得对待爱情，对待他人以及对待不同于我们已有的价值观。《无人问津的学校》给了这样一个反思和交流的突破口，给了教育以另一种形式。（李田）

象棋公园
Chess Park

艺术家：萨利赫·侯赛因、
　　　　杜奥·库达·波尼
地点：印度尼西亚 雅加达
形式和材料：壁画和象棋桌装置
时间：2009年
推荐人：史黛拉·普拉瑟塔

Artist：Saleh Husein，Duo Kuda Poni
Location：Jakarta，Indonesia
Media/Type：Mural painting and
　　　　　　chess table installation
Date：2009
Researcher：Stella Prasetya

作为2009年雅加达双年展的一部分，萨利赫·侯赛因与杜奥·库达·波尼合作开展了当地特色的艺术作品，包括与公众互动，并创造了公共空间。

空间利用已成为雅加达人民的一个问题，创造空间让人们能够通过公共活动进行互动。印度尼西亚雅加达作为国际都市，每年都是快速都市化进程的目标。随着都市化进行的快速发展，当地政府与都市空间发展步调不一致。人们必需的公共空间已被公路和商业利益空间取代。其结果是，下层立交桥成为雅加达人民利用空间的目标；尽管如此，他们仍然需要与企图利用空间获取商业利益的私营部门竞争，比如停车场。

《象棋公园》是通过艺术家与象棋社区的合作建造而成，象棋社区经常使用这一空间。坐落于南雅加达Tebet火车站附近的立交桥下，这一空间曾被视为死亡空间，而如今通过当地市民的象棋活动成为社交场所。自从象棋活动从市政处去除后，这一社区经常受到警察的突袭；

尽管这一积极活动为社区安全作出贡献。

基于相关问题，萨利赫·侯赛因与杜奥·库达·波尼观察到创建公共空间为象棋公园的潜力，2009 雅加达双年艺术节在将公园合法化的过程中给予艺术家支持。这一观察通过与当地市民沟通和讨论，持续了一个月之久。

为了批准这一选址，两列公路结构被涂上象棋子的黑白色和想起社区的名称。而且公园配备可可携带桌子，桌面上印有棋盘以及座位安排，可容纳 16 人。作为对警察突袭的应对，社区要求桌腿是可拆卸的。

萨利赫·侯赛因与杜奥·库达·波尼（Yusmario Farabi 和 Aprilia Apsari）设想通过公众参与公共艺术创造来恢复城市空间为公共空间的潜力。这是一种公共教育方式，让人意识到公共空间的概念是共享的空间。

As part of Jakarta Biennale 2009, Saleh Husein in collaboration with Duo Kuda Poni has conducted site-specific artworks that involved a public interaction along with the process in creating a public space.

Utilization of a space has been an issue for people in Jakarta to create a space that enables people to interact through numbers of public activities. As a metropolis, Jakarta, Indonesia has always been a target of rapid urbanisation every year. Within the rapid flow of urbanisation, local government has not been able to balance with the pace of urban space development. Since public spaces necessities for people has been replaced by the development of highway and commercially interest spaces. As a result underneath flyover became the target of space utilization by people in Jakarta; nevertheless they still need to compete with private sector that attempt to utilize the space as commercial interest such as parking lot.

Chess Park is created through artists' collaboration with chess community that has been using the space frequently. Situated underneath flyover near Tebet train station, South Jakarta, the space that was considered to be a dead space has become a place of interaction through chess activities by local citizen. Since the activity is unlisted in the municipal office, this community has always been targeted by sudden police raid; although this positive activity has contributed to the neighbourhood safety.

Based on the related issue, Saleh Husein and Duo Kuda Poni observed the potentiality of public space creation to become Chess Park and Jakarta Biennale 2009 art festival supported artists' intention in the process of legitimating the park. The observation took one month through the process of communication and brainstorming with local citizen.

In order to authorise the site, two column of highway structure was painted by mural painting of two black and white chess piece complemented with the chess community title. Moreover the park is equipped with portable table that has chessboard painted on the surface along with seating arrangement that was designed to accommodate 16 people. As an anticipation attempt of sudden police raid, the community requested the table leg to be detachable.

Saleh Husein and Duo Kuda Poni (Yusmario Farabi and Aprilia Apsari) envision in restoring the potentiality of city space as a public space through public engagement in public art creation. This is a means of public education in perceiving the conception of public space as a shared space.

[解读]

空间服务于大众

随着城市的急剧商业化,非利益驱使的公共空间显得如此稀少而宝贵,这些公共空间在被商业占领的大面积区域的包围下,终于争得这样的一席之地,《象棋公园》就是通过艺术家与象棋社区的合作建造而成。象棋公园坐落于火车站附近的立交桥下,这一空间曾被视为死亡空间,但是现在这个空间变成了公共艺术的空间,成为了市民交流和活动的场所。公共艺术驱使公共空间成为一个公共教育方式,也发掘了公共空间的潜力,在被消费与商业包围的城市,艺术与文化能否在这个荒漠中慢慢生长蔓延,也取决于公共空间。而对于城市的改造就是由这些一点一点的公共艺术空间中开始,并且作用于人们的观念、生活、文化和想法。公共空间只有在艺术家的引导下真正服务于大众,才成为真正的大众艺术。(李田)

Cigondewah 文化中心
Cigondewah Cultural Centre

艺术家：缇斯纳·桑扎雅
地点：印度尼西亚 万隆
形式和材料：建筑
时间：不详
推荐人：史黛拉·普拉瑟塔

Artist : Tisna Sanjaya
Location : Bandung, Indonesia
Media/Type : Building
Date : Unknown
Researcher : Stella Prasetya

Cigondewah 位于印度尼西亚的万隆市西南方，被工业化和城市化所破坏。因此，洪水，水源和土壤污染以及垃圾蔓延等环境污染问题时常发生。

缇斯纳·桑扎雅的童年就是在 Cigondewah 度过的。缇斯纳·桑扎雅还见证了 Cigondewah 从以前免费的农村共有土地，向如今以纺织厂和垃圾回收厂为主的工业化用地的巨大转变。2008 年，缇斯纳·桑扎雅决定购置一块土地，用以创建 Cigondewah 文化中心，旨在通过各种不同形式的公共环保活动来唤醒当地人们的环保意识和恢复当地的生态生机。该文化项目是此艺术家用自己的 6 件艺术作品换得 600 平方米土地，自筹资金而建立起来的。

在项目施工过程当中，大量的塑料垃圾污染了该区域地点和水源。随着地基挖掘施工的进行，缇斯纳·桑扎雅发现他所购置的土地下面被深达 2 米的垃圾塑料所污染。这些塑料垃圾后来以艺术品的方式，被桑扎雅展示在新加坡国立大学博物馆内。此次生态恢复活动由万隆科技大学艺术系的学生和 Cigondewah 当地居民共同参与。环保活动第一步就是就号召当地居民参与义务植树。由于尽管 Sanjaya 试图提高公众对如何适应和采取更为可持续性的城市生活方式等问题的关注度，他仍需要应对因现代生活方式所带来的公众对此类活动的冷漠心理。大部分生活在类似于印度尼西亚的发展中国家的人们，必定会经历由现代化进程导致的贫富差距而带来的社会宽容度和公共意识的缺失。因此，为了强化人们对"土地资源和水资源不能当做废物处理"这一观念的认知，将艺术平民化可以被视为通过知识转移这一有创意的途径来缩小社会差距的有利手段。

缇斯纳·桑扎雅把这座房子想象成一件能够还原艺术的自然价值的艺术品。而现如今房子更多是只是被赋予了商业价值。缇斯纳·桑扎雅有意识地把这项活动的宗旨定为：提醒人们艺术活动的本质为对人性的关注，而不仅仅是利益的得与失。Cigondewah 文化中心活动向人们展示了传统公用地在开发和利用方面的更多选择自由，而这种选择上的自由在如今的城市生活中早已不复存在。缇斯纳·桑扎雅把因工业化和城市化所导致的环境破坏看作是现代化的阴暗面。然而，通过他举办的文化项目，一种新的意识已经开始在世人的脑海中扎根。

Cigondewah is a place situated in southwest Bandung, Indonesia that has been violated by industrialisation and urbanisation; as a result several environmental issues have risen including flooding and pollution of soil and water and overflow garbage.

Spending his childhood times at Cigondewah, Tisna Sanjaya has witnessed a dramatic change of the area from rural setting of free communal land into a highly industrialised area that is dominated by textile factories and garbage recycling. In 2008, Tisna Sanjaya then decided to purchase a plot of land and creates Cigondewah Cutural Center as an attempt to rejuvenate the area and creating environmental awareness through various public events. The project itself was self-funded by the artist through negotiation process of exchanging six pieces of his artworks with a six hundreds square meters land.

During the building process, an enormous amount of plastic waste has polluted the land and water of the area. As the excavation foundation begun, Tisna Sanjaya found two meters depth of plastic waste contaminated the land that he purchased; the waste then presented as an artwork that he exhibited at NUS Museum, Singapore. The rejuvenation process involves collaboration with students of Art Department of Institute Technology Bandung along with people in Cigondewah. The first attempt of rejuvenation to invite people for tree plantation was questioned by the people, as they no longer recognize the value of having green land.

Although he attempt to raise public awareness on how to adopt and adapt the more sustainable modes of urban living, Tisna Sanjaya has to deal with public apathy as a consequence of modernisation that shaped individual character. The majority of society that live in a developing country such as Indonesia has to experienced the lack of tolerance and public awareness as a result of social gap that was created by modernisation. Therefore by providing art to be approachable to the society can be used as a medium to fill the social gap through creative approach of knowledge transfer; in order to revise their conception of land and river for not to be used as waste disposal.

Tisna Sanjaya envisioned the house as the artwork that is capable to return the nature value of art, in which today has overly dominated by commercial value. Tisna Sanjaya intentionally presented this project as a way of reminder action of humanity as the essence of art activities instead of loss and profit consideration. As an artwork, the house is installed by diverse medium of artworks that by Tisna Sanjaya as a respond to and in collaboration with people in Cigondewah.

The activities of Cigondewah Cultural Centre demonstrate the freedom of traditional communal uses of open spaces, which is no longer, can be found from urban area today. Tisna Sanjaya referred this environmental destruction by industrialisation and urbanisation as the dark side of modernisation. Nevertheless through his art project a new awareness has started to embedded in the society.

[解读]

身处渠沟，仰望明月

污染问题好像成为了国家和城市发展和崛起的诅咒，发达国家很早就经历了这个黑暗阶段，但是发展中国家还在阵痛之中。艺术家缇斯纳·桑扎雅从童年起就生活在这个地方，他亲眼目睹了这个地方从一个可以种植的清洁之地，一步步变成一个被垃圾覆盖被污水弄脏的地方。过去美丽的故乡消逝在城市的尽头，而城市和工业以牺牲土地和自然为代价如春笋般兴起，人们虽然对此早已察觉，但是现代化的进程和飞速发展的工业又不能够立马停下来，人们为了经济的持续繁荣，对这种泥沙俱下的发展方式已然麻木。艺术家缇斯纳·桑扎雅将《Cigondewah文化中心》想象成一件能够还原艺术的自然价值的艺术品，他发起了各种环保的公共活动，唤起人们对自然的情感和对城市污染的担心。我们如今虽然身处污浊的渠沟，但是却心向明月，而艺术家所希望的那个自然清洁的世界也会实现。（李田）

公共艺术、哀悼和恢复
Public Art，Mourning & Resilience

艺术家：多位艺术家	Artist：Multi artists
地点：黎巴嫩贝鲁特	Location：Sabra and Shatila Refugee Camps, Beirut，Lebanon
形式和材料：壁画	Media/Type：Murals
时间：2012 年	Date：2012
推荐人：纳赫拉·阿尔·塔瓦	Researcher：Nahla Al Tabbaa

1982 年，黎巴嫩贝鲁特郊边的 Sabra 和 Shatila 难民营内发生了一起使 3500 名巴勒斯坦难民殉难的凶残屠杀案。毫无征兆地，Faluga 士兵在以色列的授意下采取了人口消灭的丑恶行径，致使幸存者永远遭受着心理的创伤。30 年过去了，这片难民营依旧饱受恶劣的卫生环境、鲜有的水电供应，以及沉重的心灵创伤之苦。这个项目旨在将这段值得追悼的事件构建为一个当今形式的史实。

Susan Greene 曾做过 20 年的心灵创伤、精神恢复、记忆和创造力的工作并创建了"艺术的力量"（Art Forces）。"艺术的力量"旨在通过社区艺术项目和社会媒体激发批判性的想法和行为以及将巴以冲突的意识传递到远方的美国。

本地的艺术家们应邀进入 Sabra 和 Shatila 难民营，这个难能可贵的过程需要得到准许，成堆的文书工作才能见到限制在这些难民营内的居民。艺术家们了解到的是沉重的悲痛气息，恶劣的生活环境以及搬离这里的愿望，而他们的任务就是创造使居民受到鼓舞的作品，加强他们的集体回忆尤其是对于刚出生的新一代人。

在壁画绘画过程中，社区的参与则为艺术家和社区居民之间设立了一个相互沟通的平台。他们可以交流彼此的故事、梦想以及抱负期望，无论是个人的还是整体意义上的，并对他们的日常冲突进行反思。通常，这样的项目往往会超出人们的期望，通过壁画的创作就可以产生共鸣，尤其是当之间的关系可以有机建立时，意想不到的合作就会发生。

从Sabra和Shatila难民营的交流中流露出的主要情感主题，一个是沉默，一个是失落感。随着时间的推移，大屠杀的见证者和幸存者会感到他们的社区会在没有一个永不侵蚀的史实记载下不知不觉地沦落丢失。Art Forces要采取的方式不仅要以永恒的方式来证明这一史实，还要以难民营的建筑来影响世界观。

短语"难民营（CAMP）"仅仅用来指代可被正规称之为城镇的水泥建筑，因为这里的居民是难民。这些安置点被称为难民营的理念使这里的居民一直梦想哪一天可以逃出这里回到属于他们的家乡。这一思路被Art Forces以一种方式有效地借鉴，这种方式是壁画都是共同创建的，从而为这一特殊社区提供一种公有制的艺术，这种艺术与他们有公共联系性。

The year 1982 saw the heinous massacre of 3500 Palestinian refugees in the Sabra and Shatila camps in the outskirts of Beirut, Lebanon. Unannounced, Faluga soldiers with the allowance of Israeli soldiers wiped out a population, leaving the survivors permanently traumatized. 30 years on, the camp suffers from poor sanitation, hardly any electricity and water, and the heavy hearts of its survivors. This project aimed to frame this memorial as a current form of documentation.

Susan Greene had been working with the intersections of trauma, resilience, memory and creativity for two decades and founded Art Forces. Art Forces aims to inspire critical thinking and action through community art projects, social media, bring awareness of the Palestinian conflict, across the waters to the United States.

Regional artists were invited to enter into Sabra and Shatila camps—a rare process which requires permits, limited visitations and heaps of paperwork, to meet its inhabitants who are confined to these camps. The artists learnt of a heavy sadness, poor living conditions, and a desire for these people to move, create, feel inspired, and to reinforce their collective memory, particularly as new generations are being born.

Community participation during the mural painting set a platform to engage in a dialogue between the artists and the community. They were able to exchange stories, dreams, ambitions- personal or collective, and reflect on their daily conflict. Often, such projects tend to exceed expectations, echoing past the mural making, particularly as relationships are built organically, and unexpected collaborations occur.

The main emotional theme that emerges from the Sabra and Shatila discourse is one of silence, and a sense of loss. The witnesses and survivors of the massacre feel, as time passes, that the gravity of the manner in which they lost a community would in itself be lost without documentation that wouldn't erode. The manner in which Art Forces not only manifested that documentation in a timeless, and traditional, manner, not only brought a sense of tangibility to the saga, but also affected the worldview that the camp had of its own buildings.

The term "CAMP" is only used to describe the concrete buildings that would otherwise be validly termed as towns, because the residents are refugees. The idea of these settlements being camps evokes the constant dream that the residents may one day be able to "decamp" and return to their homelands. This train of thought was effectively addressed by Art Forces in the manner in which the murals were collectively created, and provided a public ownership of art that was publicly relevant to that particular community.

[解读]

地狱中的微光

人类文明进程里其实充满了血腥苦难战争和无可言说的哀伤，历史的滚滚车轮不断向前的过程中，人们力求避免残酷的屠杀和血腥的战争，我们向往天堂，可是我们必须一步步从地狱走出来。历史是善于遗忘的，也许几千上万的无谓死亡，带给生者无尽折磨的灾难只是历史书上轻描淡写的一句话，直至最后记得这场灾难的人都已死去。为纪念 Sabra 和 Shatila 大屠杀 30 周年，艺术的力量想用绘画让这段历史被人们永远铭记。这些为难民营做壁画的艺术家们得以看到这个阴冷混乱的人间地狱，用绚丽的色彩带给他们一点温柔，一点记忆。Mahmoud Darwish 在一幅画里写道："当殉难者永远的沉睡了我醒了，我告诉他们，'希望你们可以醒来发现在一个由渴求爱交织而成的国度里'。"有些悲剧，究其恩仇根源已太过复杂，只愿没有人为的地狱和故意的践踏。（李田）

吉卜兰·图韦尼纪念馆
Gebran Tueni Memorial

艺术家：弗拉基米尔·久罗米洛维奇　　Artist：Vladimir Djurovic
地点：黎巴嫩贝鲁特　　　　　　　　　Location：Beirut，Lebanon
形式和材料：雕刻、装置艺术　　　　　Media/Type：Sculpture/ Installation
时间：2011 年　　　　　　　　　　　　Date：2011
推荐人：纳赫拉·阿尔·塔瓦　　　　　Researcher：Nahla Al Tabbaa

吉卜兰·图韦尼纪念馆位于贝鲁特新设中心地带，是对黎巴嫩最富盛名的政治人物中吉卜兰·图韦尼（Gebran Tueni）先生一生伟业和功绩的公共献词。图韦尼先生无论是在振奋人心的演讲中还是他的书稿文件中都用其特色的语言传递着赤忱的爱国情怀，并被广大市民周知。而本纪念馆则可恰如其分地体现出上述内涵。

Djurovic 决定不用传统的纪念碑来对这位先生表达哀悼和致敬，而是将多种元素符号无缝嵌入这块景地以赋予它更多的公共意义和民众色彩。这块单条状铺设物用宽为 15cm×9cm×57 cm 的花岗岩设计而成（鉴于吉卜兰·图韦尼生于 1957 年 9 月 15 日），上面铭刻着图韦尼先生许多的名言警句和不朽誓言。该纪念碑的内侧则种植着黎巴嫩的本土植物。过往行人可以在这里拿到刻有图韦尼先生名字的鹅卵石，这些鹅卵石会被他以前供职的报社同事来不断添充，以表达他们对图韦尼先生无尽的缅怀和思念。

该艺术家醒目大胆的举措在于他始终坚持不用外在的物理形象来表现当事人（图韦尼先生），不用雕刻形式来体现他的相貌或者外形，而是用一种更新颖的方式来使公共民众与图韦尼先生的当今社会影响相互作用。该纪念物没有采用与公众互动脱节的铭记方式，而是处处体现其真正意义上的公共价值，即其实际用途就是作为一条人行道。该艺术家也想公共民众在走过、跨过、踏上该纪念物时可以与它进行无意识的互动，这便可以体现韦尼先生在整个黎巴嫩社会中的存在，无论这个社会对他的铭记是主动还是被动。

作为一处景观纪念广场，艺术家赋予了它丰富的公共色彩和图韦尼先生独特的人物角色，这种表现形式则应被人称道。代表图韦尼先生不朽名句的文字，即影响大多数民众最重要的方式，在这里用多种语言清楚列出，就算是黑夜也会用灯光照映，使人们可以无时无刻受到图韦尼先生的激励和鞭策。这也象征着图韦尼先生报社的名字"An Nahar"，是"白天"的意思，而图尼在生前则赋予了它特定的文化含义，即生命不息，希望不灭。而选择的材料则体现了一种微妙和谐而又强有力的精神；花岗岩表现出的力量和树木映衬出的柔美相互交融，其不朽的精神内涵更使这块纪念碑为人称赞。

The Gebran Tueni memorial, situated in the new central area of Beirut, is a public dedication to the life and work of one of Lebanon's most popular political figures. Tueni was known for his patriotism and for his words, either through his inspiring speeches or in the written form. The memorial fittingly highlights both of these elements in abundance.

Instead of using a traditional representation of the man as monument, Djurovic decided to invite the public into the effect of the personality through multiple symbolic dimensions, curated in a seamless landscape. A single linear column, created of granite bands 15, 9 and 57 cm wide (his birthday being 15/9/1957) lies on the floor etched with many of his most famous statements and oaths. The plantation lining the memorial are all species indigenous to Lebanon. Passers-by are invited to take pebbles engraved with his name, constantly replenished by his old colleagues at the newspaper he worked at signifying an act of constant remembrance on their part.

The boldness through which the artist has insistently created the memorial devoid of full physical representation of the subject, without sculpturally mentioning him in visage or shape, presents a relatively new manner in which the public can interact with the contemporary presence of Tueni's legacy. Instead of remaining a memory external to public interaction, every part of the memorial exhibits truly public values, namely its practical use as a walkway. The artist has also allowed for the public to interact with the memorial even subconsciously, as they walk past, over and through the memorial, it represents the subject's presence in Lebanon's society whether that society remember him actively or passively.

As a landscape memorial, particular recognition should be given to the manner in which the artist has reflected the very public personality and persona of Tueni. The font used for the recognition of Tueni's words, how the majority of the public interacted with him in his life, is clear cut, in multiple languages, and illuminated at night, always available for the public. This is also symbolic of the fact that Tueni's newspaper was named An Nahar, meaning "The Day", and particular cultural reference was made during Tueni's life of the idea of this daylight, and hope, never being extinguished. The choice of materials used provide a subtle yet resonant display; the strength of the granite, merged with the fragile beauty of the trees lining them, provides the element of timelessness that renders this memorial truly fit for praise.

[解读]

无声的致敬

人们从花岗岩的地砖上走过，吉卜兰·图韦尼先生的成就还有他的名言就在脚下提醒着人们，有些人和有些历史应当被人们永远铭记。艺术家Djurovic为吉卜兰·图韦尼先生所做的纪念碑给了我们一个新的视角，这不仅是一个纪念碑，其实更像是一处沉静的公共景观。人们可以走近他，和他产生某些互动。肃穆的黑色地砖上有各国的文字书写的他的生平，鹅卵石上也刻着他的名字，人们可以拿鹅卵石回去当作纪念，似乎这些坚硬的材质给了这个纪念碑恒久的意义，石头将永远记载着他，而人们也不会忘记。这一切看上去与的周围环境融为一体，整个纪念碑静穆而伟大，无声地向这位伟大的公民致敬，同时也是向这内战结束后的20年里，黎巴嫩人民所承受的冲突、经历过的苦难、昂扬的奋斗致敬。（李田）

计数器
[En]Counters

艺术家：艺术氧气 (NGO) 等多位艺术家	Artist : Art Oxygen, various artists
地点：印度孟买	Location : Mumbai, Indian
形式和材料：互动行为、多媒体艺术	Media/Type : Various Interventions—Multiple
时间：2010年至今	Date : 2010-on-going
推荐人：伊芙·莱米斯尔	Researcher : Eve Lemesle

《计数器》是由非营利艺术组织——艺术氧气在 2010 年发起的一个公共艺术课题。从事公共领域有双重需要：一是给观众提供接近当代艺术形式的途径，他们通常不会接触到这些艺术形式；另一个是询问新的叙述手法是如何通过艺术实践形成的，这些新的叙述手法来自影响我们生活的空间与场所的日常问题。

《计数器》是一系列的伙伴关系和合作的催化剂，这些伙伴关系与合作不仅是与印度和国际艺术家和从业者，而且是与其他学科的组织机构。自从 2010 年，这个课题已经涉及超过 50 个印度和国际的艺术家，与 10 多个亚洲和欧洲的文化机构建立了合作关系，当代艺术家的人数超过 500 000 名，并且举办了超过 1000 名初中生和高中生的讲习班。

此课题的目标是把艺术品放在日常生活中，通过在特定场地进行的一次活动，鼓励印度艺术家进行询问并探索孟买本性的尝试。艺术家把孟买定义为分裂自我和独特的灵魂。他们对梦想和绝望，无节制和过度节俭，民主与无政府状态的生活进行了对比。通过对比，每个艺术家都成为该场地的一个组成部分，融入社会动力学并且成为创建日常关系的一部分。该课题试图通过这种方式将艺术引进公共空间，方便人们之间相互影响，参与以及和艺术家及其作品进行对话。

《计数器》的各种版本已经探讨了孟买的城市发展以及鼓励改变城市中日常生活的景观创意和行动的问题。2011年版的《计数器》集中于水的问题，艺术家图沙佐治模仿城市流行的戈库尔 Asthami 节，以喷泉的形式组装了一个叠罗汉，出现在城市的高层建筑中。这个作品是对现在城市的豪华建筑以及开发商不能提供供水设施的无能的刻薄评论。2014年版的标题是这片天空是否还有爱？艺术家卡拉特用珠瑚海滩的盐印了一系列的情书，与作品的短暂性进行对比。这些情书被海浪冲走，而盐却是经常用于保存食物，而且是持久性和持续时间标志的一种物质。

现如今即将到来的计数器的版本是以"空间过渡"为主题。

[En]Counters is a public art project that was initiated by Art Oxygen, a non-profit arts organisation in 2010. The need to engage with the public sphere was two-folded: one to give accessibility to contemporary artforms to audiences which are normally not exposed to them; the other, to interrogate how new narratives stemming from everyday issues affecting the spaces and places we live in could be generated by artistic practices.

[En]Counters was the catalyst for a series of partnerships and collaborations not only with Indian and international artists and practitioners, but also with organizations operating in other disciplines. Since 2010 the program has involved more than 50 Indian and international artists, establishing more than 10 collaborations with Asian and European cultural institutions, bringing contemporary arts to more than 500,000 people and organizing workshops with more than 1,000 junior and senior students.

With the objective to place artworks amidst the drone of the everyday, the project was an attempt to encourage Indian artists to question and explore Mumbai's nature by carrying out site-specific activities. The involved artists investigated the opposing elements which define Mumbai/Bombay's divided self and unique soul; they explored the contrasts which populate its daily life, between dream and desperation, excess and parsimony, democracy and anarchy. In doing so, each of the artists became an integral part of the site, integrating into its social dynamics and forming part of the daily relations created. In this way, the project tried to bring the arts to public spaces for people to interact, engage and dialogue with artists and their works.

The various editions of [En]counters have explored issues related to Mumbai's urban development and encourages creative ideas and actions to transform the city's everyday life and landscape. In [en]counters 2011's edition dedicated to the issue of water, artist Tushar Joag assembled a human pyramid in the form of a fountain mimicking the city's popular Gokul Asthami festival, which was performed in front of high-rise building coming up in the city. The work was a caustic comment on the current spurring of luxury buildings across the city and the incapacity of developers to provide water facilities. In the 2014 edition titled "Is There Love in This Air?", artist Reena Saini Kallat stenciled a series of love letters with salt on Juhu Beach, playing with the contrast generated by the ephemerality of the work, which got washed away by the waves, and the use of salt, a material often used for conserving food and a symbol of permanence and duration.

The forthcoming edition of counters is themed Spaces Transition.

[解读]

艺术走进公众

艺术之于众人，一直犹如阳春白雪般曲高和寡。艺术源于生活何处，高于生活几何？艺术氧气试图通过计数器给出"准确的数字"，解开艺术神秘的面纱。来自世界各地的当代艺术家通过跨学科、多领域的结合，在特定场地举办艺术活动，将原本束之高阁的艺术品放置于日常生活中，以各式各样具象的、新奇的公共艺术形式引发问题的思考，进而影响我们的生活。同时，艺术家们尝试着以多样化新颖的叙述手法付诸于艺术实践，诞生出各种版本的计数器，每个都以特定的主题引发我们对于艺术介入生活问题的思考。无需严格计算艺术与生活的距离，我们只需将艺术融入生活，将生活思想具象化，如此便可消除人们与艺术之间的隔阂，使之切身感受到艺术始终存在于我们的公共空间中，触手可及。（闫丽祥）

背景：公共·艺术·生态
In Context: Public · Art · Ecology

艺术家：多位艺术家	Artist：Multiple
地点：印度新德里	Location：Delhi
形式和材料：装置艺术、调研、公共艺术	Media/Type：Installation Research Public Art
时间：2010 年至今	Date：2010-on-going
推荐人：伊芙·莱米斯尔	Researcher：Eve Lemesle

这个课题试图调查新德里市的多个生态。背景一：公共艺术生态居所项目于 2010 年 3 月开始。艺术家（来自印度、德国、日本和美国）的项目聚焦于对公共空间的干预。项目意图以居住地为基础，强调艺术家创作主题与干预对话。

课题从映射的天气模式和气候变化对当地社区的影响进行研究；交互式视频雕塑；在公路带上检查树木的意义；设计一个人们关于问题和保护问题进行交流互动的画面；构建一个自然生物的水净化系统；追寻人们的足迹以及人们从月光集市到古尔冈的基本目标。

这个项目成立多年以来，居住地课题已经有各种版本，如谈判路线版本（2011），这个版本采用结合艺术家和当地社区的研究和艺术创作的跨学科的方法，邀请艺术家、专业人士从事特定场地的特定项目，访问目前看得见的和看不见的转换模式代替访问直接环境。

背景中的食品版本（2012和2013）：公共·艺术·生态旨在当前的社会与文化中，集体环境和个人环境中，审视食品的相关性和意义，赞美我们的身体和自我之间的内在联系，并且把它作为一个基本的鼓励参与、互动与合作的仪式进行探讨。关于周围食品政治问题的关键检查，通过把食物作为艺术媒介进行使用，并且把它并入性能、艺术装置或互动活动，居住地提供了一个与周围食品相关的进行有趣对话的机会。

课题：希尔维亚温克勒和史蒂芬——PPR（乘客驱动人力车）

通过使用自行车作为德国的日常交通工具的试验，他们相信在未来城市可持续发展中，自我驱动的车辆在短距离运输中将发挥重要作用。在公共艺术生态背景中：他们在新德里用常见的公共交通方式——人力三轮车工作，同时构建了一个PPR（乘客驱动人力车）原型。车上的乘客/拉车工的等级关系转化成一个临时的合作关系，两个人用体力和逻辑知识共同向前推进人力车。艺术家通过乘客驱动人力车，把路人纳入新德里的不同街区，为后石油社会的梦想添加燃料。

The program attempts to interrogate multiple ecologies in the city of Delhi. The first In Context: public.art.ecology residency commenced at the beginning of March 2010. The artists (from India, Germany, Japan and the USA) projects focus on interventions in the public sphere. The curatorial intent underlying the residency stresses on the dialogic aspect of the artists thematic and intervention.

Projects ranged from mapping weather patterns and the effects of climate change on local communities; an interactive video sculpture; examining the signification of trees in the context of road zones; designing a tableaux that will interact with people around questions around conservation; building a natural biological water purification system; to one that traces the paths of people and their constitutive objects from Chandni Chowk to Gurgaon.

Over the years, since its inception, the residency program has had various editions such as The Negotiating Routes edition (2011), which invited artists, professionals to work on site-specific projects with an inter-disciplinary approach that combined research and art creating by artists and local communities, addressing the visible and invisible transformations currently taking place in their immediate environments.

The food edition (2012 and 2013) of In Context: Public · Art · Ecology aimed to re-examine the significance and relevance of food in our current social and cultural, collective and individual milieu, celebrate its intrinsic connection with our bodies and our selves, and explore it as a primary ritual that fosters engagement, interaction and collaboration.

Alongside a critical examination of issues surrounding the politics of 'food', the residency offered an opportunity to engage stimulating conversations around the discourse of food through employing food as artistic medium incorporating it in performance, art installations or in interactive events.

Projects: Sylvia Winkler & Stephan Koeperl—PPR (passenger propelled rickshaw)

Through their experience of using the bicycle as a daily means of transport in Germany, They were convinced that self powered vehicles are going to play a major role for short-distance transport in the sustainable cities of the future. During "In Context: Public · Art · Ecology" they worked with the use of cycle rickshaws in Delhi, which is a common mode of public transport. A prototype of a PPR (passenger propelled rickshaw) was constructed. In this vehicle the hierarchy of passenger/puller is transformed into a temporary collaborative unit where physical power and logistical knowledge are shared to bring things forward. The artists engaged the passers-by in the different neighborhoods of Delhi by riding the PPR to fuel the dream of a post-oil society.

[解读]

"绿色"的艺术

"假如地球上没有一滴干净的水，没有一口干净的空气"，我们的生存将去向何处？在这尝试提问和解决问题的并不是环境学家，而是来自印度、德国、日本和美国的艺术家。世界那么大，我们无法丈量所有的地域生态，但是紧紧环绕身边的衣、食、住、行却近在咫尺。艺术家们以居住地为基础，或将食物作为艺术媒介，融入艺术创作；或探寻社区研究与艺术创作的交流对话；或形成公共艺术与生态间的互动合作。希尔维亚温克勒和史蒂芬——PPR（乘客驱动人力车）基于未来城市的可持续发展，以自然环保的自我驱动、体力与逻辑知识的共同结合，使乘客参与到艺术家的驱动中，形成临时合作体，共同构成艺术生态的"燃料"。（闫丽祥）

"贾亚赫"（印度的荣耀）
"Jaya He"（Glory to India）

艺术家：拉杰夫·塞西
地点：孟买希瓦吉国际机场

形式和材料：空间艺术
时间：2014 年
推荐人：伊芙·莱米斯尔

Artist：Rajeev Sethi
Location：Chhatrapati Shivaji International Airport, Mumbai
Media/Type：Space Art
Date：2014
Researcher：Eve Lemesle

国际机场候机楼标题为《"贾亚赫"（印度的荣耀）》的艺术项目，是由南亚的著名舞台设计师、亚洲遗产基金会主席拉杰夫塞西策划、提供概念化和帮助设计。机场候机楼布满印度艺术品，包括传统的工艺品以及当代艺术。在芝加哥成立的建筑师团体斯基德莫尔、奥因斯和梅里尔从印度的国鸟孔雀身上得到启示创作了 2 号候机楼，并且使它成为一种艺术和设计奇观。

作为印度最大的公共艺术之一，机场有 7 000 件艺术品，分为两个不同的部分。第一部分包括特别委托的艺术作品，这个作品把城市绘制成一个分层的结构，当游客在走廊行走的时候，每个作品都会讲述一些关于城市的故事。第二部分是一个像中央脊柱曲线的一面墙。它用艺术作品的形式反映了印度的多元文化遗产、生活传统和当代艺术。

此项计划仍处于不断发展中，目前包含了 1 500 位艺术家的 7 000 件作品，展示在一个 3 000 米长的多层艺术墙上。机场除了恢复的历史文物、当地工匠和民间艺人创作的作品外，还有一些全国著名的当代艺术家受委托创造的艺术作品和公共艺术装置。

整个项目的策展聚集了一群艺术家、设计师、建筑师、历史学家、人类学家和技术管理员，从文化、美学、历史和社会等方面一起工作来诠释印度文化艺术。

位于印度的 2 号候机楼项目，受到大众的认可和喜爱，对于每年 4000 万名通过机场的乘客，这是一个包含多样性的候机楼。据报道，每个艺术作品的创作过程都由工匠通过手机进行实时记录，这些视频也将被安装在机场，以便观众更深入的了解艺术家及作品创作的整个过程。

1. 阿克沙伊《再生的孟买》

艺术家阿克沙伊使用废弃的材料，组成孟买航行图中熟悉的地区，深深吸引了人们的眼球。作品材料全部由艺术家从孟买当地收集而来，如帕雷尔、达拉维和佐巴刹跳蚤市场。阿克沙伊用废弃的按钮、微芯片和电路，在孟买的像素化图像上，创建了此件大型装置，甚至在谷歌地图上都很容易可以看见。

2. 加尔米沙哈《城市弹性共振》

加尔米沙哈创建的小型建筑结构，唤起了人们对老建筑拆除相关的怀旧和失落之情。她在 2 号候机楼的装置比她平时的作品规模更大。她花费一年的时间，做了大量的勘测图纸和模型，与策展和设计团队反复讨论最初的作品概念，并以孟买迦特克帕郊区以及周围废弃的建筑物为基础创作而成，同时孟买迦特克帕也是她生活和工作的地方。

3. 迪利普《沉默的方式》

印象中的机场始终伴随着喧嚣，但是迪利普的作品却给观众提供了片刻沉默的场所。它给观众描绘了一个精心详细的内部世界，在这个世界里，每个对象和场景尽管平凡却都赋有诗意。五种不同的场面犹如正在进行的城市建筑现场，观众的目光随着画面的展开逐步推移。画面的单色设置将各式各样的建筑风格进行强烈的对比；通过画面中的窗户看到的景观同样是不同的天空，观众的目光随着画面时而远眺高山，时而鸟瞰，永远不知道带你去的下一个安静的角落在哪。

4. 珠帕森的《希望》

在珠帕森的作品中充满瞬间的神奇时刻；图像从一个出租车或建筑物变化到伊卡洛斯。从这个 36 英尺长的透镜状印刷作品，你可以看到繁华的孟买：骑行者，上班族，装饰着不同艺术风格的建筑以及标志性黑色和黄色的菲亚特出租车。当改变观看位置时，图像就会转变到希腊神话中敢于飞翔的叛逆英雄伊卡洛斯。

The international airport terminal's art programme, titled "Jaya He" (Glory to India), has been curated, conceptualised and facilitated by Rajeev Sethi, one of South Asia's leading scenographers and Chairman of the Asian Heritage Foundation. The terminal was built in a manner to accommodate artworks from India that comprise traditional arts and crafts as well as contemporary Indian art. Chicago-based architects Skidmore, Owings and Merrill, who are behind the Burj Khalifa in Dubai drew inspiration from India's national bird, the peacock, for the architecture of T2, making it something of an art and design wonder.

One of India's largest public art initiatives, the airport houses 7000 artworks. There are two different sections. The first section comprises specially commissioned artworks that map the city as a layered narrative, where each work tells something about the city as travellers walk in the arrival corridor. The second section is a wall running like a central curvilinear spine. It contains artworks that echo India's plural cultural legacy, living traditions and contemporary artistic expressions.

Planned not to remain static, the constantly evolving programme consist of 7,000 artworks by 1,500 artists in the form of a three-kilometre long multi-storey art wall. Apart from restored historical artefacts and works by local artisans and folk artists, several nationally renowned contemporary artists were also commissioned to create art and installations.

To begin with, the curator gathered a team of designers, artists, artisans, architects, art historians, anthropologists and conservators with technicians, all working together to collect and interpret India culturally, aesthetically, historically and socially.

The project at T2 also endorses and locates India's position as dynamic and one that embraces diversity to the annual 40 million passengers that pass through the airport. The process of the creation of each work of art has been documented by providing artisans with mobile phones, said the newspaper. These videos will reportedly be installed in the airport as well, providing an insight into how and by whom the works were created.

Some of the artworks:

1. Recycled Mumbai by Akshay Rajpurkar

The aerial map of Mumbai that artist Akshay Rajpurkar has eked out of waste material draws our eyes to familiar parts of the city. The artist collected materials for the works from local places in Mumbai, like Lower Parel, Dharavi and the flea market of Chor Bazaar. Created from discarded buttons, microchips and circuits, Rajpurkar's large-scale installation builds upon the pixellated image of Mumbai as seen on Google Earth.

2. Resilient—Resonances of a City by Charmi Gada Shah

Charmi Gada Shah creates miniature building structures, which evoke the nostalgia and loss associated with the demolition of old structures. Her installation at T2 is on a much larger scale than her usual work. It took a year and a number of exploratory drawings, models, and discussions with the curatorial and design team to detail her initial concept of a work based on the abandoned buildings in and around the Mumbai suburb of Ghatkopar, where she lives and works.

3. Silent Ways by Dilip Chobisa

In the hustle and bustle of an airport that never sleeps, Dilip Chobisa's work offers the viewer a few moments of silent refuge. It gives the viewer a meticulously detailed interior world where each object and scene, however mundane, is endowed with poetic meaning. Five different mise-en-scenes disorient the viewer, much like the scene shifting that is going on with the city's architecture. His monochromatic settings throw the varied architectural styles into strong contrast; the views through the windows are equally diverse–skyrises, distant hills, an aerial view—one's eye travels up and down, never knowing where the next quiet corner may take you.

4. Hope by Baiju Parthan

Wait for that 'magical moment' in Baiju Parthan's work; that moment when the image changes from a taxis or buildings to that of Icarus. On viewing this 36 feet-long lenticular print work, one sees the bustling city of Mumbai with its bikers, commuters, Art Deco buildings and the iconic black-and-yellow Fiat taxi-cabs. As one changes one's position of viewing, the image turns to that of Icarus, an anti-hero from the Greek myth who dared to fly.

[解读]

艺术的荣耀

机场不是一座机械工业下的冰冷建筑，它是一首诗，是一首充满多种元素的诗。每一天这里都有着相聚、别离、幸福、悲伤，它是人们悲伤的起点站，也是走向幸福的终点站，一座冷冰冰的建筑是无法承担这些充沛的情感的。孟买希瓦吉国际机场2号候机楼以艺术和科技的结合，优美的孔雀造型，充满情怀的设施艺术，是整个建筑鲜活并附有人情味，完美的融合了人们的相聚和别离、幸福和悲伤。候机楼内的《贾亚赫》（印度的荣耀）艺术项目娓娓讲述了印度的故事和文化，7 000件饱含艺术家呕心血的艺术品，使机场等候的人们沉浸于艺术的故事，感受着众生百态的生命和情感，或喜悦，或悲伤，此时的个人已融入整个建筑内的氛围，融入整个生命的洪流，奔向幸福终点站。（闫丽祥）

与此同时在别处
Meanwhile | Elsewhere

艺术家：瑞克斯三人组
地点：达卡市
　　（达卡艺术峰会的一部分）
形式和材料：装置艺术
时间：2014年
推荐人：伊芙·莱米斯尔

Artist：Raqs Media Collective
Location：Dhaka（As part of Dhaka Art Summit）
Media/Type：Installation
Date：2014
Researcher：Eve Lemesle

公共艺术项目被瑞克斯三人组命名为《与此同时在别处》，最常见的公众日常的标志：道路标志和广告牌，已经转化成艺术空间和场地。

大批的广告牌和海报遍布达卡市的城市环境中。每个广告牌都显示相关的词语或短语特色一个侧面，而且可能会有相对的指向。总之，广告牌创造了一套排列和组合的心态，并通过时钟的时针和分针的动作调节。在达卡市追踪这些成对的词的路径时就像在街上跟随自然的诗。思想完成的表现就是当你在任何一个地方读到海报或广告牌时，脑海中会浮现散落在其他地方的广告牌的影子。

"阅读一种情感或一个时刻是发生在意识不同方面的事情。阅读城市的墙壁和广告牌也成为我们认可的词语，然而，当我们向上看或向外看去寻找它们的时候，它们总是让我们措手不及。我们发现自己以及我们自己之外的其他地方。"正如管理者笔记中所写。

艺术品把公众的每个工作日转变成他们每天行走车道和道路城市景观和建筑的情感体验。

在 1922 年，吉比什·巴什、莫妮卡·纳如拉和森古普塔组成瑞克三人组，并且已合作创造了巧妙深刻的作品，包括多媒体设施、个人雕塑、在线项目和演出。通过这些作品，他们严格观察城市发展以及现代化的力量形式。

The public art project, titled Meanwhile I Elsewhere by Raqs Media Collective transformed the most common public everyday signange—road signs and billboards—into spaces and sites for art.

An array of billboards and posters was scattered across the urban landscape of Dhaka. Each billboard showed a clock-face that featured words or phrases that relate to each other, possibly in counter-point. Taken together, the billboards produced a set of permutations and combinations of states of mind and being through an invocation of the actions of the hour and minute hands in a clock. Tracing the path of these word-pairs through the city of Dhaka was like following a found poem on the streets. The completion of the thought set out in any one poster or billboard required one to read its scattered shadows on other posters, other places.

"Reading a feeling or a moment is something that happens between different aspects of consciousness. Reading off the walls and billboards of cities, we become the words we think we feel, and yet, even as we look up or out to search for them, they always catch us unawares. We find ourselves, besides ourselves, meanwhile, elsewhere." An extract from the curatorial note.

The artwork transformed the every commute of the public into an emotional experience within the landscape and architecture of the city in lanes and roads that they travelled everyday.

In 1992, Jeebesh Bagchi, Monica Narula, and Shuddhabrata Sengupta formed the Raqs Media Collective and have been collaborating to produce smart, poignant works—including multimedia installations, individual sculptures, online projects, and performances—through which they critically examine identity, urban development, modernization, and forms of power.

[解读]

见或不见

城市的出现，是人类走向成熟和文明的标志，也是人类群居生活的高级形式。林立的摩天大楼，纷繁复杂的交通线路，随处可见的摩登广告，已经成为人们心中城市的标志，同时也成为重复和乏味的代表。一陈不变的建筑、虚假空洞的广告、单调麻木的表情，使得城市慢慢丧失了可读性，留给人们的只有空洞的城市景观。然而，一点的改变却又让城市大为不同，瑞克斯三人组的《与此同时在别处》，以最常见的道路标志和广告牌，转化成充满情感的艺术空间。原本每天重复单调的走路，此时犹如解密探宝般趣味横生，在寻觅的过程中静心阅读了这座城市，头脑中形成的碎片日益完整，最后拼出了整座城市的面貌。我在这里，《与此同时在别处》，见之我幸，不见我命。（闫丽祥）

菲尼克斯市
Phoenix Market City Art Project

艺术家：多位艺术家
地点：印度泰米尔纳德邦金奈
形式和材料：装置艺术
时间：2013 年
推荐人：伊芙·莱米斯尔

Artist：Multiple
Location：Chennai，Tamil Nadu India
Media/Type：Installation
Date：2013
Researcher：Eve Lemesle

菲尼克斯市是印度南部钦奈的一个购物中心。ArtC 是一个艺术基金会，致力于促进印度当代艺术由商场单纯空间发展成艺术展览空间。这个想法是通过支持艺术实践以及扩大公众与艺术的互动来推动印度当代艺术的发展。

为当代艺术建立新观众提供代用空间，通过各种媒体展示作品，推动艺术的定义和艺术品的改变，该项目通过对英国和泰米尔作品的描述和解释，让局部的人群接触艺术品。

该项目包括动画视频艺术的展览，作品《仍在移动》由威廉肯特里奇、西荻达、施坦坎普、北野武村田、怀特、娜塔丽尤尔贝里和汉斯伯格展示。开创性的声音艺术作品《在电梯中》，由小野洋子、罗宾兰波和穆尔塔扎通过一个日常电梯经历创作。该作品由戴安娜坎贝尔贝当古指导，伊芙 莱米斯尔创作。

馆内的地图告知游客每件艺术品的位置，这使游客更好地理解这个项目。在印度，人们一直努力在商店的橱窗、人行道、机场、公共设施、废弃的建筑物等空间中展示艺术。这个项目努力把艺术纳入代用空间，这已成为普通百姓日常生活的一部分。许多个人收藏家开始以容易被普通人接受的方式创作艺术。大多数的项目都是由那些身为慈善家的收藏家所拥有。

Phoenix Market City is a mall in Chennai in the southern part of India. ArtC, an art foundation dedicated to promoting contemporary art in India has converted the mall into a space for art exhibits. The idea is to promote contemporary art in India by supporting the development of artistic practices and expanding the audiences that engage with art.

Providing an alternative space that builds new audiences for contemporary art, exhibiting works from a wide variety of media that push the definition of art and changing the artworks after a duration, the program also includes description and explanation of the work in English and Tamil along with the artworks to reach out to the regional crowd.

The program includes a curated exhibition of animated video art works, Still Moves, showing works by William Kentridge, Shahzia Sikander, Jennifer Steinkamp, Takeshi Murata, Pae White, and Nathalie Djurberg and Hans Berg. In the Lifts, seminal sound art works by Yoko Ono, Scanner (Robin Rimbaud), and Mehreen Murtaza transform an everyday elevator experience. The program is curated by Diana Campbell Betancourt and produced by Eve Lemesle.

An illustrated map that informs visitors about the location of each artwork is one of the efforts to enable greater understanding about the program. In India, there have been efforts to showcase art in window displays of shops, on pavements, airports, corporate complex, abandoned buildings among other. This program is an effort towards taking art in an alternative space, which has become part of the everyday lives of the common people. A number of individual collectors have initiated these ways of making art easily accessible to the common people. Most of the projects are owned by collectors who have become philanthropists.

[解读]
空间纳入艺术

巴黎的卢浮宫、英国的大英博物馆、美国纽约大都会艺术博物馆，作为人们心中的艺术圣地，似乎只有它们才是艺术的标志，艺术似乎也只属于那里。如果艺术只是如此，那恐怕很快就只能束之高阁，以至无人问津。当代语境下的艺术，越来越多的关注日常，由高高在上的神坛走向凡间的世俗生活。艺术家以印度菲尼克斯市购物中心为实验基地，将艺术纳入商业空间，进入人们的日常生活。当漫步在充满商业氛围的购物中心时，映入眼帘的不再仅是玲琅满目的商品，充满韵味的艺术介入使整个中心成为贴近民众的"日常艺术宫"。商业独立空间内的艺术实践，拉近了民众与艺术的距离，扩大了公众与艺术的互动，促进了当代艺术的发展。（闫丽祥）

案例研究
中南美洲

CASE STUDY
Central and South America

City Dormitory	城市宿舍
Mural De Valparaíso	瓦尔帕莱索的壁画
Mural on Avenida 23 de Maio	德马约大道 23 号的壁画
Ad Trees/Urban Nature	城市自然
Pets	铁特河岸的汽水瓶
O Morro (The Hill)	坎陶广场贫民窟绘画项目
Pimp My Carroça	改装垃圾车
Luz NasVielas	Luz NasVielas 社区
City Dormitory	城市宿舍

城市宿舍
City Dormitory

艺术家：古加·费拉兹
地点：巴西里约热内卢
形式和材料：装置艺术
时间：2007 年
推荐人：彼得·莫拉雷斯

Artist : Guga Ferraz
Location : Rio de Janeiro,Brazil
Media/Type : Installation
Date : 2007
Researcher : Peter Morales

2007 年 3 月，古加·费拉兹安装了一个装置，乍一看，它像一个从里约热内卢市中心的蒂尔卡里奥卡画廊里伸出来的八层的双人床。这是一个整体性的工程，城市宿舍的安装没有借助于额外的结构接到墙上。毕竟，这堵墙看起来很脏，好像是被弃用的，而且它位于普拉查革命英雄和 Helio 公司奥气文化中心的这个名声不太好的酒店附近。

为了更全面地领会费拉兹的作品，了解这个遗址的历史背景是非常重要的。把它命名为 Praça Tiradentes，是源自一位巴西革命运动的领袖。在 18 世界末期，他致力于建立一个独立于葡萄牙的巴西共和国。随着该运动的发展，Tiradentes 被捕了，随后经历了审判，被认定为有罪，最后被公开绞死的地点正是这个以他名字命名的广场。作为一个出身寒微的人，他现在被认为是巴西的民族英雄。

Hélio Oiticica 文化中心是以一个视觉艺术家的名字来命名的。他成名于参与了巴西的新具体主义组织，以他对于色彩的大胆创新和后来他称为"穿透环境的艺术"——一个人们可以走进去并且相互影响的创作。Oiticica 的作品模糊了高、低艺术之间的区别，这个与我们非常熟知的费拉兹的作品是一样的。

《城市宿舍》创造了一个不稳定的汇合，把一个有趣的、滑稽的游乐场和一个具有重要功能的城市设施结合起来，不管他们是谁，也不管谁愿意花费时间来玩、来逛，或者住在这个安装在铁架上的像"石槽形状"的空间里。据观察，孩子们在这个结构上玩耍； 相比于叨扰邻居，流浪汉把它作为落脚的地方。正如巴西评论家鲁本斯 PileggiSá 所观察到的，每一层的木板垫使人联想到极简主义，即使它们是在纯粹地重复叠加。它们不尊重艺术史， 用即时存在实用性排挤艺术历史。

从每个层面上讲，尤其是在整个巴西和里约热内卢，这个街头艺术装置面临着房屋短缺和普遍的贫穷问题。它可被看作是艺术家普遍对贫穷这种状况的私人的对抗——艺术家在市中心的情况。但是它也有其他的意义，在这个装置安装的 4 个月间，任何一个无家可归的人，想要居住在这层层排列的空间之一，都是可行的。

这个作品也是巴西房屋流行风格的视觉缩影，在山坡和深谷中，类似于鸟巢样的形状的房屋，整个拉美洲都随处可见，每个附加的结构都随意建造在其中一个之上。这个作品也展示了一种解放的游牧生活，一个人可以在某天在这里睡一晚，也可以在另一天在那里睡一晚，等等。

古加的社会批判是尖锐的。在功能和美学之间模糊状态、城市设施和对他作品的嘲讽。出乎意料地把一个人带入思考。就如 Pileggi Sá 暗示的，一个人不要怀疑是否自己看到了艺术，而正相反有人问他在看什么，重新认识我们的周围的现实事物。

In May of 2007 Guga Ferraz installed what looked at first glance like an eight-tiered bunk bed outside the Gentil Carioca gallery in central Rio de Janeiro. The integration of the work, City Dormitory went beyond the question of bolting a structure to a wall. After all, it was a dirty, seemingly abandoned wall near hotels of dubious reputation in the vicinity of Praça Tiradentes and the Hélio Oiticica Cultural Centre.

Understanding the history of the site is important in order to fully appreciate the work of Ferraz. Praça Tiradentes is named a leading member of the Brazilian revolutionary movement whose aim was to create a Brazilian republic independent of Portugal toward the end of the eighteenth century. As events unfolded, Tiradentes was arrested, tried, found guilty and publicly hanged and quartered in the very square that now bears his name. Once a man of humble origins he is now considered a Brazilian national hero.

The Hélio Oiticica Cultural Center is named after a visual artist best known for having participated in the Brazilian Neo-Concrete group with his innovative use of color, and for what he later termed "penetrable environmental art," work one could walk into and interact with, Oiticica's work blurred the distinctions between high and low art, this is all very much the case with Ferraz as well.

City Dormitory creates an uneasy confluence between what could be a fun and ironic playground and a vital, functional urban amenity for those, regardless of who they may be, who want to spend some time playing, lounging—or—living in one of the "bunk shaped" spaces on the iron frame. Children were observed playing on the structure and homeless people did use it as a dwelling place, much to the chagrin of the neighbors. As the Brazilian critic Rubens Pileggi Sá observes, the mats on the wooden planks on each tier are suggestive of minimalism but even in the purity of their formal repetition, they mock art history by supplanting it with an immediate existential practicality.

On each and every level, this street-art installation confronts the issue of housing shortage and poverty prevalent throughout Brazil and Rio de Janeiro in particular. It can be seen as a personal protest against the artist's general state of poverty—the case of artists in the city center. But it extends to others as well, in that any homeless person who wants to inhabit one of these tiered spaces could do so for the four-month duration of the installation.

The work was also a visual reverberation of a popular housing style known as in Brazil as "birds nests" seen on the hillsides and ravines throughout Latin America, where each additional structure is haphazardly built over the one below.

The work might also suggests a type of liberating nomadism where a person might sleep one day here, another day there, and so on.

Guga's social critique is sharp. In the ambiguity between the functional and the aesthetic, the urban amenity and the sarcasm his work gets one to think in unsuspected ways. As Pileggi Sá suggests, one does not wonder if one is seeing art but rather one asks what one is looking at, a re-cognition of the reality around us.

[解读]

城市尽头

建筑的边界在哪里？居住的边界在哪里？而艺术的边界又在哪里？艺术家古加·费拉兹的作品探讨了这些边界，艺术是否必须要为艺术而艺术，艺术是否可以具有实用功能，是否可以成为对抗贫穷的工具。这些一层层裸露于建筑之外的简易双人床，成为给人们提供的一个简易居所，也许每天都有不同的人来居住，他们可能孑然一身，到处漂泊，处处为家，这个装置里每天居住着一批又一批不同的人。装置所在的地方是巴西的平民英雄 Tiradentes 被绞死的地方，艺术家将装置选在此地也许是为这个英雄感到痛惜，这片他为之付出生命的土地，在独立之后是否让平民能够居有其所，是否让人们真正成为自己的公民。这件作品成为了巴西普通人民生活现状的一个缩影，在这个国家慢慢崛起，代表博爱的基督在里约热内卢上空张开双臂，俯视着整座城市，这座上帝之城巨大的贫富差距让人们怀疑大家是否都是上帝一视同仁的子民。（李田）

瓦尔帕莱索的壁画
Mural De Valparaíso

艺术家:英迪
地点:智利瓦尔帕莱索
形式和材料:壁画
时间:2012 年
推荐人:皮特·莫拉莱斯

Artist:INTI
Location:Valparaíso, Chile
Media/Type:Mural
Date:2012
Researcher:Peter Morales

从瓦尔帕莱索老城区的中心地带——康赛浦西翁山向远处眺望大海,你的视线将被眼前不断出现的景象所吸引,一排排鳞次栉比的墙,大胆、醒目的色彩,神秘永恒的人物,以及无数象形符号。在经过了一些仔细考察后,这些都被充分肯定。这就是 INTI 的作品。

INTI 是丘克亚语中"太阳"的意思,自印加时代起,它就代表着安第斯高原的一位神。INTI 是当代姓氏为卡斯特罗的著名智利艺术家、涂鸦艺术家的笔名。西班牙的姓氏是源自拉丁语"为了防御的古罗马兵营",让人回想起古卡斯提尔王国,不断的征服和颠覆。这位艺术家的名字——INTI Castro,正象征着他所有作品的融合。

去年,巴黎市长 Bertrand Delanoë 邀请 INTI 在市中心公寓式的塔上做一幅 140 英尺的壁画。他在巴黎的画廊—— Itinerrance,承接了这个工程。自从他还是青少年的时候,他就在夜色下开始涂鸦创作,已经 32 岁的 INTI ,已经在创作这条路上走的很长了。据说,早年(14 岁就开始绘画）他是偷偷摸摸的在墙上画，或者是他自己这样认为，墙主有时候会允许他在墙上画，并且会观看。但毕竟，他的家乡瓦尔帕莱索一定程度上有一些波西米亚人，不要求什么"创作"，欣赏其墙上醒目的颜色和设计，尤其是当做得好的时候。他在早期参考拉蒙那·帕尔·旅（Ramona Parra Brigade）和既是智利街头艺术家又是巴西街头艺术家二人组的 DVE 成员。

以前哥伦布风格为基础，通过不同文化元素的混合，INTI 已经建立了一个与众不同的街头艺术风格。该风格在他玻利维亚的之行后，定义更加完全。

INTI 如是说："在奥鲁罗狂欢节，舞者从非常少的题材中创造自己的演出服装。我惊讶于他们丰富的创造力，深深地爱上她们扮演的角色。我喜爱 kusillo 人物形象，一个来自玻利维亚高地的小丑，我已经把他们转化并运用到我的作品中。我的作品可以反映出拉美人是什么样的，其中关键点就是汇合，西班牙文化和前哥伦布文化的强强混合。"

从那个时期起，几乎他所有的作品中都会出现 kusillo 形象，他们身着马普撤人或玛雅人的装扮，常常装饰有饰品以展示渴望，这种做法让人不禁想起另一个安第斯山的传统角色。Ekeku 是另一位看护家园的神，代表着丰收、肥沃和幸福。INTI 经常把 Ekeko 运用到他的作品中，他把辣椒、玉米、头骨、娃娃、铃铛和乐器和 ekeko 作为一个整体。

INTI 还曾经常在阿根廷、玻利维亚、巴西、德国、黎巴嫩、挪威、秘鲁、波兰、西班牙、瑞典和美国进行工作。所有这些国外的关注给他带来了国家的认可。在 2012 年，他在家乡瓦尔帕莱索创作了一幅 50 米长的壁画，在大白天创作而不像以前在夜色中进行；有一大群助手跟着他，并且带了大量的脚手架系统。这个壁画的主要元素似乎也是 kusillo 的分解。

智利理事会全国文化宫拉斯阿特斯创建了《瓦尔帕莱索壁画》，现在是展现这座城市性格的标志化建筑，尤其是它的历史悠久的城区被联合国教科文组织列为世界遗产。

As one looks out to sea from Cerro Concepción in the heart of the Historic Quarter of Valparaiso one's eye is captivated by an image that appears and reappears, weaving a series of walls together with bright, bold colors, a mysterious timeless figure and a myriad of symbolic elements that are recognizable yet bear some careful consideration to fully appreciate. This is the work of INTI.

Inti is the Quechua word for the sun, considered a god in the Andean highlands since the times of the Inca. INTI is the assumed name of the now world-renowned Chilean muralist and graffiti artist whose last name is Castro. The Spanish surname is derived from the Latin Castrum for fortification, and is redolent of Castile, the conquest and the re-conquest. The artist's very name, INTI Castro, is emblematic of the syncretism that is at the heart of his work.

Last year the mayor of Paris, Bertrand Delanoë, invited INTI to paint a 140-foot mural on an apartment complex tower in the center of the city. The artist's Parisian gallery, Itinerrance, managed the project. At thirty-two INTI has come a long way since his teenage years, when he tended to do his graffiti work under cover of night.

It is told that in his early years (he has been painting since he was fourteen years old), he had been working on a wall surreptitiously, or so he thought, only to discover that the owner of the wall had been looking on with approval for some time. But after all, his native Valparaiso has a certain bohemian disregard for the "establishment" and appreciates bold colors and designs on its walls, especially when done well.

His early references were the Ramona Parra Brigade, and the DVE Crew collective both Chilean street-art crews as well as the Brazilian street-art duo Os Gemeos.

INTI has since created a distinctive style of street art by mixing various cultural elements over a basic pre-Columbian base. This style came to be more fully defined after a trip to Bolivia.

As INTI tells it,

At the Oruro Carnival, the dancers create their outfits out of very little. I was struck by the richness of their creativity and fell for the characters. I appropriated the kusillo figure, a buffoon from the highlands of Bolivia that I have been transforming until I have made him my own. My painting reflects what Latin America is, where syncretism is the key and the mixture of Spanish and pre-Columbian cultures is strong.

From that time the kusillo figure appears in most of INTI's work, dressed in Mapuche or Maya outfits and often adorned with representations of desired things a practice that is reminiscent of yet another Andean traditional character. The ekeko is a house god of abundance, fertility and happiness and INTI often places on his figures collars of peppers, corn, skulls, dolls, bells and musical instruments as one would with the ekeko.

INTI has also worked in Argentina, Bolivia, Brazil, Germany, Lebanon, Norway, Peru, Poland, Spain, Sweden, and the US. All this international attention has brought him national recognition. In 2012 he created a 50 meter-long mural in his native Valparaiso, no longer working clandestinely but in full daylight; with a crew of assistants and a substantial scaffolding system. The main element of this mural is a seemingly disassembled kusillo.

Chile's Consejo de Cultura y las Artes funded the Valparaiso mural now iconic of the character of the city of Valparaiso, especially its Historic Quarter a UNESCO World Heritage site.

[解读]
海边的神秘风光

于智利狭长曲折的海岸上，殖民文化与传统文化在这里交融，神秘的壁画掩藏着神秘而古老的故事，诉说着南美洲、智利或者是瓦尔帕莱索的过去和未来。智利像很多美洲国家一样从被殖民国家逐步成为一个独立的国家，所以智利的文化是十分多元的，他既保存了西班牙的一部分文化也存留着美洲本土的文化，而艺术家本人的名字INTI Castro 就是两种文化的结合，INTI 是丘克亚语中太阳的意思，Castro 西班牙的姓氏源自拉丁语"为了防御的古罗马兵营"。这幅画在公寓墙壁上的大幅壁画充满了瓦尔帕莱索特有的风格，成为了这座城市的显著标志，奇幻而神秘的色彩让墙壁从周围的环境里跳脱出来，与灰色的公寓形成对比，远远望去宛如童话世界的一部分。代表着丰收、肥沃和幸福的神明挂满了辣椒、玉米、头骨、娃娃、铃铛和乐器，他们守护、祝福着这座城市。（李田）

德马约大道 23 号的壁画
Mural on Avenida 23 de Maio

艺术家：奥斯兄弟
地点：巴西圣保罗
形式和材料：壁画
时间：2008 年
研究员：皮特·莫拉莱斯

Artist : Os Gemeos
Location : São Paulo, Brazil
Media/Type : Mural
Date : 2008
Researcher : Peter Morales

这面长 680 米的墙上全是卡通图案和身着各色服装的角色，这些都是由世界闻名的奥斯兄弟创作的。

试着来描述其中的一段：在一端，我们可以看到一个小孩，身穿小丑风格的紧身衣，五颜六色的袜子，而且是运动着的，看起来像是他全身都有动物图形的纹身。

他睡时，脸色红润，却怪诞、愤怒，这些景象勉强由观众的想像和思维整合在一起。一丛生长在该壁画前面的真实灌木，似乎马上要燃烧，因为它后面的壁画上画有火焰，其后有一个人，这个人抽着雪茄，冒着烟，保守谨慎。但是他抽的不是雪茄，是一个长的、带有面的东西，这样似乎说的通，因为这个谨慎的人左手持有一个燃着的树枝，而它反过来被树根抓住拼命向着一个小房子的方向。

紧接着有一个坐着的人物，他撕掉面具来揭示这个面具其实是由许多的面具连接在一起的，最后所有的面具以各种各样的方法打开来揭露面孔底下究竟是什么——然而是另一个面具,这个面具包含有许多带有面具的人物,他们玩着巴拉克拉法帽,和由五颜六色不同图案的织物做成的头巾。

他们的服装风格不像最初坐着的人物那样。相应的,这后面有一块地,一个有着一头绿色披肩秀发的女子的圆脸庞,位于这块地的中间,从她的头部发出的腐朽的树干和树桩,被当做基底和脚手架,用来休息以及悬挂各种活动中剩余的人物,这些活动如梦如幻。如此继续,从复杂性来说,它真是一个惊为天人的艺术作品,运用了多种颜色,社会解读以及仅仅是很普通的乐趣——对现实的二次创作。

随着孩子们在圣保罗逐渐长大,Gustavo 和 Otavio Pandolfo 发现了一个独特的玩耍方法,通过艺术表达来进行沟通。当 20 世纪 80 年代嘻哈文化被引入到巴西时,他们的家庭鼓励孩子们的艺术尝试这一做法,被他们进行了定义。

黄色色调的人物象征他们的工作来自一个共同的梦想。由于梦想和梦魇享有同样的范围,所以这些人物既滑稽又忧郁。

这些人物是属于这个世界的,他们展现了希望和愿望,以及现实和侮辱,这些都是有着超过 1100 万人口的大都市中居民的日常经历。

壁画本身所包含的人物,以及圣保罗的市民,还可以经历任意官僚机构的不公正和不尊重——现代化大都市的现实。2002 年,奥斯葛妙思在相同的位置绘画了较早的壁画,后来在 2008 年年初,圣保罗市把整面墙作为清洁城市项目全部要重新喷涂。奥斯兄弟迅速作出反应,重新绘制了 680 米高的墙,这一次与艺术家 Nina Pandolfo, Nunca, Finok 和 Zefix 进行合作,又一次没有借助圣保罗市的财政支持。到 2008 年 12 月,历经 150 小时的工作,Avenida 23 de Maio 又一次被这些进行着陌生又有趣活动的人物所覆盖,被这些丰富多彩的舞台造型所点缀。这是一次上述艺术家的共同协作的成果,每个艺术家负责一部分。截至作者写这篇文章时,圣保罗已经把这个作品树立了起来。

The 680 meter-long wall is populated by cartoonish, patched-up-clothes-wearing characters that have made the twin brothers, known as OS GEMEOS, world famous. An attempt to describe the piece might go like this: at one end we have a baby wearing a harlequin style onesie, multicolored stockings and sporting what look like tattoos of animals all over his body. He sleeps under a grotesque, angry red face barely kept together by the will of the viewer. An actual bush growing in front of the mural appears to be burning because of the flames painted behind it is followed by an image of a cigar-smoking, conservatively-suited man—the man. But it's not a cigar that he smokes, its a log, with a face, which makes sense because the suited man is also holding a burning tree in his left hand that in turn is holding on for dear life by its roots to a small house. Next follows a seated figure who has pulled off his mask to reveal that the mask is a series of masks hinged together and eventually they all swing all the way open to reveal what is behind the face—yet another mask itself containing a multitude of small masked figures sporting balaclavas and head scarves made of multicolored fabric of varying patterns. The sartorial style is not unlike that of the original seated figure. This in turn is followed by a land that has at its center the round head of a girl with long green tresses, dead trunks and tree stumps emanating from her head serve as the substrate and scaffold upon which rest and hang a plethora of figures engaged in a variety of activities that can best be understood in dreams. And so it goes, it is a truly amazing work of art in its complexity, use of color, social commentary and just plain fun—a creative reimagining of reality.

As children growing up in São Paulo, Gustavo and Otavio Pandolfo developed a unique way of playing and communicating through artistic expression. Their family encouraged their artistic endeavors, which they later defined when hip-hop culture was introduced to Brazil in the 1980's.

The yellow-hued figures that are emblematic of their work come from a shared oneiric source. The figures are simultaneously funny and melancholic since dreams and nightmares share the same domain. The figures are also of this world and suggest the hopes and aspirations as well as the realities and indignities that are the daily experience of the inhabitants of a metropolis with a population of more that 11 million.

The murals themselves, like the figures they hold and like the people of São Paulo, can also experience the injustices and disrespect of an arbitrary bureaucracy—the reality of life in a modern metropolis. OSGEMEOS had painted an earlier mural on that same location in 2002, then early in 2008, the city of São Paulo painted over the wall as part of its Clean City Project. In prompt response, OSGEMEOS repainted the 680 meter wall, this time in partnership with the artists Nina Pandolfo, Nunca, Finok and Zefix, once more without the financial support of the City of São Paulo. By December 2008, after 150 hours of work the Avenida 23 de Maio was again covered by a series of colorful tableaux of funny figures conducting strange and intriguing activities. It was a collaborative effort given that each artist mentioned above was responsible for a portion of the whole yet hewing to a cohesive unit. The city as of this writing has let the work stand.

[解读]
超现实梦魇

绘画的历史从最早的雪维洞穴的野牛壁画源起，是石器时代的人们祈祷和希望的表达，后来教堂里的湿壁画传达着虔诚的宗教的希冀。如今，城市里墙面上的绚烂的壁画是艺术家对这个城市、对未来的幻想。奥斯兄弟在圣保罗创作的壁画十分具有超现实主义的意味，小丑、火焰、面具、帽子和无比怪诞的场景让人们不禁猜测隐藏在其中的种种奥义，里面的人物滑稽可笑，但是又十分忧郁，这也许是艺术家自我内心的真实写照。用树木来点燃雪茄的资本家，去掉自己的面具里面仍隐藏着无数面具的小男孩，都是艺术家对这个世界看法的抽象表达，我们透过这些热闹纷繁的壁画的外观看到了什么，也许不是狂欢，而是一种对命运的忧虑，对真实和假象的探索。（李田）

城市自然
Ad Trees / Urban Nature

艺术家：BijaRi 集团
地点：巴西圣保罗
时间：2008 年
推荐人：皮特·莫拉雷斯

Artist：Grupo BijaRi
Location：São Paulo，Brazil
Date：2008
Researcher：Peter Morales

2007 年，为回应过多的视觉污染，圣保罗市制定了一个以"清洁城市"法律著称的新政策。这个政策消除了整个城市布告板上的宣传标识和广告的展示。随着标识和广告被拆掉，广告牌基础设施被遗弃，广告公司走向破产。这些建筑物为有进取心的艺术家和城市设计师创造了机会。巴西的集团 BijaRi 和生于多米尼加共和国 / 荷兰建筑师 José Subero 在 2008 年把这个项目称为《城市自然》。

《城市自然》打算挪用这些被遗弃的建筑物来建立垂直花园，也以广告树木而闻名，从而创造了绿色地带，城市自然。当协作团队看到它时，一个围绕着文化和环境交集的对话便发生了，而且营销系统和技巧应用到"市场"，就好像城市自然能够被探索。这个计划寻求通过对起初对城市污染负责的实体系统和技术拨款来促进一个健康、可持续的城市环境。在企业中鼓励赞助广告树，甚至用这些建筑物提供免费的无线网络——是这些企业以提升自己的方式作出的一种改变。

尽管通过《城市自然》项目，新形成的绿色地带起初只代表 3%～7% 的大都市地区，但主要的目标是创造一个视觉语言，这个视觉语言可以促进曾经以广告过多著称的城市恢复成一个绿色城市，以及超越"清洁城市"法起初的不断追求更清洁、更绿色和更健康的城市环境的思想。

《城市自然》所声明的目标是：

1. 增加城市中的绿色地带，因此增加氧气产量和二氧化碳滞留。

2. 通过创造鸟类、昆虫、附生植物等的栖息地来促进微生态系统的生物多样性。

3. 呈现一个创意城市形象，尽管城市密度较高，但对这个城市呼吸来说是可能的。

4. 鼓励通过增强的集体意识来评价的变成城市社会文化方面的艺术生态实践。

5. 赚取这个城市的碳信用额。

圣保罗，曾经是一个雨林盆地，被现代化进程转变为一个水泥丛林。现在，圣保罗开始作为一个向更绿色的未来促进环境和文化转变的城市，重新定位自己。出现在各个地方的小块片的雨林开拓了他们通往那个未来道路的正当的位置要点。

In 2007 in response to excessive visual pollution, the city of Sáo Paulo instituted a new policy that came to be known as the "Clean City" law. The policy eliminated the display of any publicity logos and advertisement on billboards throughout the city. As the logos and ads were removed billboard infrastructure was abandoned wen ad firms went bankrupt. These structures created an opportunity for enterprising artists and urban designers. Brazil's Grupo BijaRi and Dominican born/Dutch architect José Subero conceived the project known as Urban Nature in 2008.

Urban Nature proposed the appropriation of these abandoned structures to install vertical gardens, also known as Ad Trees, and thus creating green zones, Urban Nature. As the collaborative team saw it, a dialogue could thus take place around the intersection of culture and the environment, and the use of marketing systems and techniques to "market," as it were—Urban Nature, could be explored. The project sought to promote a healthy, sustainable urban environment by appropriating the systems and techniques of the very entities responsible for urban pollution in the first place. Corporations where encouraged to sponsor Ad Trees, and even to use the structures to provide free WiFi —a change in the way these corporations promoted themselves.

Although the newly formed green zones through the Urban Nature project originally represented only 3%—7% of the metropolitan region, the principal objective was to create a visual discourse that promotes the idea that a city once known for its plethora of ads can be revitalized as a green city and that there is continuity in pursuit of a cleaner, greener and healthier urban environment beyond the initial effects of the "Clean City" law.

The stated goals of Urban Nature are:

1. Increase in the green zone of the city and consequently an increase in O_2 production and CO_2 retention.

2. Promote biodiversity in the micro-ecosystems by creating habitat for birds, insects, epiphytes, etc.

3. Present an image of a creative city where notwithstanding the urban density, it is possible for the city to breathe.

4. Stimulate artistic-ecological practices that become a socio-cultural aspect of the city that is valued through an enhanced collective consciousness.

5. Earn carbon credits for the City.

São Paulo, once a rainforest basin, was transformed into a concrete jungle by modernization. Now, São Paulo begins to reposition itself as a city that promotes environmental and cultural transformation toward a greener future. The bits of rainforest that appear here and there to reclaim their rightful place point the way to that future.

[解读]

从城市中走出来的森林

圣保罗这座城市本是雨林盆地，它慢慢从森林里走出来，而森林从城市的地盘里渐渐消退，自然和绿色成为了这座城市中难得看到的景色，我们在钢筋混凝土的现代都市里迷惘寻找，除了纷繁复杂的广告牌和被商业和资本彻底笼罩的城市，绿色和纯净显得那么稀缺。将草木以广告的形式安放在广告牌里，从最商业的地方生长出绿色来，看到这样的"广告"人们怎会不产生一丝愉悦和放松。森林又逐步以这种形式回到这座原本属于丛林的城市，同时也宣传了这座城市的草木，圣保罗的"清洁城市"给了我们一种思路，植物以这样的方式取代着广告，也是对视觉环境和生态环境的双重优化，城市与植被应当如何相处，他们不是相互占有和驱逐的关系，而是如何结合在一起展现出一个城市最大的魅力。（李田）

铁特河岸的汽水瓶
Pets

艺术家：爱德华多·史路尔 Artist：Eduardo Srur
地点：巴西圣保罗 Location：São Paulo，Brazil
形式和材料：公共装置 Media/Type：Public Installation
时间：2008 年 Date：2008
推荐人：加布埃拉·里贝罗 Researcher：Gabriela Ribeiro

铁特河全长 1 100 千米，沐浴着圣保罗市 62 个城区。铁特河在巴西的历史上具有非常重要的意义。早在 18 世纪，它曾作为巴西早期拓荒者"班代兰蒂斯"们的行进路线。这些冒险家利用铁特河到达圣保罗州的内部，进一步扩大了巴西的领土。在旅途中，敢为人先的"班代兰蒂斯"们建造了几个城市。所以正是因为铁特河，圣保罗市才得以诞生。

在随后的几年中，人们在圣保罗大都会区的铁特河段上航行或是练习水上运动。但是从 50 年代开始，这种情况发生了变化。随着圣保罗市人口的无序增加和工业的增长，河水开始因为城市市民活动的扩张和工业污水的排放而遭到污染。艺术家爱德华多·史路尔将城市和城市公共空间作为他自己的画廊。不知情的行人甚至都不知道他是谁，但是肯定已经被艺术家在大都市的一些非常便利或是非常不寻常的地点介入所影响。

在 2008 年 3 月的一个早上，由于艺术家的介入，铁特河出现了一个不寻常的场景：20 个巨大的丰富多彩的瓶子（汽水瓶）"装饰"河流。

圣保罗人（圣保罗公民）和城市的常客以前能看到并感觉到被污染的河水的气味，在污染物中总能看到很多塑料瓶。而作为一个例行常规，对于生活在这种情况下的人们而言，一切都变得司空见惯。但是如果你认识到艺术家的作品在视觉上重新激活了河流的话，也正有了这种艺术家吸引人们视线的干预，让他们真正开始注意始终弥漫在城市的污染，毕竟铁特河流经了圣保罗的大部分地区。

20个由乙烯基制成的巨大的瓶子，充气后固定在河面上的混凝土浮式平台上。这个项目持续两个月之久，共有超过60万人看到了这件作品。作为一个教育项目，还挑选了3000名公立学校的儿童和教师坐船前往河边，近距观看这件作品。由于目前的条件限制他们只能在河岸观看作品，所以这是一次近距离感受铁特河的难得的机会。

在展示结束时，充气瓶子的塑料材料被循环使用，用于制作和这个项目非常一致的另一个目的。应Srur先生的请求，艺术家和巴西设计师Jum Nakao设计并制作了带有塑料材料的背包，随机捐赠给学校。

因此，可以得出结论，这件作品所传递的信息远远超出了环境问题的范围：它提出对外观进行回收，使公民能够重新思考他们城市的形象。塑料饮料瓶项目除了在铁特河进行展示外，它也陈列在瓜拉皮兰加水库和圣保罗布的拉干萨保利斯塔市。

The Tietê River has 1100 km of extension long and it bathes 62 municipal regions of São Paulo. The River has a great importance in the history of the country, which served as a route for pioneers called "Bandeirantes" in the eighteenth century. These adventurers, who expanded the Brazilian territory, used the Tietê to reach the interior of the state of São Paulo. During the journey, the pioneers "Bandeirantes" founded several cities. It was because of the Tietê River that the city of São Paulo was born.

In the following years, it has been widely used for navigation and even to practice water sports, mainly in the metropolitan region of São Paulo. It was from the 50's that this situation has changed. With disordered population and industrial growth of the city of São Paulo, the river began to receive domestic and industrial sewage in the stretch of the city, leaving it polluted and contaminated.The artist Eduardo Srur makes the city and the urban public spaces as his own gallery. An unsuspecting pedestrian might not even know who he is, but it certainly has been impacted by some of their interventions in the metropolis, which are always in places very accessible and very unusual.

And it was through the intervention of the artist that one morning in March 2008, the river Tietê dawned with an unusual scene: 20 giant colorful bottles with PET (plastic soda bottles) format, "decorating" the river flow.

Paulistanos (São Paulo citizens) and frequent visitors of the city were used to see and feel the smell of the polluted river, and in the middle of this pollution is always possible to see many plastic bottles. And as an usual routine, for those who live with this scenario, everything becomes unnoticed. With this intervention the artist piqued the gaze of the people so that they pay attention to the pollution that always pervades the city, after all, the Tietê river runs through a very extensive area of São Paulo. However if you realize that the artist work visually reactivated the river flow.

The 20 huge bottles made of vinyl, inflated and fixed in floating platform remained for two months on the river concrete, and it was seen by over 60 million people. And also featured an educational project that took 3000 children and teachers from public schools to visit the river by boat and see the work, an unique opportunity to embark on the Tietê river, which by their current conditions, are restricted accesses to the area within the riverbanks.

At the end of exposure, the plastic material of the inflatable bottle was reused for a purpose very consistent with the proposed work. Upon the Mr. Srur request, artist and Brazilian designer Jum Nakao designed and produced backpacks made with the pads of the work to be donated to the schools that made the ride.

Therefore, it is concluded that message of the work goes beyond the environmental issue: it proposes recycling the look, bringing citizens to rethink their city. The PETS project beyond the exposed Tietê River, it was also on display at Guarapiranga and in the city of Bragança Paulista, São Paulo.

[解读]

用艺术呼吁人们关心环境

环境问题是现代城市转型中无可避免的一个社会问题，这关系到城市居民的切身利益。当一个城市中出现的环境问题变得司空见惯的时候，这个城市居民的生活质量必然下降。当艺术家在介入圣保罗市铁特河环境污染的问题时，以其独有的艺术特质和文化包容，将这一问题不断放大，用巨大的塑料瓶符号不断强化人们的视觉感受，去震撼每一个市民的心灵，从而引发他们的思考。这个持续时间长和影响范围广泛的公共艺术案例会不断消除现代社会发展给环境带来的不利影响，有可能在一定层面上刺激政府着手解决环境生态问题，创造城市居民和城市生活环境的生态和谐。针对环保问题的呼吁性质的公共艺术作品已经成为当下公共艺术的社会功能和审美的价值取向之一。（于奇赫）

坎陶广场贫民窟绘画项目
O Morro（The Hill）

艺术家：杰隆·库哈斯、德雷·尔翰
地点：巴西圣保罗
形式和材料：绘画
时间：2010 年
推荐人：加布埃拉·里贝罗

Artist：Jeroen Koolhaas，Dre Urhahn
Location：São Paulo，Brazil
Media/Type：Painting
Date：2010
Researcher：Gabriela Ribeiro

2005 年，一个荷兰人计划拍摄一部关于巴西里约热内卢贫民窟嘻哈文化的电影。在整个录制过程中，他被当地的艰苦条件所震惊，但也被当地人的创造力和乐观精神所鼓舞。由此他创立了"贫民窟绘画"，目的是推动巴西社区贫民窟的艺术产出，从而改变生活在贫民窟人民的日常生活环境，增强他们的自豪感。

在这样的思想指导下，为了项目运行筹集资金，艺术家们通过官方网站 www.favelapainting.com 收到了来自个人、机构和商业组织共计 10 万美金的捐款，达到了预期的目标。最终这个荷兰人成功地筹集到了 20 万美金来启动这个项目。

2010 年初，他回到里约热内卢。起初，为圣玛尔塔 25 位居民提供培训，使他们提高专业绘画技能，在课程结束时甚至会颁发证书。他们做的事情得到了社区的帮助，更重要的是，这些人在为这个项目服务的同时，可以领到薪酬，提高自己的生活水平。用这些筹集的资金，这个荷兰人还提供了项目需要的所有材料。

经过了一个月，在贫民窟的底部已有7000平方英尺的面积被画作所覆盖，34间房屋被涂上颜色，穿过了整个坎陶广场。作品中有的画面线条生动鲜艳，与半音音符相得益彰，时而又白色缎带横穿，远处看像一道道彩虹。这样的视觉画面与贫民窟其他颜色单调的房子形成鲜明对比，引起了大家的兴趣，更多人想加入进来，使整个地区像坎陶广场一样五彩斑斓。

当这些艺术品日渐增加，贫民窟的潜在吸引力已经超过了巴西的耶稣像。站在圣玛尔塔的最高处，自然地倾斜望去，可以看到整个城市的全景。这样的创意吸引了来自国内国外的公司。两位艺术家成功地实现了自己目标：让你用最好的理由来赞美巴西贫民窟。今天，这个社区有自己的博客，叫做"圣玛尔塔贫民窟之旅"，发布旅游公告，描述旅游攻略，导游都是当地居民。它已经吸引了大量的访客，其中不乏政界名人和娱乐明星。

In 2005 a Dutch duo met to produce a film about the hip hop culture of the favelas in Rio de Janeiro, Brazil. During the recordings, though shocked by the local conditions, were inspired by the creativity and optimism of the people. From this experience created the "Favela Painting," a project whose goal is to produce art in slums driven by Brazilian communities in order to transform the everyday environment and instill the pride in these people.

From this idea three projects, "Menino com Pipa," "Rio Cruzeiro" (both performed in Vila Cruzeiro), and as indicated in this case study, "Praça Cantão" were held to be the most recent and significant work of artists in Brazil and involves ideas that go beyond art, bringing community participation and the occupation of a large area in Santa Marta.

Santa Marta is a slum located in Morro Dona Marta (one landform located between the neighborhoods of Botafogo, Flamengo and Laranjeira). The site is famous for had been the scneario of the music movieclip "They Do not Care About Us" by Michael Jackson in 1996. Some years later, it was back in the spotlight of the news for being the first community to receive the Pacifying Police Unit of Rio de Janeiro.

To finance the project, the artists received donations from individuals, institutions and businesses made through the official website of the foundation: www.favelapainting.com. Finally, the campaign for raising $ 100 000 achieved the goal, with the price of the US currency during the campaign, the duo managed around US $ 200 000 to undertake the project.

With the money raised, in early 2010 the artist returned to Rio de Janeiro. Starting the process to conduct a training course for 25 residents of Santa Marta, which enabled the development of skills in painting, bringing a professional opportunity, even giving them a diploma at the end of the course. Everything that these people could also help in the painting from the community, or even their own homes—is also important to note that these people were paid what allowed them to improve their standard of living while they were involved in this project. With the money the duo also managed to acquire all the necessary material for the execution of work.

For a month were painted over 7000 square feet on the bottom of the favela, coloring 34 homes and occupying the entire length of the Praça Cantão. The art work consisted of paintings with vivid stripes that make a matching tone following a chromatic scale, and sometimes interrupted by a white band, resembling a rainbow to be seen from afar. The visual contrast to the monochromatic tone than other houses in the slum had, leading residents in the will of the expansion project, so that the whole place could be as colorful as the Praça Cantão.

With the collective work of art, the slum that already has a tourism potential by staying ahead of Cristo Redentor, and with a large tilt angle, and the highest part provides an insider's view of the city. The initiative attracted the interest of major companies from national and international communication scale. The two artists managed to achieve its main goal: to make you speak of Brazilian favelas for the best reasons. Today the community has a blog called "Favela Santa Marta Tour" which discloses and describes the guided tours, the guides are locals themselves. Always getting a lot of visitors, including worthies politics and entertainment.

[解读]

艺术拯救贫穷

巴西贫民窟一直是犯罪、吸毒、走私的基地和"天堂",同时又最有可能成为传染病肆虐的源头,这里深刻反映了巴西发展过程中的社会疾病。可是两个荷兰画家却让贫民窟的吸引力超过了巴西的耶稣像,足以看出艺术的魅力。艺术不仅可以为人们带来物质上的财富,也能带来精神上的慰藉。为房子粉刷的居民可以领到薪酬,提高自己的生活水平。"圣玛尔塔贫民窟之旅"让当地居民成为导游,增加了收入。贫民窟中的人们总是洋溢着一种乐观的精神。对于艺术家来说,贫民窟中同样存在希望。贫民窟中的壁画是震撼,幸福从来都不是需要靠物质来衡量。贫穷需要艺术来拯救,除了无奈的挣扎之外,全社会必须一同行动起来,共同创造美好的明天。(于奇赫)

改装垃圾车
Pimp My Carroça

艺术家：蒂亚戈·蒙达诺
地点：巴西圣保罗
形式和材料：互动行为、装置

时间：2012年
推荐人：加布里埃拉·里贝罗

Artist：Thiago Mundano
Location：São Paulo，Brazil
Media/Type：Various Interventions, Public Installation
Date：2012
Researcher：Gabriela Ribeiro

2012年，蒂亚戈在为城市墙壁涂画时，注意到小路右边的一辆垃圾收集车阻碍了交通。蒂亚戈并没有心生厌恶，而是研究并发现这种行为通常与环卫工人社会地位边缘化导致的社会歧视联系在一起。然而，在循环系统效率很低的现代城市，垃圾车是必须要存在的基础设施。比如说圣保罗每天会产生17 000吨垃圾，这里只有1%能够循环利用；但是90%的再循环物质由垃圾车和环卫工人收集起来，他们对环境保护和社会发展做出了巨大的贡献。

通过观察，蒂亚戈慢慢接近这些工人，听他们讲自己的故事，参观他们所负责的区域，慢慢喜欢上了他们。所以，他创立了《Pimp My Carroça》项目（MTV"改装车"就来源与此）。

之前，为了完成项目，蒂亚戈登陆到一个名为"Catarse"的集资网站，向公众筹集资金。在集资正式开始的前一天，实际预约人数已经超过了38 200，甚至在活动结束后，仍然有人捐款。5月18日，筹集到来自

792 个人的近 6 400 000 兰特现金。

随着这个艺术和社会项目的实现，2012 年 6 月初，联合国可持续发展大会召开前，在圣保罗城的"Virada Sustentável"，被选出的工人赢得了改装货车的机会，由专业的涂鸦艺术家完成。在此过程中，工人们和他们的家人必须接受体检、理发，还得到了食物和心理治疗。当这天要结束的时候，在市中心示威游行，号召政府认同、建设并保持这些循环组织。

艺术家觉得他们的作品是不成功的，不是因为自身服务，而是因为他们所拥有的条件。他认为他的项目并没有真正解决这个问题，但是能够通过其他艺术家和普通大众将它的影响扩大。《Pimp My Carroça》项目资金筹集入口仍然开放，捐赠者能够为他们自己的城市采取行动。

然而，这个项目通过艺术行动来认同这样一份非正式的工作，提高了公众的社会意识，既谴责又尊敬这些人的工作，因为他们的权利和工作条件需要改善和提高，需将巴西城市环境机构明朗化，每个人应该给这些工人最基本的尊重。

In the year 2012, while painting the city walls, the artist Thiago Mundano noticed a cart garbage collection that followed by the right lane, disrupting the transit of cars. Instead of seeing a nuisance, researched and found that this activity "extra-official" is commonly linked to marginalization and has a low status in society. However, in cities where recycling system is not efficient, its role is fundamental and necessary. For example, in the city of São Paulo exclusive, 17,000 tons of waste are generated per day, and only 1% is recycled; 90% of this recycled material is collected by them, doing what they generate an activity of great environmental and social importance.

From this observation, Mundano was approaching these workers, has been listening to their stories, visited places where they attended and was falling in love for their cause. So the artist created the project "Pimp My Carroça" (an allusion to the popular MTV show "Pimp My Ride").

Before, to accomplish the project, Mundano signed in Catarse [catarse.me], a website of collective funding, so could receive donations from the public. On the penultimate day officially open to donations, 38,200 actual stipulated to carry out the action had already been exceeded. Even after the end of term contributions kept coming. On May 18th, there were nearly R$ 6 400 000 in cash, donated by 792 people.

With the realization of the artistic and social project "Pimp my Carroça", in early June 2012, days before the RIO + 20 (Rio de Janeiro

- RJ) and during the "Virada Sustentável" in the city of São Paulo, the selected workers earned a reform in his wagon with art made by professional graffiti artists, including phrases chosen by the public. During this process, the collectors and their families have undergone a medical and eye exams, along with massages. Also received haircuts, food and psychological therapy. At the end of the day, a manifastation took place in the city center, calling on the government to recognize, establish and maintain recycling cooperatives.

The artist does not believe their recycling work is worthy, not because of the service itself, but because of the conditions in which they operate. He also believes that his project does not bring a definitive solution to the problem, but try to multiply it to have a wider reach, either by other artists or the general population. For this, in the "Pimp My Carroça" fundraising site is still open and the donor can take action for their own city, and rewarding with items for donation receiving a kit to be able to customize ("Pimp") the carriage of a collector with paints and safety items.

However, the intent of recognizing the work through an artistic action gives the population increased level of social awareness about such an informal profession, denouncing and honoring the work of these people, for their rights and conditions can be improved by giving them the visibility being the Brazilian cities of environmental agents, and working population in the respect that everyone should give these workers.

[解读]

艺术家的姿态

艺术家是否就是超越常人，总是一副在闪光灯下高高在上的样子呢？蒂亚戈就给出了很好的答案：艺术家应该是平易近人的。社会的发展往往是不平衡的，环卫工人社会地位低下，是一个付出很多但回报较少的群体，但是关于一个城市的整洁和人们的生活。他们处于城市底层，容易被普通大众所忽视，对环卫工人等普通体力劳动者存在着偏见和歧视，缺乏对其人格及其劳动成果的应有尊重。在这个案例中，艺术家充当一种粘合剂的媒介，呼吁政府对于弱势群体的关注，弥补环卫工人与社会的裂痕，用最为温和却十分有利的行为——艺术，用装饰斑斓色彩与美丽图案的垃圾收集车，来获得更多的社会支持和关注他们的身心健康，涂鸦艺术让整个冷漠的城市变得亲密而有趣起来。而对于每一个参加了"拉客"的清洁工，他们不再是城市隐形的英雄，而是微笑地拉着自己美丽的小板车，心中充满自尊。（于奇赫）

Luz NasVielas 社区

艺术家：博阿·米斯图拉团队	Artist：Boa Mistura
地点：巴西圣保罗	Location：São Paulo，Brazil
形式和材料：互动行为、涂鸦	Media/Type：Various Interventions、Painting
时间：2012 年至今	Date：2012-on-going
推荐人：加布里埃拉·里贝罗	Researcher：Gabriela Ribeiro

由来自西班牙的 Arkoh、Derko、Pahg、Purone 和 Rdick 组成的博阿·米斯图拉团队成立于 2001 年。他们尊重各自的多样性，以团队的身份去创作和环游世界，从而体会世界各个角落不同的生活方式。

对于他们来说，艺术是制造改变的工具和灵感的一种表现形式，《Luz nasVielas》项目是在圣保罗最大的城市维拉巴西兰迪亚的贫民窟完成，并取得了巨大成功。

在当地主人贡萨尔维斯家里，艺术家和组织者们受到了热情款待，他们花时间跟当地居民交流并了解当地社区，分析研究由狭窄的街道和蜿蜒的小巷组成的城市结构。然而，当体验过社区生活并且熟悉了城市街道网中的每一条小路后，他们在维拉巴西兰迪亚创造了独一无二的城市美学。

这个创造就是在涂抹颜色后配上白色文字 "Beleza" "Firmeza" "Amor" "Doçura" "Orgulho" 和 "Genialidade"。这些词语动机明确，语气强烈（大致意思为美丽、坚定、爱、骄傲和温暖），充满普世价值，但是真正执行起来却并非如此。这种想法旨在在几乎得不到政府资助的底层地区，用积极的词汇凝聚社区。

为了完成画作，他们请来当地的孩子们帮忙，他们教孩子们怎样采用一种新视角涂抹。正因为孩子们是在自己居住的地方尽情涂画，所以这样激动而有趣的过程让他们很享受。整个过程耗时一个月并于2012年完成。

这个项目的创作力量来自当地社区不断增加的积极意愿，给他们的生活注入了不同的元素。比如贫民窟环境中缺失的美与消极的社会因素总是出现在这些人的身边，但是生活中偶尔出现的的美好事物会让他们变得更加乐观。

尽管这个项目吸引了一小部分游客，更多的是激发和唤醒了这里的居民。事实上，它们是一项内在的工程，就像是住在那里的人们很亲近的东西。但是对普通大众来说，它们只是圣保罗 ViradaSustentável 于同年6月4、5日才正式对外亮相的一道景观。

维拉巴西兰迪亚的涂画区虽然只是当地孩子们共同完成的一个项目，但是它更是一种朴素纯真的祝愿，让孩子们相信他们能够得到更好的教育与带有完善基础设施的居住地。这个维拉巴西兰蒂亚的项目吸引了巴西媒体的关注，他们有更积极的理由来到这里，一睹这些标新立异的艺术家，进而了解当地居民的心愿和态度，引起社会大众对这一地区所存在的问题的关注，也可以以某种方式给政府施压，在这一地区引进投资项目。

Created in 2001, the collective Boa Mistura is formed by Spanish Arkoh, Derko, Pahg, Purone and Rdick. Harmonizing the diversity of styles of each of its members, the group travels the world creating jobs with ever greater proportions and intervening in the most varied ways of life existing in various places on the planet.

To the collective, the art is as a tool for change and a form of inspiration within that thought, "Luz nas Vielas" project implemented in the favela Vila Brasilândia, one of the largest city of São Paulo, has achieved its success.

Hosted by Gonçalves family where they felt welcomed, artists and organizers have had time to get to know the neighborhood, study and analyze the narrow streets and winding alleys that make up a kind of urban fabric. And especially, have had several conversations with the locals. When they experience the community and understand the paths that occur in existing alleys network in place, the collective planned interventions aimed unique aesthetics and desires of Vila Brasilândia.

The interventions consisted of paintings with words of motivation in white on strong and contrasting tones with the landscape of slums, the words they chose collectively with the locals were "Beleza" "Firmeza" "Amor" "Doçura" "Orgulho" and "Genialidade" (roughly translated as Beauty, Firmness, Love, Pride and Geniality). These words are universal values and designed to be applied in everywhere, but the reality check is not for real. The idea was to integrate the community with positive words in degraded regions that receive little government support.

To accomplish this painting, the artists relied on the help of children living in the Vila Brasilândia, who received a training to learn to paint and apply a perspective that is required in the work. In addition to enjoying an exciting and fun activity because they can paint the walls where they live. The whole process took about a month, being held in 2012.

The strength of this project is based on local impact, the rising in positive community desires, and put in their life different elements which are usually in their daily lives, such as the absence of the beauty of the slum environment and the negative social facts that surround these people, which sometimes excel the existing good things in their lives, always renewing optimism.

Despite attracting a few visits to the Vila Brasilândia, the art works much more to motivate residents and actually it works as an inner project, almost like something intimate for those who actually live there. But for the general public was exposed due to "Luz nas Vielas" in "Virada Sustentável" of São Paulo, in the 4th and 5th of June of the same year of the execution of the work.

The painting areas in Vila Brasilândia, even though it is an action involved with children in the community, it would be a certain naivety to believe that this would provide better education or other areas of basic services. But this action brought Vila Brasilândia the Brazilian media for positive reasons in relation to the gaze of the artists and the attitude of its residents, trying to call attention to the problems existing in the place and somehow pressure the government to bring investment to the region.

[解读]
艺术治愈心灵

　　公共艺术应该就是用最简洁、最直白的表现方式，去唤醒社会的良知。艺术家的社会责任感应该以一种什么样的形式呈现，我想这个公共艺术的案例就给了我们很好的答案。在这个案例里，材料与形式十分简单，近似于一种原始的涂抹，但是引发了人们对于这个破败社区的广泛关注。艺术只有走进大众，才有永久的生命力。博阿·米斯图拉团队把一种乐观向上的积极的生活态度，融化在了颜色和文字里，同时也深深地印刻在孩子们的脑海中。虽然在一段时间内孩子们仍然会生活在贫民窟里，但是这件真、善、美的公共艺术作品会持续抚慰孩子们的。艺术家通过创造劳动去感染、教化他人、造福社会，也是具有良知和道义感的一种表现。同时在奉献爱心、回馈社会的同时，自身的知名度、美誉度也会得到提升。（于奇赫）

城市宿舍
City Dormitory

艺术家：古加·费拉兹
地点：巴西里约热内卢
形式和材料：互动行为、装置
时间：2007 年
推荐人：加百布里埃拉·里贝罗

Artist：Guga Ferraz
Location：Rio de Janeiro，Brazil
Media/Type：Various Interventions Public Installation
Date：2007
Researcher：Gabriela Ribeiro

2007 年 4 月，在里约热热内卢嘈杂的市中心，古加·费拉兹现场将八张床叠加在一起，制作了他的作品《Cidade Dormitório》。这件作品借鉴了建筑物或高塔的设计理念，对称地排列在上面（每一层有八层楼高），放置于基拉甸奇宫 Gentil Carioca 画廊两侧。

通过这种形式，艺术家给大众提供了一间一旁是木条、床上是毯子的铁架结构的公共城市设施。人们可以在这样的一件小屋子里工作、学习和打发时间。落成典礼当天，馆内为观众提供了很大的活动区域，但之后被流浪汉、乞丐、酒鬼和普通观众所占据，不知为什么这些人被请到这里，它的职能开始慢慢发生改变。

任何无家可归的人若是想要获得一个屋子的使用权，必须首先为它服务四个月，并做一些基本的工作。因为这就是来这里工作的重要目的——学会合理使用空间。然而最终只有有三个人获得了有限的使用权，他们分别是男妓、前科犯和流浪汉。因为其他无人居住的"楼层"也遭到不同程度的破坏，他们决定住在这里，完成诸如整理和清洁床铺的工作。
负责顶层的是流浪汉路易斯，每天按常规一直在同一个区域做着同样的
工作，像在家乡 Irajá 一样舒适。他每天戴着天主念珠祷告，把自己的
空间安排妥帖，把自己极少的随身物品整理在一个盒子里，像一个小陈
列室一样。不久之后，他离开了这里去找了一份兼职工作，他仍然知道
怎样维护和保持房间的干净整洁，他还建议艺术家们在铁架上铺一层覆
盖物来阻挡生锈，如果出现锈迹，他会清理干净。但是这份工作要求有
固定的时间，最终他没能得到。

这些作品所呈现的美学是模仿当下流行的富人建筑。然而它也涉及社会
经济和政治问题，特别是在里约的住房赤字，超出了作品的表层含义。
这些作品表现出对城市中心贫困现状的抗议。在一种很微妙的情况下，
它们展现出穷人与现实的妥协和冲突，一种被社会和政府所忽视的阶层
仍然得不到关爱与帮助的境遇。这样的作品虽然充满尖酸微妙感，却为
需要工作、申请居住权或使用权的人们提供了场所。

In April 2007, amid the chaos of the city center of Rio de Janeiro, the artist Guga Ferraz made his installation site-oriented "Cidade Dormitório", consisting of eight beds stacked on the side of the gallery "Gentil Carioca" located on the Praça Tiradentes, the objects were stacked symmetrically (a bunk of eight floors height), referring to the idea of building or tower.

The artist work through this, offered an urban facility to anyone who wants to spend some time—or do any other type of occupation—lounging on one of the "rooms" of this iron frame with wooden crates and sinker to the wall. And with mats on all floors. At the inauguration was a kind of playground for a broader audience, but from the second day and in the following months, the work was being occupied by the homeless, beggars, drunks, bystanders in general, people who somehow were requiring too this space and began to turn a kind of route.

Any homeless who wanted to get hold of one of the "rooms" can do it for four months, which was the period in which the work remain in place. Because that was the intent of the work: the use of occupation. And some people, three in total, actually took possession of what had become, even in a limited time, their dwellings. Among them a former male prostitute, an ex-convict and homeless. Even these three people who decided to live there, who managed to retain some of the characteristics of the work to completion, such as mattresses and clean bedding, because the other "floors" which nobody lived, suffered depredations.

The resident homeless who occupied the "Top Floor" was Luis, who remained in the same place, he was keeping a routine in their daily lives, as comfortable as his hometown in Irajá. Every day, he prayed with a catholic chaplet, arranged his space folding the sheets and keeping his few belongings in a box that served as a cabinet. Soon after he left to try to get a part time job, and also he knew how to handle equipment maintenance and suggested that the artist put a product on the iron so there would be no rusting, and if that happened he would do the job. But as the work had a certain time frame, unfortunately it did not happen.

The aesthetics of the work is currently one of mimesis of popular buildings for people who can afford. However it refers to socioeconomic and political problems that go beyond the appearance of the work, as the housing deficit that prevails in the country in general and in Rio de Janeiro in particular. Showing that the work is a protest against the artist present state of poverty in the city center. This artwork demonstrates a commitment and clash with reality when dealing with a very delicate situation, the homeless, which is often ignored by society that has better living conditions and goverment agencies depending on their posture management, do not give them proper care and structure. In this site-oriented the artist Works with the acidity even in subtle tone while offering a welcoming structure for those who needed the work, applying for the residence time of the work or even an everyday utilitarian function.

[解读]

艺术——医治社会的良药

公共艺术家的目光应该是敏锐的，善于发现社会中存在的问题。或者说是社会的一个批判者，其目的是让社会发展的越来越好，而不是给社会发展唱赞美诗。《Cidade Dormitório》可以看作是一个城市贫富差距的纪念碑，是一个拷问人性的耻辱柱，政府的财政税收永远无法抹平富人与穷人的鸿沟。穷人虽然是社会最底层的人，往往那个被认为是无知与粗鲁的代名词，但是我们看到如果无家可归的人能拥有一个自己的独立空间的话，他们还是能够像普通人一样知礼有节的。中国也面临这样的问题。大城市的房价连年攀升，使得很多年轻人变成"蚁族""房奴"和"北漂"，即使奋斗一生也买不了一套理想的房子。这件作品发人深省，有很浓厚的社会批判色彩，是一件令人印象深刻的公共艺术作品。（于奇赫）

案例研究
欧洲

CASE STUDY
Europe

Flagpole	旗杆
Otwock	奥特沃茨克小镇
Partizaning	Partizaning
Search: Other Spaces	搜索：其他空间
Fundamental Frequency	基本频率
The Monument to a New Monument	从纪念碑到新纪念碑
Urban Dreams	城市梦想
Hidden Forest	隐藏的森林
Oxygenator	氧合器
PARK : birihtimal (PARK : a possibility)	公园：一种可能
Evening Echo	夜晚的回声

旗杆
Flagpole

艺术家：卡米拉·史捷诺科恩	Artist : Kamila Szejnoch
地点：波兰卢布林	Location : Lublin，Poland
时间：2010—2012年	Date : 2010—2012
形式和材料：雕塑	Media/Type : Sculpture
研究员：利西亚·普洛科·朋科	Researcher: Lesya Prokopenko

卡米拉·史捷诺科恩制造公共空间艺术，特别是装置、雕塑和城市设计（华沙、奥斯陆、贝尔格莱德、埃里温、弗里堡、卢布林）。

"最近我喜欢谈论的话题，大多数情况下是历史和意识形态，它们影响公共认同。""我已经正试图通过添加一些新的元素和对比来处理，更新、奚落或仅仅是提醒过时的层，来区分原始的风格、功能。当然，我并不表现出我是一个客观的旁观者，因为在某种程度上，我自己就是现实的一个产品，我居住在其中。"评论这个艺术家。她的大部分作品有两个主要的元素：一个元素是历史的、传统的，已经定型或者自称一型的；另一种则更现代的、开放的，具有玩乐性也相互影响着。

第一个《旗杆》在立陶宛广场卢布林创造——一个位于波兰东部的城市，在这里举行国家仪式和其他特殊事件。这个广场有历史重要性，这里放置有四个纪念碑（16世纪、18世纪和20世纪的历史事件）。每个纪念碑都讲述了一段至今还能唤起波兰人尊敬的历史和事件。同时该广场的故事也是一个令人茫然的历史混杂，人们可能不仅被他的数量众多的故事所惊讶到，而且故事常常会突然转折，改变方向。该项目试图说明这种情况。当时的想法是把这个永久不变树立的旗杆替换成一个新的旗杆——一个纠结的旗杆。这个"旗杆"放置于官方的地方应当是一个笑话，但同时也是对这个地方的一个诊断：因为一个旗帜象征一个积极的想法，旗杆是一个支撑物——在新的形状下，它可以指国家的状况和不断吹拂在这里的"历史之风"。

经过进一步研究，卡米拉·史捷诺科恩提出了"旗杆"与同时代大事件之间关系的象征概念。在 2012 年，她建立了欧洲旗杆，在波兰和乌克兰 2012 年欧洲杯的大背景下安装的雕塑。它们解决文化问题和通常被大的商业运动事件而掩盖的地理政治问题。其中一个《旗杆》临时安装在波兰 Wrocław 的"Oławka"运动场，另一个成为国际公共艺术节日的主要项目的一部分，乌克兰 2012 年的基辅雕塑工程。在基辅永久安装旗杆的另一个想法正在进行，就像现在"旗杆"已经变成一个社会共同体之间的交流工具，为了文化价值和人权，他们经历了一系列抗议来自政府权利和腐败的压力。

对于 Szejnoch，旗杆本身不仅展示了民族骄傲，而且彰显了国家机构和规草制度。《旗杆》的雕刻是扭曲和转动的，就好像这些规则是灵活性的，建议是可改变的，嘲笑固定的身份，是一个有远见的民族身份的溶解。

Kamila Szejnoch (Poland) deals with art in public space—specifically installations, sculpture and urban designs (Warsaw, Oslo, Belgrade, Yerevan, Fribourg, Lublin).

Recently my favourite topics have been in most cases history and ideology, the elements which influence the public identity. "I have been trying to approach, update, ridicule or just remind outdated layers of the past by adding something new and contrasting - distinct from their original style or function. Of course I do not behave as an objective spectator as I am myself to a certain extent a product of reality I live in", comments the artist. Most of her works consist of two main elements. One is historical, traditional, already fixed or closed in its form, second one more contemporary, open, playful, and interactive.

The first "Flagpole" was created in Lithuanian Square in Lublin—a city in eastern Poland. There are state ceremonies held and other occasional events. The square is of a historical importance, there are four monuments placed (of the 16th, 18th, and 20th century historical events). Each monument represents a part of Polish history and events which still evoke respect. At the same time the story of the square is a fascinating historical mishmash, which may surprise not only with too big a number of stories but also take different directions and sudden turns. The project attempts to illustrate this situation. The idea was to replace a permanently standing flag pole with a new one—a tangled one. The "Flagpole" is supposed to be a joke in an official place, but a diagnosis of the place as well: Since a flag symbolises a positive idea, a flagpole is a supporting construction—in a new shape it can refer to the state's condition and 'historical winds' blowing there constantly.

Further on, Kamila Szejnoch has developed the symbolic concept of the "Flagpole" in its relation to contemporary events. In 2012 she created the Euro Flagpole, sculpture installed within the context of the Euro Cup 2012 in Poland and Ukraine, addressing cultural and geopolitical issues often hidden within big commercial sport events. One "Flagpole" was temporarily installed in "Oławka" Stadium in Wrocław (Poland). The second one became a part of the Main Project within the international festival for public art, Kyiv Sculpture Project 2012 (Ukraine). A further idea to install the object in Kyiv permanently is in progress, as now the "Flagpole" has become a communication tool for the community that has experienced a series of protest against the pressure of authorities and corruption—for liberal values and human rights.

For Szejnoch the flagpole itself represents not only national pride but the institutions of state and their rules and regulations. The sculpture of the "Flagpole" is twisted and turned as a suggestion of flexibility and change for these rules, mocks the fixed identities, and is a visionary of dissolution of national identities.

[解读]

公共认同

旗帜自古以来一直被作为是一个团体的精神象征，从古代的部落图腾到如今的各个国家的国旗，"旗杆"作为一个纯粹的精神产物，往往带给人们神圣感和崇拜感。而此时的"旗杆"是历史的诉说和记录者，让人们想起波兰那段尘封的历史，同时"旗杆"是扭曲的转动的，像被某种无形而巨大的力量扭动弯曲，随风飞扬。让人们反思"旗杆"背后所代表的规则和骄傲，权威和精神，是永恒的还是变化的，是固定的还是旋转的是艺术家想要讨论的，过去被认为绝对权威的东西如今有了更多值得商讨的余地。Kamila Szejnoch 的一系列旗杆雕塑不仅是为了记录某个事件，或者象征着什么，而是成为了公众精神的一个连接点，面对一个不再笔直挺立着的"旗杆"，公众应该何去何从。（李田）

奥特沃茨克小镇
Otwock

艺术家：米罗斯拉夫·巴勒卡	Artist：Miroslaw Balka
地点：波兰奥特沃茨克小镇	Location：Otwock，Poland
时间：2011 年	Date：2011
推荐人：莱西亚·普洛科朋科	Researcher：Lesya Prokopenko

《奥特沃茨克小镇》是一个试图通过艺术，从艺术角度发现城镇中存在的问题的项目。从其自身来看，作为一个"完全艺术品"，它使波兰奥特沃茨克小镇成为了汇集国际艺术家们的平台。该项目在 2011 年启动，发起人是波兰艺术家米罗斯拉夫·巴勒卡、策展人 Magda Materna 和 Kasia Redzisz。

奥特沃茨克是一个位于华沙西南 24 千米的小镇，在 Świder 河沿岸，有多达 44 000 名的常住人口。1989 年之后的私有化和建设热潮导致了城市极为混乱。城市如今的建筑非常的杂乱。在该市中心，20 世纪 60 年代建设起来的商业大厦成为了废弃多年的废墟。一个未建成的钢筋混凝土公园时刻提醒人们当地的景观状态。

这个项目的意义是双重的。"奥特沃茨克"是 Bałka 的故乡，他把他的家改造成了工作室。这个工作室反过来激发了他在艺术和空间之间的灵感，就是在这个地方，他的灵感被激发了出来。该项目邀请了艺术家和作家对这个城市背后的意义进行改造和创作。他们的作品和他们的双于揭开了这些被隐藏的意识、事实和其他相关的情况。《奥特沃茨克小镇》项目是一个在这片土地上进行思想创造和意识描述的练习。被邀请的艺术家、管理人和作家利用他们能够发现的背景进行创作。他们的成就存在于既定的历史和未来之间，在个人和官方的语境之间，在空间功能之间变换。

这个想法是对于特定场域的概念和场域回应最好的描述。这个作品以难以捉摸的艺术姿态与手法，背景或事件，揭露了新的含义，并塑造了松散的联络网。通过互补，他们提出问题，并给出导致变化的解释和建议，以及考虑实际的新方法。"奥特沃茨克"是一个需要花费时间去和艺术交流的地方，满载承诺与想象力。迄今为止，在这里工作过的艺术家有：Lara Almarcegui，Tacita Dean，Anna Molska，Charlotte Moth，Luc Tuymans，Anna Waliszewska 和 Jos de Gruyter i Harald Thys。该项目背后是一个广泛的公共项目，包括艺术与城市研究。它能够让当地政府和公众研究所参与，并支持艺术研究干预该镇的发展。

Otwock is a project revolving around a place perceived through the lens of art—with the aim to look at the town in question from the angle of art. It is both a "total artwork" in its own right, and a platform for international artistic participation in the life of a small Polish town Otwock. The project was founded in 2011 by established Polish artist Mirosław Bałka, together with curators Magda Materna and Kasia Redzisz.

Otwock is a town located 24 kilometers south-east of Warsaw, on the river Świder. Its current population is estimated at 44 thousand inhabitants. The privatization and the construction boom that followed after 1989 resulted in great urban chaos. The present-day architecture seems accidental and chaotic. In the city center, commercial pavilions erected in the 1960s were turned into a construction site, abandoned now for years. Reinforced concrete foundations of an unfinished retail park remain a bitter reminder of the current condition of local landscape.

The place has a twofold meaning. Otwock is the hometown of Bałka, who has transformed his family house into a studio. The studio, in turn, has inspired a reflection on the relations between art and the space in which it is created. The artists and authors invited to the project are here to create in the found context, their works and gestures to uncover layers of hidden senses, facts and associations. Otwock is an exercise in creating a subjective description of the place. Invited artists, curators and writers work with the found context. Their contributions are suspended between history presence and future, between personal and official narratives, between changing functions of space.

The adopted method is best described by the notions of site specific and site responsive. The works, the elusive artistic gestures, the texts or events uncover new layers of meanings and form a loose grid of associations and connections. By mutually complementing each other, they pose questions and make proposals leading to changing interpretations and new ways of considering the local. Otwock is an attempt to spend time with art in the place, approached with commitment and imagination. The artists who have thus far worked in Otwock include Lara Almarcegui, Tacita Dean, Anna Molska, Charlotte Moth, Luc Tuymans, Anna Waliszewska, Jos de Gruyter i Harald Thys.

The project is followed by an extensive public programme, containing art and urban research. It has been able to engage the local municipality and public institutions to support the development of artistic research and interventions in the town.

[解读]

想象力网络

一个废弃的小镇，一段掩藏的历史，一个被遗忘的时代，我们生活在此刻，而过去的时光则疾驰而去。时光的流逝被这个小镇记录了下来，波兰近代十分动荡的政治和经济变化也表现在这个波兰首都华沙附近的小镇上，他成为了那个时代特有的产物，艺术家试图在一个城市被遗弃的遗骸里找寻一些问题的印记，是苏联模式的经济向私有化市场经济转型过程中的巨大浪费和损耗。这个小镇被**巴勒卡**改造成一个公共的艺术空间，这不是一件孤立的艺术品，它是延展开来的，它作为一个整体的场域，不是新建成的，而是旧的、厚重的。所有的物品共同构成了一个联络网，给了很多艺术家一起创作艺术品的机会，并且赋予这个场域以新的含义，挖掘它内部的新的想象力。《奥特沃茨克》从一个被时代遗忘的弃儿，成为了无数想象和承诺的承载地。（李田）

Partizaning

艺术家：多位艺术家
地点：莫斯科
时间：2011 年
推荐人：莱西亚·普洛科朋科

Artist：Multi
Location：Moscow
Date：2011
Researcher：Lesya Prokopenko

《Partizaning》倡议是一项由莫斯科的艺术家和研究者于 2011 年创建的项目，有着自己的同名博客，以国际跨学科现象命名。该项目崇尚自由表达，旨在重新思考城市行为和重新构建景观和社会结构思路。通常这样类似项目的完成都有市民的参与，他们匿名提出意见，而不是抱怨这样的产品不是艺术——这样消除城市现实的设施和艺术品之间的差异。

《Partizaning》是一场运动、一个集体、一个网址，记录了许多来自世界各地的以艺术为基础的行为、干预措施以及城市重新规划的实例。这些艺术家声称，越来越多共享的社会政治现实和不满，可战略性地解决以艺术为基础的"partizaning"策略。

目标是把艺术开发为一个对于包容城市再生和社会活动的实用工具。它的发展面临着当代俄罗斯的文化、政治和社会危机。《Partizaning》提供了艺术的扮演重塑公共空间、全球范围内城市和人的互动角色中的作用的文件记录和分析，同时也展示了他们自己对于俄罗斯首都的公共空间的看法。

部分《Partizaning》团队的项目已经在 2012 年春季实施，包括艺术家和市民共同参与的现场的人行道，即使交通和行人运动需要城市人行横道，在这里却没有涉及；在公共空间里的特殊板凳；在莫斯科一些有问题的街区设置附有说明的"邮箱"，以供市民递交他们对于公共空间转化的建议和要求。

由《Partizaning》创建的斑马线显然公开说明了艺术干预的游击天性，比官方城市斑马线的线条稍稍细点，这凸显了一个主要的信息：自己动手。在一些情况下，DIY 的斑马线实际上被官方替代了。

喷成明亮橙色的木质长椅被放在莫斯科周围的几个地点。上面的标签是这样写的："这个长椅是镇上人为镇上人做的。我们没有为它付钱，也没有给它打广告。这个城市是属于人民自己的，我们可以用双手把它建设得更加美好。请随意使用这些长椅，它做工精良，不会弄脏您的衣服。在制作过程中，费用全部用在了装船、喷绘、酸奶、和马芬饼干上。"这些长椅之中的一个长椅的命运让人费解，将会被莫斯科电台的在线电波覆盖。因为一些原因，公园里的长椅被重新喷成灰色的，最基本的假设是，市级技术人员发现，橙色与邻国的革命有着含糊不清的关系。

在"街区艺术分区"的工作标题下，邮箱项目在 Troparyovo 街区开工。它包括，市民可以向邮箱提交关于如何改进这个街区的建议和意见。题词是这样写的："请把在您的街区没实现的想法写下来，并留下。我们会制作一张标有问题和角角落落的地图，尽量充实它，或者找到一个比我们做得更好的人。这个项目的信息是'市民服务市民'。完全基于我们的热情，为了这个项目，我们收集想法、意见和建议，并设法完成而且我们不收取费用。因此，请记住，很多困扰你的东西其实可以用您自己的双手来解决。"

邮箱项目仍在进行中。基于 2014 年春季采集的雅罗斯拉夫地区的数据，莫斯科青年多功能中心推出讲座和讨论节目来讨论街区的变化，与街头艺术家、活动家和城市规划专家进行研讨会。

因此活动家希望当地居民参与到城市变化的过程中。

Partizaning initiative is a project by a group of the group of Moscow artists and researchers, founded in 2011, along with the eponymous blog, named after the international interdisciplinary phenomenon. The project popularizes the ideas of free expression and actions aimed at rethinking and restructuring of the urban landscape and society through socially oriented urban intervention. Often these projects are implemented jointly with ordinary citizens, who spoke anonymously, not claiming their product to be art—but rather eliminating the difference between practical urban facilities and artworks.

Partizaning is a movement, a collective, and a website, documenting examples of art-based activism, interventions, and urban replanning from around the world. The artist claim that increasingly shared sociopolitical realities and dissatisfaction can be strategically addressed using art-based 'partizaning' tactics.

The aim is to explore the role of art as a practical tool for inclusive city regeneration and social activism. It evolves out of the cultural, political and social crisis facing contemporary Russia. 'Partizaning' provides documentation and analysis of the role of art in reshaping public spaces, cities and human interactions—globally, as well presents their own interventions in the public space of the Russian capital.

Some of the projects implemented by Partizaning group by spring 2012 include artist- and citizen-made pedestrian crossings in spots, where municipal crossings were not issued, although the traffic and pedestrian movement required these; special benches in public spaces; 'mailboxes' in several problematic Moscow districts with instructions for citizens to submit their recommendations and request about public space transformations.

Slightly thinner than white stripes on the official municipal crossings, the ones created by 'Partizaning' obviously and openly demonstrate the artistic guerilla nature of the intervention, highlighting the main message: do it yourself. On several occasions, the d-i-y crossings were indeed replaced with the official alternatives.

The benches painted bright orange were placed in several places around Moscow. The text on the label states: "This bench is made for by townspeople of the townspeople. We have not been paid for it, and we are not advertising anything. This city belongs to its people and we can do it better ourselves—with our own hands. Feel free to the bench as you wish. It was made from quality materials, painted, but not getting you dirty. During manufacture, the money was spent entirely on board, paint, yogurt, and muffins." Further puzzling destiny of one of the benches was covered online by Echo of Moscow radio station. The bench in a park for some reason was repainted gray – the basic assumption is that the municipal technical workers found the color orange ambiguous because of its connotations with revolutions in the neighboring countries.

The mailbox project under the working title "street art district" was launched in Troparyovo district. It included the system of boxes in which citizens could submit their complaints or suggestions for improvement of the district. The inscription reads: "Please, write what is missing in your area and drop a message in the box. We will make a map of issues and concerns, and we will try to fulfill them, or find someone who can do it better than ourselves. The program's message is 'citizens for the citizens'. Within the program we collect opinions, complaints, and suggestions, and try to fulfill them, being based solely on our mutual enthusiasm. We are not being paid for it. Therefore please remember that much of what is bothering you can be fixed with your own hands."

The mailbox project is ongoing. Based on data collected Yaroslavsky district in spring 2014, Moscow Youth multifunctional center launched the lectures and discussion programme to discuss district changes, to conduct workshops with street artists, activists and urbanists. Thus activists want to involve local residents in the process of urban change.

[解读]

城市再生

当我们讨论艺术的作用时候，从柏拉图、黑格尔和托尔斯泰明显道德论的美学到康德和克罗齐纯粹形式主义的美学，而在这里艺术成为了连接大众和城市重建的工具，它深入到城市生活中，潜移默化地改变了人们的生活。如今艺术不是高高在上的奢侈品，而是越来越走向大众服务大众的艺术，市民自己动手制作的斑马线和长椅，让公共服务设施重新充满了温度，让市民亲自参与到城市的变化中去。艺术家成为了这场变革的引导者，艺术不是一个孤立的作品，而成为了一个温暖的联结者。曾经的农耕时代，我们的生活用品大多是自己制作的，那些让我们惊叹的美丽彩陶在那时也只是劳动者制作的实用工具，而工业社会和完备的社会机制让我们距离生活越来越远。艺术让城市更加完整起来，让城市再生。（李田）

搜索：其他空间
Search: Other Spaces

艺术家：基辅艺术家和
　　　沃瓦·沃罗特尼沃夫
地点：乌克兰基辅
时间：2012 年
推荐人：莱西亚·普洛科朋科

Artist : Kyiv artist and Vova Vorotniov
Location : Kyiv，Ukraine
Date : 2012
Researcher : Lesya Prokopenko

由基辅艺术家和策展人沃瓦. 沃罗特尼沃夫一起合作的项目《搜索：其他空间》，涉及历史公共部份和破坏公共空间之间的关系，发生在 2012 年 4—5 月。

这个城市的标志建筑——大力士喷泉亭，展示了当地的巴洛克风格的传统。包括大力士在内，沃瓦·沃罗特尼沃夫在 2 周时间内，完成对展览的大力士喷泉亭的象征性的改变，他使用一些日常物品（一个背包、运动鞋、网球、面料、彩色纸等）作为材料，来创作物体和拼凑雕像——暗示当今的生产过剩和过度消费。第二阶段是项目的介绍，以及在同一个空间的讲座，旨在提出研究的目标和目前的文件材料。

实际上沃瓦·沃罗特尼沃夫利用大力士所做的是一个明确的证明，他证明个人的战术可以构建一个公共艺术作品，即使在最后它获得了机构的授权。大家都熟知，沃瓦·沃罗特尼沃夫作为基辅 Muralissimo 涂鸦艺术节的艺术总监，他至少有 10 年都在创作个人非机构授权的概念壁画，并在波兰和乌克兰安装。这恰是在公众的视线中重新建某一个地方的例子，即使这突如其来的功能转变是在短时间内确定的。

此外这个项目的选址有一个特殊的城市意义：大理石喷泉亭的位置就在基辅的老城区，Podil 的建筑遗产，目前因为私人房地产的开发和市政当局的城市规划而快要绝迹。17—18 世纪巴洛克和新古典主义的建筑，没有得到适当的恢复，或许 / 已经进行了商业装修而使其历史价值大打折扣。此外，一些提到的控制限制了公众使用这些建筑的权利。当沃瓦·沃罗特尼沃夫把达达主义活动应用到这些遗址中的一个上，它提供了人们关注的艺术范畴内一个新的部分展馆。在某种程度上，通过把流行音乐和荒诞主义引进公共空间，艺术活动防止它沉浸在偏执的中立性和社会潜意识的破坏性。

作为大力士工程的一部分，沃瓦·沃罗特尼沃夫制了一系列的塑料制品来连接特定地点和街头艺术，在它们之中，大力士喷泉既是一个公共的展览空间，又是一个照相馆。开放给路人，让他们自由观看，对这个地方这样干预的结果进行记录和说明，特别是一般意义上的公共空间存在的问题。"我也想在我们这个时代里的巴洛克式，过度生产和客户产品的功能性之间画一条平行线"，该艺术家这样么说。该项目的高潮是一个公开演讲。在演讲中，他试图勾勒出项目的目标和吸引艺术家的一系列问题。

The project by Kyiv artist and curator Vova Vorotniov involved interaction with a historical public site and subversion of public space within the programme «SEARCH: Other Spaces» and took place on May 4—May 15, 2012.

The Samson Fountain pavilion, a symbol of the city, is representative of the local baroque tradition. Within Samson, Vova Vorotniov during two weeks made symbolic interventions in the exhibition space of the Samson pavilion, using everyday things (a backpack, sneakers, tennis balls, fabric, coloured paper, etc.) as supply material to create objects and sculptural collages that hinted at overproduction and the consumer excess of today. The second stage was the presentation of the project and a lecture in the same space that aimed to outline the goals of the research and present the documented material.

What indeed Vova Vorotniov implemented within Samson, was an explicit proof that individual tactics can construct a public artwork, even if in the end it does get an institutional authorization. Known for his curatorial work as an artistic director of Muralissimo Kiev graffiti festival, Vova has been creating personal non-institutional projects for at least a decade –conceptual murals and installations in Ukraine and Poland. This is the exact case in point of re-establishing a certain place in the public eye, even if this sudden functional shift was transient in time.

Moreover, the project's location had a particular civic meaning: the architectural legacy of Podil, the old part of Kyiv where Samson pavilion is situated, is currently endangered by private property developers and city municipality. Baroque and neoclassical 17-18th century constructions do not receive proper restoration and/or become subjected to commercial renovation, which depreciates their historical value. In addition, many of mentioned manipulations limit public access to these constructions. When Vova Vorotniov conducts his Dadaist activities around one of such urban sites, it provides the pavilion with a new portion of attention within the artistic context. In a way, by bringing pop and absurd into the public space, artistic activities keep it from sinking in paranoid neutrality and destructive social automatism.

As part of the Samson project, Vorotniov made a series of spontaneous plastic interventions bordering on site-specific and street art in which the Samson fountain was both a public exhibition space and photo studio. Open to the spontaneous contemplation of passersby, the results of the intervention were documented and illustrated aspects of this place, specifically, and problems of the public space, in general. "I also wanted to draw a parallel between baroque excess, overproduction and the functionality of consumer products in our time", says the artist. The culmination of the project was a public lecture in which he tried to outline the project's goals and the range of issues that led the artist to it.

The project offered new functions for the Samson pavilion—social (lectures) and expositional (exhibition with documentation). Working openly in a public space, the artist's goal was to become an added component to the place in the perception of the random spectator, clearly illustrating the practice of artistic invasion and thus expanding the traditionally ambivalent, respectfully alienated perception of the street by the average person. This location also inspired several improvisational actions. For example, "Vedel's House" —for an hour he played the music of the 18th century Ukrainian composer Artemiy Vedel on a ghetto blaster —a symbol of street youth (sub) culture.

A result of the Samson project was the understanding of the practical opportunities and need to overcome the alienation of Kyiv's modern public space through local, contextual, and, most importantly, systematic interventions aimed at redefining street objects, at exploiting their functional resources, at their actualization and realization.

[解读]

陈旧中的新

基辅作为乌克兰的首都，有着 1500 年的历史，被誉为"俄罗斯众城之母"，陈旧与创新、传统与现代、遗迹与重生在这里共同存在着，艺术有时候并不是摧毁旧的东西，而是让新的东西在旧的物体上演绎。艺术家将废弃物组成的雕塑与旧的巴洛克建筑放在一起，让人们反思过去与现代，过剩的装饰和过剩的生产到底给了我们什么。重建与改造是 Vova 一直在做的事情，历史公共空间的重新塑造成为艺术对生活的一种干预，人们可以观看、参与，甚至改变一个地方的面貌。Vova 用塑料所做的一系列这样的街头艺术，让路人也参与进来，那些被观看了几百年而被人们忽略的古老建筑似乎也焕发了新的力量，这突如其来的变化给观者的视觉和观念带来洗礼，人们有时候会认为当代艺术是反传统的，其实当代艺术往往是赋予传统以新的活力和想法。（李田）

基本频率
Fundamental Frequency

艺术家：安科·瓦伦特
地点：意大利那不勒斯
形式和材料：雕塑
时间：2011 年至今
推荐人：茱茜·乔克拉

Artist：Bianco-Valente
Location：Naples，Italy
Media/Type：Sculpture
Date：2011—Now
Researcher：Giusy Checola

《基本频率》是一个永久性的大型雕塑，在 2011 年迁入位于那不勒斯附近的波蒂奇维拉马斯科洛公园。另外建立在一个 3 000 千平方米的地表的完全废弃和停止使用的 18 世纪的建筑也得到了恢复。事实上，基本频率不是来自维苏威火山地震波传感和扩散的一个复杂系统，而是根据波蒂奇的维苏威火山的天文台与音乐家马里奥马索罗（又名马斯）的合作构想出来的。

维苏威火山是全球公认的最危险的活火山之一。从 1944 年开始，这个火山戏剧性地变得不再活跃了。但历史记录表明，即使在 130 年后，它仍然可以再次活跃。这个火山作为那不勒斯及其周围居民历史和生活的象征性图标，与他们有牢固的联系。从城市的每一个角落都能感觉到火山的壮观，但人们似乎没有意识到其内在的生命来自地球的核心。生活在那不勒斯 24 个直辖市和 3 个区的将近 80 万居民生活在维苏威火山脚下，由于火山爆发的可能性很高，所以这个区域已被标记为"红色区域"。一旦火山爆发，后果将是毁灭性的。全世界都记得公元 79 年火山喷发时，滚烫的岩浆淹没了庞贝古城和埃尔克雷诺城市的著名遗迹，至今却成为了意大利主要的旅游点。然而，尽管政府的经济措施努力让居民们搬到别的地方去，但是居民还是不想离开火山。当然政府只是要求居民搬走，并没有想法设法建立一个更具创新的监控系统来创造更多的逃生通道，所以部分火山脚下的人到现在还是没有搬走。

《基本频率》大型雕塑由钢和发光灯构成。通过维苏威亚诺（火山天文台）监控网络的各种传感器进行测量，获得的数据来自维苏威火山的物态变化，随时调节成为一个合理的方案，然后通过由大约30个花园扬声器构成的音频环绕声系统传遍整个公园。除了良好的装备，雕塑安装在公园，并俯瞰整个别墅，与连接火山口到公园的假想线垂直，产生了一个复杂的关系结构。无论是公园内还是周围的建筑物，不断合理的改变让每个人满心欢喜，就像微小的物态变化的直接表达，聆听它们的实时"呼吸"。

因此，艺术品在整个范围内创造了一种居民、游客和火山地质之间不寻常的关系。事实上，这是第一次，维苏威亚诺已获准使用每日记录的关于内部火山活动的数据，目的是使公民变得敏感并且使他们具有风险意识。同时，作品改变了居民对于火山的看法，与火山的关系变得更加亲密、真实，而不只是图像上显示的那样。此外，由于艺术品的真实性，从事雕塑作品的专业人士（建筑师、设计师和工程师）和公共管理之间形成了一个之前不存在的网络。

Frequenza Fondamentale (Fundamental Frequency) is apparently a permanent large scale sculpture, settled in 2011 at Villa Mascolo's park a Portici in the neighbours of Naples, a XVIII century building recently restored after several years of complete disuse and wreckage, developed on a 3 000 square meters surface. As a matter of fact, Frequenza Fondamentale is instead a sophisticate system of sismic-waves sensoring and diffusion, coming from the Vesuvio volcano. It was conceived with the collaboration of the Vesuvius's Observatory of Portici and the musician Mario Masullo (aka Mass).

Vesuvius is an effusive active volcano, one of the most dangerous and analyzed volcanos worldwide and it has been quiescent since 1944. But its history demonstrates that it can wake up even after 130 years and counting. Naples' and surroundings inhabitants have an indissoluble tie with the volcano as a symbolic icon of their history and life, whose imposingness is felt from every point of the city, but they seem to have little consciousness of its inner life, coming from the Earth's viscera. At the foot of Vesuvius live almost 800 000 persons, organized in 24 municipalities and 3 districts of Naples city, that have been declared "red zone" for its high risk. In case of eruption the aftermath would be devastating: all the world remembers the great eruption in 79 D.C., of which today remain in the famous remnants of the lava-submerged cities of Pompei and Ercolano, amongst the main touristic sites in Italy. However the inhabitants don't want to «separate» from the volcano and despite the regional administration economical incentives to make them move somewhere else, the "vesuvian" people have in large part stayed, asking instead to establish an even more innovative monitoring system and to create more escape lanes.

Frequenza Fondamentale, made by steel and electroluminescent lights, uses the datas coming from the Vesuvius' changes of state, measured by various sensors of the monitoring network of the Osservatorio Vesuviano (Vesuvian Observatory), in order to modulate, in real time, a sound scenario to be spread across the park by an audio surround system that consists of about thirty garden speakers. In addition to the sound installation, the sculpture has been installed in the portion of the park overlooking the front of the villa, orthogonal to the imaginary line that links the volcano's crater to the park, generating a complex relational structure. The sound changes continuously for everyone to be delighted, both inside the park and also from the surrounding buildings, as direct expression of micro-changes of state of the volcano, to listen to his real time "breathing".

The artwork has created an unusual relationship between residents and visitors and the volcano's geologic reality, hence with the whole territory. For the first time, infact, the Osservatorio Vesuviano has given permission to use the datas which have been daily recorded about the inner volcano's activities, with the aim to sensitize citizens and stimulate their awareness of the risks. At the same time the artwork has transformed the perception of the inhabitants regarding the volcano, making their relationship more intimate, physical and not just iconographical. Moreover, thanks to the realization of the artwork, a network arouse among professionals that worked at the sculpture production (architects, designers and engeneers) and public administration, that was non-existing before.

[解读]

艺术需要科技的依托

维苏威火山与在这里居住的居民产生了很深的情感。政府只是单方面地强调居民的安全，却忽视了居民同火山之间产生的一种地域文化的认同。与其要求居民搬离火山地区，还不如建立一个监控系统来指引居民在火山将要爆发的时候逃离，安科瓦伦特却很好地解决了这个问题，他运用雕塑艺术的表现手法，赋予火山监测仪器一种美的形式，散发出科技的气息。艺术不单纯是天马行空的想象、华丽的颜色与繁复的图案，有时候艺术家也需要成为"半个科学家"，既要注重美的表达，也要充满富有逻辑的、科学的思考，将艺术与科技完美统一在一起。我们常说 21 世纪是一个"大数据"的时代，人们往往难以接受成百上千的数字信息，如何将获得的数据转化为视觉上的直观感受，既易于理解，又方便记忆，是艺术家和设计师需要思考的一个问题。（于奇赫）

从纪念碑到新纪念碑
The Monument to a New Monument

艺术家：詹娜・卡德罗娃 Artist : Zhanna Kadyrova
地点：乌克兰基辅 Location : Ukraine
形式和材料：雕塑 Media/Type : Sculpture
时间：2012 年 Date : 2012
推荐人：莱西亚・普洛科朋科 Researcher : Lesya Prokopenko

詹娜・卡德罗娃，生于 1981 年，一直广泛致力于乌克兰的公共艺术，旨在反思和改造苏联时期留下的城市遗迹和公用空间艺术的传统。

正如这位艺术家在对她自己工作的评价中所说的：独立之后、超过 20 年的时间，乌克兰社会已经经历并且还在经历一个尖锐的冲突，这个冲突关于历史遗迹的方方面面的理解和接受（或不接受）。这些差异中最明显的一个指标表现在和历史纪念碑之间的关系上。应该把纪念上个历史时期英雄的纪念碑废除吗？因为这些英雄的丰功伟绩在当代乌克兰的发展中是受到质疑的？应当用这些在先前的历史中被封杀的、所从事的活动被从历史记忆中抹去的人们的纪念碑来代替吗？

纪念碑新碑是一个白色瓷砖雕塑，与真人的高度相当，看起来像在典礼前夕被一张白色的纸张所覆盖的纪念碑。白色的瓷砖无疑显示了人体的轮廓，"新英雄"的身份却难以识别。具有比喻涵义的白片可遮住任何事情和任何人。这个雕塑品评了现代乌克兰的历史和政治游戏以及激发了每一个旁观者饱满的精神世界。观众开始能够看到他们愿意看到的纪念碑。

安装纪念碑新碑由詹娜·卡德罗娃在2009年乌克兰的Shargorod提出。2007年举办的建筑学盛会，这是詹娜·卡德罗娃第一次来Shargorod，在这里她展示了该作品的首个版本。她愿意为这个小镇考法艺术潜能，詹娜·卡德罗娃提出了全市建设永久性的工作，看看人们会作何反应。她和市议会达成共识：在列宁大街，给纪念碑建造一个相应的小广场——作为城市设施的一部分，带有小径和长椅，切实重构苏联解体后的公共空间的结构。接下来的几年间，准备工作就绪，该作品于2009年启动，该镇上的社区和市政当局都积极参与。

当地人在空间安排工作上做出了贡献，调整和完善自己家乡的基础设施。在第55届威尼斯双年展上，这个作品进一步展示了一个关于乌克兰建筑的视频，但实际上纪念碑是詹娜·卡德罗娃的一部分，她经常在Shargorod创建并展出自己的作品。同时，Zhanna已经获得了该镇镇长的官方致谢，并且已经从社区收集反馈意见。

这件作品展示了非传统对象在不变的环境中促进变化的能力，同时保留了作品本身的概念价值。在各种不同的媒介、不同的时间展出，安装的主要点仍作为Shargorod的参考保留。

Zhanna Kadyrova (born in 1981) has been working extensively with public art in Ukraine, aiming at reconsidering and transforming the heritage of soviet urban planning and traditional notions of art in public spaces.

As the artist states in her comment to the work, for more than twenty years after independence, the Ukrainian society has been experiencing a sharp conflict of interpretations and (non) acceptance of various aspects of its historical heritage. One of the clearest indicators of such differences is shown in relation to the historical monuments. Should monuments to the heroes of previous eras, whose beneficial effects on the development of Ukraine today are under question, be dismantled? Should monuments to the people whose activities were previously censored and blotted out of historical memory, be placed instead?

The Monument to a New Monument is sculpture of white tile in full human height that looks like a monument covered with a white sheet on the eve of the opening ceremony. The tiled white mass undoubtedly shows the outlines of the human body, but the identity of the "new hero" is unrecognizable. The metaphorical white sheet can be concealing anything anyone. The sculpture comments on the historical and political games of modern Ukraine, and activates the psycho-mental fullness of every beholder. Spectators begin to see in the monument that they are willing and able to see.

The installation Monument to a New Monument by Zhanna Kadyrova was presented in Shargorod, Ukraine in 2009. Zhanna Kadyrova came to Shargorod for the first time during the architecture festival held in 2007, where she presented the first version of the work. Willing to explore the potential of art for the small town, Zhanna suggested its municipality to construct the permanent work, and see the way people will react. She has agreed with the city council that a new small square should be arranged for the Monument—as a part of the big urban installation, with paths and benches around it in Lenin Street, practically reconsidering the structure of post-soviet public space. For the next couple of years, the preparations have been taking place, and the work was launched in 2009, with the active participation of the town's community and municipality. The locals contributed to the work on space arrangement, adjusting and improving the infrastructure of their hometown. The work was further presented as a video hologram within the Ukrainian pavilion at the 55th Venice Biennale, but the actual Monument is a part of Shargorod, where the artist regularly presents and creates her pieces. Zhanna has received the official acknowledgement from the town's mayor as well, and has been collecting feedback from the community.

The piece demonstrates capacity of non-traditional objects to catalyze changes in quite immobile environments, simultaneously preserving the conceptual value of the work itself. Presented over the time in a variety of media, the installation's main focus remains on the reference to Shargorod.

[解读]

历史的面纱

近代的东欧历史充满了种种无法言说的动荡、误解、激进和衰颓，乌克兰便是其中的一个集中体现。苏联时期的种种遗留在这个大帝国分崩离析之后，如何处理公共空间里曾经树立的英雄纪念碑，当历史风向彻底改变的时候，我们如何去思考和定义黑与白、正义与邪恶、英雄与贼寇呢？如此尖锐的冲突应当如何化解？也许在短时间之内我们是迷茫的，不知道去信奉什么或者摧毁什么，詹娜·卡德罗娃的纪念新碑给了这个问题一种反思和答案，用白色的瓷片制作一个幕布遮盖住纪念碑，艺术家给历史蒙上面纱，他是神秘的、未知的，我们无法定义或者评判。一切都隐藏在幕布之下，一切都等待被揭开，历史给我们的是真实或者假象，幕布下是什么我们难以揣摩。这个未知的纪念碑放在公园里，给这个曾经动荡的国家的人们抚慰和反思。（李田）

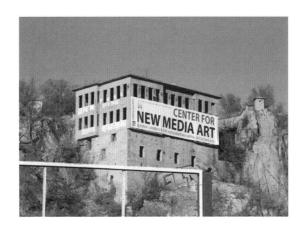

城市梦想
Urban Dreams

艺术家：多位艺术家

地点：保加利亚普罗夫迪夫
形式和材料：装置艺术、
　　表演和行动
时间：2012年
推荐人：茱茜·乔克拉

Artist：Emil Mirazchiev，Sevdalina
　　kochevska，Veronika Tzekova，
　　Vladislav kostadinor，HR—Stamenov
Location：Plovdiv，Bulgaria
Media/Type：Istallations,
　　Performances and Actions
Date：2012
Researcher：Giusy Checola

《城市梦想》是在 2012 年 10 月 20 日—11 月 17 日，由若干艺术家在普罗夫迪夫探讨公共和半公共空间问题时进行的一个项目。因为它的目的是使艺术与居民相互作用，从而使居民在日常生活中感知城市的一些问题。所以该项目包含了涉及艺术家和市民的装置、表演和行动。

普罗夫迪夫作为巴尔干半岛、希腊和土耳其之间的一个历史、文化和商业的十字路口，一直保持着重要的历史地位。但如今，大部分公共空间被遗弃或只建设到一半。考虑到人们对于当代文化的一种需求，这里丰富的文化遗产既没有被很好地保存，也没有完整地留给公民。同时，城市错过了用建设文化基础设施来举办文化活动的时机。

基于人们希望到更吸引人的地方生活，并且希望在当代视角下恢复和保存过去的历史和留下的遗产，这个项目得到了欧洲委员会、保加利亚歌德学院、普罗夫迪夫市、罗伯特博世基命会和2019年普罗夫迪夫基命会的支持，与国际网络亚特兰蒂斯进行合作，旨在恢复不友好地方的关系并且发展更多的文化空间。

|CON|临时图书馆由来自8 1/2工作室的保加利亚建筑师弗拉迪斯拉夫科斯塔迪诺夫设计，这是十六世纪以来在普罗夫迪夫古代巴斯的半公共空间建立的一个独特建筑纪念碑，为自柏林墙倒塌的近20年来的艺术城市研究建立了"当代记忆"。这个图书馆由木头建造而成，收藏了书籍、视频和杂志，并为旅客提供了舒适的地方。它是围绕一个中心按照曲线建造，科斯塔迪诺夫写道，就像"爬到屋顶和天空"一样。

城市的一座大型体育场已经被废弃了20多年，现在已经通过位点的安装和维罗妮卡的团队的表演得到了复兴，被命名为"Fooootballll"。这是一个前所未有的四支球队的足球比赛（马里查河队、火车头队、斯巴达克队、波特为队），这场比赛有四个门，但只有一个球。通过增加一个和现有足球相同的足球，足球场才得以创建，但是为了增加难度以及解决方案和游戏策略的可能性，足球场设置在垂直方向。每个参与者（运动员和观众）面对挑衅时，都应该反思与确立的社会行为相关的足球场规则和他们的职位意义，因而干预其内部机制。

最后，在其他干预下，作为艺术家和策展人的埃米尔，通过在老城区的一所废弃的房子中建立电影与摄影博物馆，在保加利亚工厂建立现代美术博物馆和建筑博物馆，在科斯莫斯电影空间建立无限艺术中心，在老城区的另一个摇摇欲坠的房子中建立新媒体中心，填缺空间内的各种文化中心的顺序符号空缺，他称之为"平行的废城"。

大规模的公民参与到这次行动中，表明了他们需要更多的物质文化空间，以及他们对于不同的视觉和精神环境需求的日益增长。建立新的博物馆的行动强调了当代艺术博物馆的缺乏对于国家当代认同的严重性已经成为一个问题，保加利亚其他著名的艺术家也面临这个问题。通过在1995年赢得在古代巴斯建立当代艺术中心的权利，以及认识到博物馆需要一个涉及城市梦想中的公民集体意义，今天的艺术给予一个鼓吹"共有社会"问题的回答。后来，由于这个项目作为2019年欧洲文化资本在普罗夫迪夫的选举中发挥了基础性作用，所以这个过程甚至扩大到国际水平，包括项目中普罗夫迪夫对其他建筑的艺术性重建。最后，通过整体反思城市中的主角、建设过程中的动力与建筑本体，艺术的干预措施打破了现有城市的功能模型。

The project Urban Dreams is composed by several artistic interventions exploring public and semi-public spaces, which took place from 20 October to 17 November 2012 in Plovdiv. The project produced installations, performances and actions involving both artists and citizens, as it aimed at making art interact with the problems of the city as perceived by its inhabitants in daily life.

As an historical, cultural and commercial crossroad between Balkans, Greece and Turkey, Plovdiv has always been holding an important historical role, but nowadays the most of the public spaces are abandoned or unfinished, and the rich cultural heritage is not neither preserved nor transmitted to the citizens, taking care of the relation to contemporary cultural needs. In the meanwhile the city misses cultural infrastructure dedicated to cultural events.

Produced with the support of the cultural program of the European Commission, the Goethe Institute Bulgaria, the City of Plovdiv, the Robert Bosch Foundation and the Foundation Plovdiv 2019, in partnership with the international network Atlantis, Urban Dreams aimed at developing more spaces for culture by revitalizing unfriendly places, based on the desires of people for a more attractive place to live and for preserving and proposing the past and the heritage under a contemporary sight.

|CON|Temporary Library, designed by Bulgarian architect Vladislav Kostadinov from Studio 8 1/2 , has been installed a in the semi-public space of the Ancient Bath in Plovdiv, an unique architectural monument from the XVI century, building a kind of "contemporary memory" of the artistic urban research from the last 20 years, since the fall of the Berlin wall. The library, made of wood coming from buildings construction, hosted books, videos and magazines and provided comfortable places for visitors. It was developed as curves developing around a center like "climbing to the dome and the sky" as Kostadinov wrote.

The big stadium of the city, disused for more than 20 years, has been revitalised through a site-specific installation and the group performance by Veronika Cekova, entitled Fooootballll, an unprecedent football game played between four teams (Maritsa, Lokomotiv, Spartak, Botev) with four doors, with only one ball. The football field has been created by adding a second one, identical to the existing one, but set up in a perpendicular orientation in order to increase the difficulties but also the possibilities for solutions and game strategies. Each participant (players and audience) was stimulated to deal with the provocation to rethink the meaning of the field's rules and their position related to the established social behavior, and thus to interfere with its inner mechanism.

Finally, among other interventions, artist and curator Emil Mirazchiev, initiated sequential symbolic openings of various cultural centers in spaces that he calls "parallel abandoned city", by installing the Museum of Cinema and Photography in an abandoned house in the Old Town; the Museum of Contemporary Art and the Museum of Architecture in the factories of Bulgartabac; the Center for UNLIMITED Art in the KOSMOS cinema space; and the Center for New Media in another crumbling house in the Old Town.

The massive participation of citizens to the actions demonstrated their need for physycal cultural spaces and the growing demand for a different visual and spiritual environment. The actions of placing new museums underlined the lack of Museums for Contemporary Art as a problem regarding the contemporary identity of the country, faced also by other wellknown Bulgarian artists. Art Today gave to the problem a communitarian response, by winning in 1995 the right to place and establish the Center for Contemporary Art in the Ancient Bath and by recognizing a collective meaning for the need of a museum, involving citizens in the project Urban Dreams. Later on, this process even extended its fram on the international level, since the project played a fundamental role for the election of Plovdiv as European Capital of Culture 2019, including in the program the artistic regeneration of other buildings in Plovdiv. Finally, the artistic interventions helped to shatter the existing functions' models that are present in the city by rethinking collectively the role and the dynamics of construction buildings and architectures.

[解读]
艺术连接古今

我们目前的现代化的城市，就是一个鳞次栉比的钢筋玻璃水泥建筑与电线、柏油马路的集合体。城市"千篇一律"的面貌不断地被复制，旧建筑不断被推倒、重建，认为这就代表着进步与现代化。对于一座历史古城来说，如何把历史与现实相连接，让人们不知不觉地走入历史，融入历史，将历史性与实用性统一起来，《城市梦想》的内容就十分丰富。这个项目中艺术家和市民都参与进来，形式包括装置、表演和行动，满足了人们对于当代文化的一种需求。兴建的博物馆、美术馆和艺术中心也让人们在闲暇之余找到了一种让精神"栖息"的地方。中国是一个拥有悠久历史的文明古国，可是目前的建筑、服饰与礼仪等方面都和过去相对发生了很大的改变。这种基于不同历史时期所产生的断裂，需要依靠公共艺术去弥补。（于奇赫）

隐藏的森林
Hidden Forest

艺术家：马泰奥·碧桃丝、
　　　　丹尼尔·科洛尼亚、斯蒂法诺·拉斐
地点：意大利米兰
形式和材料：互动行为、装置艺术

时间：2009 年
推荐人：茱茜·乔克拉

Artist：Matteo Balduzzi,
Daniele Cologna, Stefano Laffi
Location：Milan, Italy
Media/Type：Various Interventions, Installation

Date：2009
Researcher：Giusy Checola

2007 年，建筑师和策展人马泰奥·碧桃丝，社会学家和汉学家丹尼尔科洛尼亚和福利研究员斯蒂法诺拉斐构思了《福雷斯塔纳斯卡斯》这个项目，目的是为了在具有显著社会和文化多样性的米兰腹地重新定义圣朱利亚诺镇的共同身份。

始于 2009 年的《福雷斯塔纳斯卡斯》项目是受米兰省委托，作为对地方文化政策革新需求的响应，目的是在米兰的外围实现文化的干预，并且这种干预涉及从都市文化生活中分离的公民。在圣朱利亚诺，当地政府有建立与年轻人对话的特殊需求，之前他们还不习惯在邻居身上花时间，因此也就越来越没意识到他们领土的问题。同时，该项目旨在响应在与不同的社会群体建立真实的联系方面的一种当代政治语言能力缺乏。

《福雷斯塔纳斯卡斯》计划在时间和空间上考虑圣朱利亚诺发展的社

会和城市结构，这解决了20世纪50年代以来的大规模移民到来的问题。为了在社会、政治、公共问题和私人问题方面创建一个跨时代的对话，并且建立一个不同的领土范围，碧桃丝、科洛尼亚和拉斐在意大利故事叙述的基础上创建第一个公共艺术项目。不久，在社交方面融入的艺术主要有形式方面和艺术家授权的特点。

该项目的主要部分是围绕空间和通信的相关设备制成的。两个集装箱放在城市的五个区，由于不同社会群体生活在那里，每一个集装箱都对应其历史。两个集装箱开设了故事吧，成为居民集会和公开讨论的地方，也是年轻人有效的办公地点。在项目持续期间，地区临时博物馆是展览场地，光博物馆是由市民创建，也是为市民创建，这里有图像和文本以创建一个集体传记。公共空间散布着1 500个广告牌，并与当地报纸长期合作，《市民报和米兰体育报》允许其作为一个扩展的项目空间使用，事实上作为报纸的特别补充，已经印刷了33 000份，作为刺激共同假想的生产的开放过程。在项目结束时，由一个大博物馆组装集装箱，一个集体展览已经在一个公共空间召开。

此外，这个项目还创建了一个收集了80个公民的故事以及1 600多个熟悉的人物的在线档案。这项工作是由与此相关的年轻居民收集和恢复的，后来从居住者的角度与审视城市历史的角度出发，辨认和推广这些公民与人物，集装箱成为大约6 000名参与者的第二个家。正是因为这些原因，它被称为"市民博物馆"。

在开始《福雷斯塔纳斯卡斯》项目之前，圣朱利亚诺被认为是一座"宿舍城"。该项目过早结束后，由于市议会严重的政治危机，市民写抗议信，发表在当地报纸并接受市民询问，项目刺激了文化的需要。然后相关的年轻人构成了一个协会，如今他们与地方行政合作，创造圣朱利亚诺和米兰之间的联系，成为该地区的文化和金融中心。因为有关居民日常生活不同方面更深入的历史研究，《福雷斯塔纳斯卡斯》已经取代了该地区推测的历史真相。

In 2007, architect and curator Matteo Balduzzi, sociologist and sinologist Daniele Cologna and welfare reserscher Stefano Laffi conceived the project Foresta Nascosta, with the aim at redefining the common identity of the town San Giuliano, in the hinterland of Milan, characterized by a prominent social and cultural diversity.

Commissioned by the Province of Milan, Foresta Nascosta started in 2009 as a response to the need for renovation of local cultural policies, aiming at realizing qualitative cultural interventions in the periphery of Milan and at involving citizens usually detached from cultural life of the metropolis. In San Giuliano the local administration had the specific need to create a dialogue with young people, unused to spend time in their neighbours and therefore more and more unaware about their territory. At the same time, the project aimed at responding to the lack of ability of the contemporary political language at establishing a real connection with different social groups, by avoiding any ghettization form based on generational aspects.

Foresta Nascosta has been planned in time and space considering the social and urban structure of San Giuliano development, that dealt with the massive arrival of immigrants since the 50s. In order to activate a trans-generational dialogue about social, political, public and private issues and to establish a different narration of the territory, Balduzzi, Cologna and Laffi created one of the first public art project based on storytelling in Italy, in a time when socially engaged art was mainly characterized by the formal aspect and the autoriality of the artist.

The main part of the project, was produced by spatial, relational and communication devices. Two containers were set up in five areas of the city, each one with a correspondence with the main layers of its history since its existence and with different social groups living there. The two containers began the Bar of the Stories, that functioned as place for meetings and open discussions of the inhabitants and operative office for the young people, and the Quarter Temporary Museum, which was the venue for exhibitions during the duration of the project, conceived as light museums, made by citizens for citizens, where images and texts were set up in order to create a collective biography. The 1500 billboards spread in the public space and the long term collaboration with local newspapers (Il cittadino and La Gazzetta del Sud Milano) allowed their use as an extended space of the project, the special supplements of the newspapers infact has been printend and distributed in 33 000 copies, as an open process stimulating the production of a common imaginary. At the end of the project, a collective exhibition has been realized in a public space which resulted from assembling the containers in one big museum.

Moreover has been created an on-line archive, collecting 80 citizens' tales and more then 1600 familiar images, that have been collected and restored by the young residents involved, which later identified and spreaded them as the history of the city from the point of viewof its inhabitants. The containers became a kind of "second home" for the around 6000 participants. For these reasons, it has been called "the Museum of Citizens".

Before starting Foresta Nascosta, the city of San Giuliano was considered as "dormitory-town". After the premature ending of the project, due to a serious political crisis of the City Council, the citizens wrote letters of protest published on local newspapers and obtained a public query, because the project activated a need for culture. Then the young people involved constituted an association that still nowadays collaborates with the local admnistration and creates a link between San Giuliano and Milan, the cultural and financial center of the region. Foresta Nascosta had replaced the presumed historical truth of the place with its deeper history, made by collective aspects of the citizens' daily life.

[解读]

艺术需要多学科的支撑

这个案例中的策展团队十分多元，三个人的身份分别是建筑师、社会学家和汉学家。社会和文化的多元说明艺术学要综合历史学、人类学和社会学等人文学科，进行交叉研究和实验。在这个案例中，空间的连接依靠通信完成。两个集装箱成为了市民们沟通的特殊物质载体，这个空间不断地被人们的情感所填满。隐性的森林项目是一个动态的、跨层次的、关系性的概念体系，是一个社会互动的过程，是艺术促使赋权行为发生的一个案例。它源于个体对城市文化的内在需求，在这个意义上，赋能就是通过提升强烈的个人效能意识，以增强个体达成目标的动机，它是一个让个体感受到能自己控制局面的过程，所以最后的年轻人组成了一个协会来促进圣朱利亚诺和米兰的互动。（于奇赫）

氧合器
Oxygenator

艺术家：约翰娜·拉吉克瓦斯卡
地点：波兰华沙
形式和材料：装置
时间：2007 年
推荐人：茱茜·乔克拉

Artist：Johanna Rajkowska
Location：Warsaw, Poland
Media/Type：Public Installation
Date：2007
Researcher：Giusy Checola

《氧合器》是在 2007 年 7 月 20 日—11 月 20 日，由约翰娜·拉吉克瓦斯卡安置在波兰首都华沙格日博夫斯基广场的一个当代公共艺术装置。

在第二次世界大战之前，格日博夫斯基广场是仅次于纽约的最大的犹太人生活社区。在 2007 年，格日博夫斯基广场还只是位于广场的办公室、五金商店和 Prozna 街古犹太区之间的一个空间间隙，其公共领域由一些不相连的、暂时的社会、宗教和建筑空间组成。现在人们居住在高层公寓楼中，而这些公寓楼则代表了资本主义自由市场的住宅和金融建筑。这些高层被居民视为不协调的，并且被一个空的非物质的空间所取代，这个空间需要被其他某种过去的图像以某种方式填满。约翰娜·瑞克瓦斯卡试图在这些高层之间创造一个对话，作为居民共同的空间。

为了不受统治公约的影响，艺术家创造了一块具有隐喻意义和物理意义的理想地方，以促进地区和空间作为理想场所的恢复。暂时作为一个独特的创造其他概念和物质空间的日常模式，不限制居民之间潜在的相互作用。这个装置是由一个圆形的、被植物和观众席围绕的、从地上挖出的并且装有含氧水的池塘构成。通过创造一个能够正式提醒自然洗礼的地方，它引起了自然和建筑景观的变化，在这个地方艺术家推翻仪式条款，自悬浮空间到公有空间的通道被激活后，通过一个标志性的姿态，从自然到人，不是"浸泡"水中，而是"浮出水面"。

这个项目取得了巨大的成功，并且受到了曾参与该项目日常开发的观众的热烈欢迎。居民甚至要求地方当局永久性保留这件公共艺术装置。之后，市政当局为广场的恢复组织了官方建筑竞赛。因此，即使约翰娜·瑞克瓦斯卡没有赢得比赛，她也启动了地方本身的非物质和物质转换，正如华沙居民所期望的，在那之后氧合器重新恢复了新生命。此外，因为它已被文化和社会的多元化的愈合空气所"氧化"，充满氧气，所以作品给予地方质的内涵。该项目提出了关于记忆、社会文化想象和建筑设施之间的关系问题。提问城市的权利，反对沙龙式的并且富有魅力的城市环境与忽视城区如格日博夫斯基广场之间的二元论，正如之前约翰娜·瑞克瓦斯卡的干预，策展人卡娅帕勒解释道。由于自发的公众反应和其自身的社会和文化影响，氧合器在波兰公共艺术史上是一个具有里程碑意义的项目，带来了与市政当局不寻常的合作，并且成为最近更多的城市活动家运动的一个先例。

Oxygenator by Joanna Rajkowska is a temporary public installation set up in Grzybowski Square in Warsaw, Poland, between 20 July and 20 November 2007.

The square was part of the former Jewish ghetto where, before the World War II, the biggest Jewish community (after New York) used to live. In 2007 the Grzybowski Square was a kind of insterstitial space whose public dimension was composed by un-connected temporal, social, religious and architectural layers, placed between corporate offices and hardware stores of the square and the ancient Jewish ghetto of Próżna street. Nowadays people leave in rab high-rise apartment blocks, to which recently have been added residencial and financial buildings representing the capitalist free-market architecture. These layers were perceived as incompatibles by residents and gave place to an empty immaterial space that needed to be filled in some way, by something else, something going over the image of the past. Johanna Rajkowska attempted creating a space of dialogue among these layers as fields to rise a common space for inhabitants.

In order to be exempt from the dominant conventions, the artist created an ideal place, an enclave of fresh air playing with its metaphorical and physical meaning to promote the revitalisation of the area and reinassance of the space as ideal place, suspending the everyday patterns as a unique to create a conceptual and material other space, opened to a potential interaction among residents to be finally claimed by the inhabitants. The installation is made by a circular pond escavated in the land with oxygenated water, surrounded by the plants and the seats for the audience. It produced an alteration of the natural and archietctural landscape, by creating a place that formally reminds a natural baptismal font of which the artist overturned the ritual terms, since the passage from a suspended space to a communitarian place has been activated through a simbolic gesture not of "immersion" but of "surfacing" of the purified water, from the nature to the people.

The project obtained a huge success and an enthusiastic welcoming by the audience, who has been involved in the daily development of the project. The residents have even demanded to local authorities to made it permanent, afterwhile the municipality organized an official architectural competition for the revitalisation of the square. Thus, even if Johanna Raikowska didn't won the competition, she activated the immaterial and material transformation of the place itself, that after Oxygenator reborn again with a new life, in accordance with Warsaw inhabitants wishes. Moreover the artwork introduced a qualitative connotation of the place, since it has been "oxygenized", enriched by oxygene, thus by the healing air of cultural and social pluralism. The project raised questions about the relation between memory, social-cultural imaginary and architectural infrastructure, questioning the right to the city for wholeness, in opposition to the dualism between the "salon-like and glamorous" urban environment and the neglected urban areas such as Grzybowski Square as before Johanna Raikowska's intervention, as curator Kaja Pawelek explains. Oxygenator, due to the spontaneous public reaction and its social and cultural impact, has been a milestone project in the public art history in Poland bringing unusual collaborations with the municipality and has been a precedent for the development of more recent urban activist movements.

[解读]

城市需要艺术

在世界城市化的进程中，钢筋混凝土所构成的建筑已经让人们产生了视觉疲劳，而城市规划往往也是混乱的，许多公共空间被侵占、挤压。人们在忙碌的、被赋予多重含义的城市空间中渐渐迷失自我。城市需要艺术，需要一个让人们放松、娱乐的地方。有多种含混不清的语境叠加下的格日博夫斯基广场，《氧合器》的出现梳理了空间，让城市能自由呼吸，人们坐在这里心情一定是好的。水的喷薄而出具有宗教洗礼的意味，它引起了自然和建筑景观的变化，经过洗礼后的城市"浮出了水面"。虽然《氧合器》因为种种原因不能被保留下来，但是它的存在告诉了人们现状是可以改变的，这引发了人们对于自身生活空间的关注，所以《氧合器》在波兰公共艺术史上是一个具有里程碑意义的项目。（于奇赫）

公园：一种可能
PARK: birihtimal（PARK: a possibility）

艺术家：坎·阿尔泰
地点：土耳其伊斯坦布尔
形式和材料：装置艺术、
　生活行为、出版
时间：2010 年
推荐人：茱茜·乔克拉

Artist：Can Altay
Location：Istanbul，Turkey
Media/Type：Installation，live action，publicaton
Date：2010
Researcher：Giusy Checola

2010 年，在 Nişantaşi Cumhuriyet 公园（伊斯坦布尔靠近欧洲的一边），艺术家、建筑师兼调查人坎·阿尔泰开展了名为《公园：一种可能》的公共艺术项目来质疑公共空间与私人界限的具体意义。三年后，这对大规模公共空间维护运动——盖齐抵抗运动的兴起起到根本性作用，这一运动在当地和全球范围以及物质和非物质层面上均有其意义。这股构成 Nişantaşi Cumhuriyet 公园的社会、文化和环境微系统的力量似乎以较小的规模重构了整座城市的生态系统，其内部复杂而矛盾的关系就如人造自然与军事基地的关系一样。

在埃尔多安政府的统治下，负责社会住宅建设和服务的住宅开发局 TOKI 将其议程转向新自由主义；另外，得益于城市重建过程的中心化，整个建筑业也得到了促进和发展。这一过程起始于伊斯坦布尔的边缘地区，其后延伸至城市中心以及整个土耳其地区。在此框架下，公园（伊斯坦布尔仅存的一点绿地）作为用于重建的宝贵土地，成为更具吸引力的区域。

该项目面积约300平方米,在市民、住户以及公园游玩者面前直接开展。由于公园位于伊斯坦布尔市中心,吸引了众多人前来。他们在空间中与项目的互动一起成为围观共同需求的一种途径。随着该项目以及设备的拆卸,同时发行了一份出版物。然后,对该项目几个想法的进一步讨论也同时出现在一些书籍及报纸文章中。

将捐助者、观众、"创意阶层"以及曾居住和工作在公园里的人调动到一起的结果可以用一部复调作品来形容,即一部关于公共空间、其意义以及其社会、文化及金融含义的作品。艺术家 Nils Norman 质疑都市规划和城市振兴的功效;Ceren Oykut 通过她日常都市生活中的书法作品追踪伊斯坦布尔的都市变迁景象;集体主义者 SinekSekiz 质疑通过在都市果园种植蔬菜的做法来唤醒人们对营养和食物政治的意识;Future Anecdotes 设计工作室提供了该项目的标示牌和出版物,而且召集了萨班哲大学视觉艺术系的一群年轻学生组成公园团体,与艺术家共同开发该项目。

三年之后,人们了解了《公园:一种可能》项目的初衷以及它的智力和力量为:该项目只是设想以不同方式拥有公共空间持久进程的一部分,将来自不同社会、文化和宗教背景的人们聚集到一个共享空间。该项目的参与者不仅是一个集体,而更像一个社区,为创造"公共"的新形象而努力,在盖齐活动期间,他们甚至与以残酷著称的土耳其警察对抗,形成了史无前例的抵抗形式和新的抵抗象征。

On 2010 in Nişantaşi Cumhuriyet Park, on the European side of Istanbul, artist, architect and researcher Can Altay run the public art project PARK: bir ihtimal (PARK: a possibility), questioning the sense of private boundaries in the public space, that three years later will be playing a fundamental role in the upraising of a massive mouvement for the defence of public space, the Gezi resistance, which acted both on a local and a global dimension, as on material and immaterial level. The forces that compose the social, cultural and environmental micro-system in Nişantaşi Cumhuriyet Park seemed to reproduce in smaller scale the whole ecosystem of the city, with its inner complex and contraddictory relations, like the one existing between the construction of the artificial nature and the proximity of a military base.

Under the Erdogan government, the Housing Development Authority (TOKI), being an institution for the contruction of social houses and services, turned its agenda into a neoliberal one, by facilitating the entire construction industry thanks to the centralization of the process for redevelopment of the city. This process started in the periphery of Istanbul to be later extended to the city center and in the whole Turkish territory. In this frame, the parks, the few green areas that still exist in Istanbul, are quiet attractive as valuable land to be redeveloped.

By providing a structure covering around 300 square meters, the project acted within the immediacy of the presence of citizens and inhabitants as like as the visitors of the park, that is always crowded because of its location in Istanbul city center. Their interaction with the space and the project constitutes a way to gather around a common need. A publication has been distributed along with the project and after the dismantling of the installation. Then further discussions on several ideas stimulated by the project were developed in several books and newspaper articles.

The results of the involvement of contributors, the audience, the "creative classes" and the people that used to live and work in the park, has been a kind of polyphonic work on public space, on its meanings and its social, cultural and financial denomination. Artist Nils Norman challenged the efficacy of urban planning and the practices of city regeneration; Ceren Oykut, traced the landscape of the urban transformation of Istanbul, through her caligraphic transcription of the city daily life; the collective SinekSekiz questioned the awareness for nutrition and politics of food, by growing vegetables in the urban orchard; the design studio Future Anecdotes, moreover providing the signage and the project's publication, collected a group of young students from Sabanci University Visual Arts Department under the title PARK Collective, who collaborated with the artist to develop the project.

What the project PARK: bir ihtimal activated became visible in all its intellectual and physical force three years later. The project has been part of the long-lasting process to imagine owning public space differently, bringing people from different social, cultural and religious backgrounds to share spaces and ideas. The people involved in the project acted more as community rather than as collectivity in order to create new images of the "publicness" that during Gezi events they defended even in front of the brutality that characterized the Turkish police in those days, by generating unprecedent forms of resistance and a new iconography of the protest.

According to the artist, the project has been mentioned during and after the Gezi events as one of the moments able to feed how the imagination of the park resulted in the actual occupation of Gezi Park.

[解读]

公共空间与私人空间

哈贝马斯强调了公共空间的重要性，但是每个人都有自己的空间和隐私，所以又有了私人区域之说。一个人不可能完全沉浸在私人生活中，他对公共活动的拒绝和对公共生活的冷漠也对公共生活构成影响，同时，他的这种态度也影响着私人生活的心态和形态。阿伦特认为，每个人的存在都展现在两种生活领域中，一个是私人生活领域，另一个是公共生活领域。2013年土耳其抗议运动被抗议者称为"占领盖齐运动"。抗议运动最初是约50名环保人士在土耳其最大城市伊斯坦布尔搭建帐篷营地，抗议政府将塔克西姆盖齐公园征收改造成购物中心。土耳其警察逮捕抗议者，焚烧他们的帐篷。之后土耳其工会举行全国大罢工，演变成了一场席卷土耳其全国的重要抵抗运动。所以公共生活领域影响着每个人。个人的生活志向、抱负和希望都是在公共生活领域中形成的，个人的才能，不管是现实化的，还是潜在的，都是在公共生活中获得发展的具体可能性。（于奇赫）

夜晚的回声
Evening Echo

艺术家：麦迪·利奇　　　　　Artist：Maddie Leach
地点：爱尔兰科克　　　　　　Location：Cork Ireland
形式和材料：装置　　　　　　Media/Type：Installation Art
时间：2011 年　　　　　　　　Date：2011
推荐人：布鲁斯·菲利普斯　　Researcher：Bruce Phillips

麦迪利奇的作品《夜晚的回声》是对爱尔兰科克市沙洛姆公园的回应，质疑关于集体记忆的持久性。沙洛姆公园于 1989 年开放，这个名字与科克萎缩的犹太社区有着紧密的联系，犹太人自 19 世纪起就居住在周边地区。作品开幕式上还包括了由科克煤气公司提供的煤气灯照明，以此来对馈赠的公园土地表示感谢。由于年久失修，公园又在 2003 年进行"升级"，其中包括一个新的六件套电灯。

2011 年，利奇安装了三个附加的配套灯具，形成 9 的序列来与光明烛台的蜡烛数目进行统一，同时和犹太人的传统构建成一个概念框架，并与科克城议会建立合作伙伴关系，利奇的《夜晚的回声》第九盏灯每年只在光明节最后一晚的日落时点亮一次，只点亮 30 分钟。不过，公众对这一短暂的年度事件的遵守仍然是一个不确定的周期性事件。作品的持续相关性是科克未来社区记忆和怀念的代表。该作品还依赖于科克城尊重与艺术家的合同，以确保灯具照明每年一次，一直到永远——这是一个简单的要求，就是后勤和技术给予方便。

这短暂而永久性的公共艺术假定生命是一个缓慢而安静的流逝过程，强调那些与多数基本原则不匹配的文化视角，试图重新唤起几乎被遗忘的纪念。通过这样，利奇的作品唤起了那些城市现存居民日益遗忘的关于已萎缩社区的记忆。

然而，将利奇的这种动机视为创造持久的社会变革将是一个严重误读。作为一种社会参与，利奇仅仅建立了一种参与的可能性，但并不代表这个提议或者假定已经被采纳了。科克的免费社区报纸每年一次通过一系列广告宣传使得《夜晚的回声》为公众所知。

在这里，艺术家没有试图制造一个积极的公众情况。可替换地，利奇建立的概念框架，优先考虑以这样的方式亲身参与，其允许艺术家自身命题被忽略或由社区抛弃。这种方法允许社会参与以各种程度发生。为此，利奇制定沟通的一种形式，分享特定的地点和社会的共识。这些交际贡献通过一个参与者的自由意志参与，使得建立一组关系成为可能。这里的例子是举足轻重的，因为它减少了艺术家作为一个假设的位置操纵者的可能性，而这个位置也不是社会的最佳利益。进一步假定，时间，地点和社区设置了有意义的参与内容。在这里，交往行为的战略形式和象征意义挥之不去的形式平衡了参与和观看的思考。在这样做时，有意识地决定抵制观看者假定的解放，虽然社会参与———一个不与任何人有联系的风险，但是同时有可能产生深远的联系的形成潜力的命题。

《夜晚的回声》 真正建立了社会和个人的深刻的连接，在随后几年中出现的当地新闻报道和事件可以证明。这是一个强大的成果，挑战政治权力的关系，是如此频繁告知公祭，最终以一种发自内心的和智能的方式服务于特定的社区。

Responding to the site of Shalom Park in Cork Ireland, Maddie Leach's work Evening Echo (2011) questions permanence in relation to collective remembrance. Shalom Park was opened in 1989 and its name marks a connection to Cork's dwindling Jewish community who had lived in the surrounding area since the 19th Century. The opening ceremony also included the illumination of a gas lamp—a gesture acknowledging the gifting of the park's land by the Cork Gas Company. After falling into disrepair for some years, the park was again 'upgraded' in 2003 and included a new suite of six electrically powered lamps.

In 2011, Leach installed three additional matching lamps, completing a sequence of nine to correlate with the number of candles on the Hanukkah candelabrum. Connecting a conceptual framework to Jewish tradition and a contract partnership with Cork City Council, Leach's work proposes that the ninth lamp to be lit for only 30 minutes once a year, at sunset, on the last night of Hanukkah. However, the public observance of this fleeting annual occasion remains an uncertain cyclical event. The work's continued relevance is partially contingent on future communities in Cork to remember and enact remembrance. The work is also dependent upon the city of Cork to honour the contract with the artist to ensure that the lamp illuminates once a year in perpetuity—a simple request that is logistically and technically complex to facilitate.

This ephemeral but permanent public work posits the notion that life is a slow and quiet passing and that this is heightened for those whose cultural perspective is in a direct mismatch with the fundamental tenets of the majority. Leach's intervention essentially attempts to reactivate a memorial mostly forgotten. In doing so, she calls on the problematic question of empathetic remembrance for a dwindling community whose memory is fading amongst the city's current inhabitants.

However, to mistake this motivation as an attempt at creating lasting social change would be a grave misreading. As a type of social engagement, Leach sets up the possibility for participation but does not assume or presuppose that the offer is taken up. The annual occasion is made known to the public through a series of advertisements in Cork's free community newspaper the Evening Echo.

Here, the artist makes no attempt to fabricate a positive public situation. Alternatively, Leach establishes a conceptual framework that prioritises the possibility for physically present participation in such a way that allows for the artist's own proposition to be ignored or discarded by the community. This approach allows social engagement to take place with various levels of criticality. To do so, Leach enacted a form of communication that partakes in the shared understanding of a specific location and community. These communicative contributions allow the potential for a set of relations to be established through a participant's own free will to engage. The example here is pivotal because it reduces the possibility for the artist to act as a manipulative agent in presupposing from an assumptive position what is or isn't in the community's best interests. It further posits a dramaturgical situation where time, place and the community set the context for meaningful engagement. Here, strategic forms of communicative action and lingering forms of symbolic significance balance contemplation with participation and spectatorship. In doing so, there is a conscious decision to resist the presupposed emancipation of the viewer though social engagement—a proposition that risks not connecting with anyone but at the same time has the potential for deeply profound connections to be formed.

Evening Echo did indeed established deeply personal and profound connections for the community as evidenced through local news reports and events that have occurred in subsequent years. This is a powerful work that challenges political power relationships that so often inform public memorial and in the end serves a specific community in a heartfelt and intelligent manner.

[解读]

夜晚的回声

麦迪利奇的这组灯具装置作品内涵深刻、发人深省，它设置在爱尔兰科克市沙洛姆公园，该组装置总共设有九盏灯，灯具和灯光设置和犹太人的传统构建相结合。在大多数时间里只亮八盏灯，第九个灯每年只在光明节最后一晚的日落时点亮一次，只点亮 30 分钟。作品的持续相关性与科克这个萎缩的犹太社区未来的社区记忆有关，试图重新唤起几乎被遗忘的记忆。艺术家并没有制造一个积极的公众情况，因此将利奇的这种动机视为创造持久的社会变革是一种严重误读，他仅仅建立了一种参与的可能性，这种方法允许社会参与以各种程度发生。利奇的这组装置作品真正建立了社会和个人的深刻的连接，是一个挑战政治权力的关系的强大成果，这种如此频繁的公祭，以一种发自内心的方式作用于社区，不断唤醒人们的记忆。（祁雪峰）

案例研究
东亚 东南亚

CASE STUDY
East Asia, Southeast Asia

Eternal Recurrence & Scattered Light	永恒再现和散射的光源
Rainbow Nest	彩虹巢
Art "as" Environment —A Cultural Action "at" the Plum Tree Creek	树梅坑溪环境艺术行动
Green Aesthetics	绿色美学
Corn Time	玉米田时间
Mirage: Disused Public Property "in" Taiwan	海市蜃楼—— 台湾闲置公共设施摄影计划
Daily Houses	日常民房
Floating Island	浮岛
Living Room	客厅
My Town Market	我的城市市场
Sun Self Hotel	太阳自助旅馆
Time Travel Museum	时间旅行博物馆
Bonn Phum Nov Bo-Ding	白色建筑中的乡村节日
31st Century Museum of Contemporary Spirit	当代精神的第 31 世纪博物馆
Archipelago Cinema	群岛电影院

永恒再现和散射的光源
Eternal Recurrence & Scattered Light

艺术家：吉姆·坎贝尔
地点：中国香港特别行政区
形式和材料：装置
时间：2014 年
推荐人：里奥·谭

Artist：Jim Campbell
Location：Hong Kong，China
Media/Type：Installation Art
Date：2014
Researcher：Leon Tan

"转瞬即逝的光"是由香港艺术发展委员会筹资，于 2014 年 9 月举办的第四次大型互动媒体艺术展览活动。这个项目由新媒体先锋杰弗里·肖和香港城市大学的莫里斯·贝纳永教授，包括两个由旧金山艺术家吉姆·坎贝尔创作的壮观的公共工程。这是此项目基于梁美萍（香港）参与的艺术作品的"应用"，以及另一个由杨嘉辉（香港）主导的美术馆展示和表演活动。坎贝尔的项目是这次展示的主题。

第一项是题为《永恒再现》的新艺术项目（9月10日—10月8日），这件灯光作品被装置在横跨环球贸易广场正对面。艺术家所面对的挑战是构思一个适合于环球贸易广场三个细长形状外部表面的形象，同时也要兼顾其多样化的公众性。坎贝尔选定了人体形态的形式，创作出了游泳者滑行的剪影，以发光的形式穿梭建筑表面。艺术家解释道，"在这部作品中能够很清楚地看到人们不同形式的运动，以游泳者在水中动作的流畅性与这所建筑形状所造成的不受重力限制里上下起伏"。

《永恒再现》这个项目在规模上令人印象深刻。环球贸易广场经测量有484米高,是香港特别行政区目前最高的建筑,这意味着这幅作品对于这个人口密集的大城市而言,具有广阔的受众范围。另一方面,作品的定位,由于一个完整的形态,从视觉上为更多人所接受。在商业城市极富戏剧性的天际线和高山,大海的地平线映衬下,产生的移动形象更富有魅力。当正在游泳的人体被清晰辨认出来的时候,他们的无个性特征(和相关的抽象概念),将他们与周围出现在模拟和数字化广告牌上面漂亮的人物形象区别开来。该装置所传达出来的这种在视觉上跨越极大的想法,成为非常受欢迎的元素,从而达到了缓解经济社会公共空间节奏紧张的效果。

坎贝尔的第二个贡献是《散射的光源》,这是一个最初被纽约曼哈顿麦迪逊广场公园保护机构所委任的短暂的公共装置艺术。《散射的光源》于9月11日—10月1日在爱丁堡广场(位于香港中心的一个公共广场),由2000个被排列在不同高度,并且以不同速率跳动的灯泡组成一个"3D矩阵"。这个灯光矩阵的内部为观众渲染了一个迷人的氛围,有时候会唤起人们的乡愁(它被想象为"对电灯泡的敬意"),有时候也会引起人们的嬉闹和惊奇。从远处看,它有着单调的外表,成为一个2000个灯泡像素的低分辨率人体动画。

《永恒冉现》和《散射的光源》共同刺激了香港特别行政区公众与非市场导向的媒体体验,这些作品包含了吉姆·坎贝尔在城市大学校园艺术馆25年的广泛调查。

Fleeting Light was the "4th large-scale interactive media arts exhibition" financed by Hong Kong's Art Development Council in September 2014. The program was curated by new media pioneers Professors Jeffrey Shaw and Maurice Benayoun of City University (HK), and included two ambitious public projects by San-Francisco based artist Jim Campbell, an "app" based participatory work by Leung Mee Ping (Hong Kong), and a gallery show and performance by Samson Young (Hong Kong). Campbell's projects are the subject of this report. The first was a new commission entitled Eternal Recurrence, a light work installed across the facade of the International Commerce Centre (ICC) between 10 September and 8 October. The challenge for the artist was to conceive of images that suited the elongated shape of the ICC building's 3 exterior surfaces, which would also engage an extremely diverse public. Campbell decided on the human form, creating silhouettes (shadows) of swimmers gliding across the lit lengths of the skyscraper. The artist explained, "It was clear in looking at different forms of human movement for this work that the smoothness of swimmers in water worked well with the inherently gravity defying up down motion forced by the shape of the building".

Eternal Recurrence was impressive in scale; the ICC building measures 484 meters and is currently the tallest building in Hong Kong, China.

This meant that the work enjoyed wide audience reach, being visible across large and densely populated stretches of the city. It was also pitched well, given that most people are visually drawn to the human body as a gestalt. Set against a dramatic skyline of commercial and residential towers, mountains and the sea, the resulting moving-images were both captivating and hypnotic. While the swimming bodies were clearly recognizable as such, their anonymity (and relative abstraction) differentiated them from surrounding images of "beautiful people" on analog and digital billboards. The visual idea of smooth movement across infinite length conveyed in the installation provided welcome respite from the barrage of advertising characteristic of the finance capital's public spaces.

Campbell's second contribution was Scattered Light, an ephemeral public installation originally commissioned by the Madison Square Park Conservancy for Madison Square Park in Manhattan, New York. In its Hong Kong incarnation, Scattered Light once again consisted of a "3D-matrix" of 2000 light bulbs arrayed at different heights and "pulsing" at different rates, this time in Edinburgh Place (a public square in central Hong Kong) between 11 September and 1 October. The "interior'of the light matrix provided an enchanting atmosphere for audiences, at times evoking nostalgia (it was conceived as "an homage to the light bulb"), at times, playfulness and wonder. At a distance, it acquired the appearance of flatness, becoming a low resolution image composed of 2000 light bulb-pixels, animated by the throughput of human bodies.

Eternal Recurrence and Scattered Light worked well together to stimulate the publics of Hong Kong, China, with non-market-oriented media experiences. They were accompanied by The Jim Campbell Experience, an extensive survey at City University's campus gallery, covering 25 years of the artist's career.

[解读]

永恒的瞬间

面向所有的人群，在最高的建筑上展示自己的作品，艺术家要考虑的不光是空间的适宜性，更要考虑受众人群的多样性。艺术家吉姆·坎贝尔以优秀的作品——《永恒再现和散射的光源》，圆满完成了这份答卷。《永恒再现》以游泳者滑行的人体形态，以发光的形式穿梭建筑表面，从视觉上满足了人群的需求，同时与周边环境完美结合在一起，动静结合的艺术形象更加富有魅力。从视觉上缓解了经济社会公共空间节奏紧张的效果。《散射的光源》以 2 000 个不同高度排列跳动的灯泡，组成一个壮观的 "3D 矩阵"，作品外观虽然单调，却为人们营造了一种迷人的氛围。灯光或许在明灭之间转瞬即逝，但艺术家耗费数十年心血的结晶散发着永恒的光芒。（闫丽祥）

彩虹巢
Rainbow Nest

艺术家：堀内纪子
地点：日本北海道
形式和材料：交互式纤维艺术

时间：2000年
推荐人：里奥·谭

Artist：Thomas Hirschhorn
Location：Hokkaido, Japan
Media/Type：Interactive Fiber Installation

Date：2000
Researcher：Leon Tan

《彩虹巢》是日本艺术家堀内纪子最著名的交互式纤维作品。它是北海道地区发展局为给孩子们在自然环境中建立一个玩耍空间，委托堀内纪子和高野景观规划机构，以及构造设计事务所的艺术家一起密切合作设计的。该项目规模宏大，用钩针编织的多重色彩的网，被安置在一个有草皮覆盖的混凝土圆屋顶的多个部分（尺寸规模为26m×20m×10m），有一部分被埋在了地下。这些网是用杜邦尼龙6-6（一种尺寸为840旦尼尔的粗纤维）手工编织而成，染色工作是由堀内纪子在其加拿大的基地进行，然后运输到札幌进行组装，经过严格的结构分析(由构造设计事务所完成)以确保孩子们在上面玩耍的安全性。《彩虹巢》于2000年在日本札幌的滝野铃兰山坡国家公园完成并安装。报道称许许多多的孩子翘首以待上去玩耍，以至于不得不将他们分成85个组，每个组又有15分钟的游戏时间限制。

《彩虹巢》无疑是非常壮观的，它的组织形式和充满生气的色彩转变了札幌景观的视觉体验。它也具有突出的功能，编织的网（必须）可以有效的支撑孩子们和成年人的游戏活动。堀内纪子和她的工程师丈夫查尔斯，一起对广泛的原型进行设计研究工作。他们在工作室里面创作出了一个有实际尺寸的木制混凝土圆屋顶模型，为的是能够在审美观和功能之间寻找一个很好的平衡，尤其是不能像那些用工厂生产的零部件进行组装的典型游乐场。《彩虹巢》经过大量耐心和重复的人工劳动完成，在这13个月的生产期间，有时艺术家每天要花费10个小时的时间来编织纤维。

堀内纪子在观察艺术馆中攀爬的孩子们时，产生了设计像《彩虹巢》这样的游乐器械的兴趣。孩子们的这种令她意想不到的行为，吸引她关注纱线和纤维之间存在的交互可能性，在这一点上她把她的实践从艺术馆和博物馆领域转到了公共空间领域。尽管在像《彩虹巢》这样的游乐结构里面玩耍存在一定程度的危险性，但是艺术家声明"一点点的危险对于孩子们的经历是很重要的"。她的直觉力已经被儿童发育文学所证实，尤其是将学习和适当的挑战（风险）联系起来的研究。堀内纪子也非常关注过去十年来社会科技的发展，还有孩子们（在日本或者其他地方的）花费越来越多的时间独自坐在显示屏前面的趋势。她说孩子们应该和别人一起玩，而不是自己一个人玩。她的游乐器械是她关注问题的一种表达，她的作品对显示屏来说，提供了一个迷人而且社会参与的替换物。虽然有一些人也许会以"盲目迷恋成就"的名义批判她的实践，但有一点是非常清楚的，那就是这位艺术家自己正在决心，通过她的作品来明确有力的表达：非机械化和非电子化的生活对人类身体和精神的重要性。

Rainbow Nest is the most famous of Japanese artist Toshiko Horiuchi-MacAdam's interactive fiber installations. It was developed by the artist in close collaboration with Takano Landscape Planning (TLP) and TIS Partners as part of a brief from the Hokkaido Regional Development Bureau for a children's play "space" set in a natural environment. Ambitious in scale, the multicolored crocheted nets were housed in multiple sections of a turf-covered concrete dome (dimensions 26m×20m×10m) partially submerged in the ground. The nets were hand knitted out of DuPont Nylon 6-6 (a yarn with a size of 840 denier) and color-dyed by Horiuchi-MacAdam from her base in Canada, and shipped to Sapporo for assembly, after stringent structural analysis (by TIS Partners) to ensure that they would be safe for children to play on. Rainbow Nest was completed and installed in 2000 at the Takino Suzuran Hillside National Park in Sapporo, Japan. There were reportedly "so many children waiting to try it out" that they had to be divided into groups of 85 and given a time limit of 15 minutes per group.

Rainbow Nest is undoubtedly spectacular, transforming visual experiences of the Sapporo landscape with its organic forms and vibrant colors. It is also eminently functional; the knitted nets (have to) effectively support the play activities of children and adults. Horiuchi-MacAdam worked with her engineer husband Charles MacAdam on extensive prototyping for the final design, creating maquettes including an actual scale wooden replica of the concrete dome in the studio, in order to arrive at a commendable balance between aesthetics and function. Notably, unlike typical playgrounds, which are assembled from factory produced components, Rainbow Nest resulted from patient and repetitive manual labor. At times during the 13-month production process, the artist would spend up to 10 hours a day braiding fibers together. Horiuchi-MacAdam's interest in designing "playscapes" such as Rainbow Nest came from observing children climbing into her work in a gallery setting. Their unexpected actions drew her attention to the interactive possibilities of yarn and fiber, at which point she shifted her practice beyond the gallery and museum sphere into public space. While there is a degree of danger in playing within a structure like Rainbow Nest, the artist has stated, "A little bit of danger is important for children to experience." Her intuition is borne out by the literature on childhood development, particularly studies linking learning with appropriate challenges (and risks). Horiuchi-MacAdam is also concerned by techno-social developments of past decades, and the tendency of children (in Japan and elsewhere) to spend increasing amounts of time alone in front of digital screens. "Kids should play together with others, not just on their own," she says. Her playscapes are an expression of her concern insofar as they provide an alluring and socially engaged alternative to screen time. While some might criticize her practice as a "fetishization of effort", it is clear that the artist herself is committed to articulating the importance of the non-mechanized, non-digital dimensions of life to the human body and mind through her works.

[解读]
虚拟世界与现实世界

随着科技的发展、社会的进步，人们越来越依赖于高科技电子产品，沉浸于虚拟的电子世界，孩子们也花费越来越多的时间，独自坐在显示屏前，而不是和别人一起玩。过度的虚拟世界已经逐渐代替了现实世界，成为孩子们游戏的天堂，这将给孩子身心的健康发展带来重大伤害。日本艺术家堀内纪子清楚地意识到这些问题，致力于以纤维创造出不同于充满机械零件的游乐场，为孩子提供一个迷人的替换物，给予孩子们柔软、舒适、安全的游乐环境，同时提供和众人一起玩耍的游乐空间。她倾心打造的《彩虹巢》，以优美的组织形式和充满生气的色彩，将孩子的目光由虚拟世界吸引到现实游戏中，强调了非机械化和非电子化的生活对人类身体和精神的重要性。（闫丽祥）

树梅坑溪环境艺术行动
Art "as" Environment—A Cultural Action "at" the Plum Tree Creek

艺术家：吴玛悧	Artist：Mali Wu
地点：中国台湾地区树梅坑溪	Location：Taiwan Plum Creek，China
形式和材料：互动行为	Media/Type：Various Interventions
时间：2009—2012年	Date：2009—2012
推荐人：熊鹏翥	Researcher：Hsiung Peng-Chu

《以水连结破碎的土地》的策展人吴玛悧以树梅坑溪中"水"的元素，试图整合过度都市化后环境生活质量失衡的竹围地区。树梅坑溪原本是人们赖以生存、不可或缺的水源，但随着工厂进驻、大楼林立，溪流的存在受到压迫，人们贪恋都市化后带来的便捷，却渐渐遗忘树梅坑溪曾带给竹围社区的单纯美好。

《树梅坑溪环境艺术行动》项目表面上藉着带领居民重新认识树梅坑溪，实际上是要激发人去思考人和环境、人和土地、人和人之间的疏离感。透过跨领域的合作和各式各样的活动参与，使居民透过实践来反思周遭环境。《树梅坑溪早餐会》于每个月选出时令的蔬果让大家享用，一来可以增进对彼此的了解，再者，也通过对水和环境的讨论来唤起社区意识。行动也向下扩及孩子间，《我校门前有条溪》和《在地绿生活——与植物有染》邀请艺术家进入校园，鼓励学童以多重感官来认识家园，并运用当地自然资源让学生体验绿生活。《村落的形状——流动博物馆》利用流动临时作品的设置,提倡都市生活中的村落感,强调田野园地的回归、废弃物再利用、手工的温度和交换的情感流动。《社区剧场》则邀请地方大人小孩参与,以肢体展演的方式,表达出对竹围的记忆。

吴玛悧长期致力于通过文化行动带起民众对公共议题的关注,而《树梅坑溪环境艺术行动》即是成功表现的案例。它不仅使居民重新意识到树梅坑溪一带环境的重要性,也引起政府的注意,反省现代都市化对地方可能带来的冲击或破坏。在为期一年半的过程中,民众从冷漠和疑惑,渐渐转成积极投入,计划的影响深入民心,造就超乎预期的广大回响。计划从艺术出发,居民的向心力在计划中被启发,牵起在地民众和艺术家的连接,为地方打造一个量身订作的活动内涵,开阔了艺术的视野。

Link the Broken Land with Water , with the water element of the Plum Tree Creek, the planner Mali Wu tries to integrate the Zhuwei area, which lost it imbalance of the quality of life due to the excessive urbanization. The Plum Tree Creek was an indispensable source of water to people. But while the factories came and a lot of buildings were built, the Plum Tree Creek was faced to pressure. People enjoy the convince brought by urbanization, but forget the beauty of the Plum Tree Creek.

Art "as" Environment—A Cultural Action "at" the Plum Tree Creek is going to lead the residents to know more about the Plum Tree Creek on surface, but actually, it's going to inspire the people to think about the sense of alienation between people and environment, people and the land, person and person. Through interdisciplinary cooperation's and various activities, leads the residents to reflect on environment through practice. Plum Tree Creek Breakfast picks out the seasonal fruits and vegetables for people to enjoy monthly. On the one hand, it can improve understanding of each other. On the other hand, the discussion about water and environments can evoke people's sense of community. The activities would also extend to the children. There's a River in front of My School and Living on the Green Land—Being Related to the Plants invite the artists into the schools, and encourage the children to know more about their homes with multiple senses. Also, they let the students to experience life with the local natural resources. "The Shape of Villages—Flow Museum depends on the setting of the temporary work, advocate the village paten in city life. They emphasize the return of field garden, the recycling of west, the temperature of manual works, the emotions of exchanging. Community Theatre invite local adults and children to express their memories of Zhuwei in the way of physical performance.

Mali Wu has been involving in inspiring people to focus on public issues through cultural actions for a long time. Art "as" Environment—A Cultural Action "at" the Plum Tree Creek is the case of success. It not only lets the residents know the importance of environment around the Plum Tree Creek, but also attracts the attention of the government. The government began to reflect the impact of the modern urbanization to local area. In the half and one year's period, the indifference and doubts from the residents turned into positive investment gradually. The project is slowly filtering into people's minds and makes greater impact than expected. Staring form art, the project inspired residents' centripetal force, and connected the local residents and artists. It created a tailor activity content for local and widened the art's field of vision.

[解读]

都市的回归

越来越快的都市化脚步，裹挟着人们一味地向前，每天穿梭于鳞次栉比的摩天大楼，已经忘了有多久没有伫足欣赏自然，没有停留亲近土地，没有停下来倾听彼此的心声。太快的都市化脚步，造成人和环境、人和土地、人和人之间日益的疏离，初生于自然的晶莹心灵，蒙上现代都市的厚厚尘埃，变得日益沉重，失去活力。与其被都市化进程的脚步裹挟着跟跄向前，不如停下脚步倾听自然的声音，回归初始的宁静。"上善若水"，以水连结破碎的土地，将失去自然活力的土地重新唤起生机。《树梅坑溪环境艺术行动》便是为了洗涤尘埃蒙蔽的心灵。通过早餐会、自然艺术学校、流动博物馆、社区剧场等实践活动，唤起人们对都市化脚步和周遭环境的思考，使都市停下脚步，重归宁静自然的美好。（闫丽祥）

绿色美学
Green Aesthetics

艺术家：柯濬彦、温柔的结构艺术团队、水内贵英
地点：中国台湾地区暨南大学

形式和材料：装置艺术
时间：2013 年
推荐人：熊鹏翥

Artist：Chun-Yan Ker, Sanfte Strukturen, Takahide Mizuuchi
Location：Chi Nan University, Taiwan, China

Media/Type：Installation
Date：2013
Researcher：Hsiung, Peng-Chu

台湾暨南大学坐落于中国台湾地区正中央，是个四面环山、名副其实的山城，拥抱着各式各样的自然景观与资产，以回归自然为核心、以艺术为手段的一系列作品与活动就此展开。沿着荫间小路，学校里的公交车亭探出头来迎接着，中国台湾地区艺术家柯濬彦的《切片／错视》如随风摇曳的草儿，轻盈地站在路边，静静陪伴着等待公交车的人们。《活柳亭》由德国艺术团队 Sanfte Strukturen 带领国际志工与学校师生，将在地的柳树固定于钢筋结构上，任由柳树滋长，形成最佳的天然庇荫。日籍艺术家水内贵英的那道《彩虹》划过草坪，它的出奇不意和不受控制，完全无法被掌握，人们只能盼望惊喜，仅在阳光和时间恰好之时，待它挥洒光芒。水内另外搜集埔里的土壤，以夯土的技法制作土墙，夯出象征生命力的《生命之树》。墙，最终会倾倒，但它夹带的种子终会破壤而出，展现新的力量。

《绿色美学》这一公共艺术计划呼应绿色大学的精神,根据当地的环境设计出属于台湾暨南大学的公共艺术。通过各种活动创作,凝聚了附近居民与学校师生,也制造出让大家能共同拥有的艺术空间。它使人们开始注意到这个自然环境,各种人与人、人与自然和人与艺术的互动逐渐出现在周遭。人和自然之间的关系被强化,在相处的过程中,我们会发现庸庸碌碌过日子的同时,有生命凋零,却也有新生绽放,人只不过是大自然中的沧海一粟。整个计划进行下来,不仅创作了艺术品,更激发了人对自然的崇敬之心,随着持续的互动而传承。

本公共艺术计划之所以独特,不仅在于其因地制宜的艺术作品,更因为它唤起了人对自然关怀的意识,将想法实践并扩展出去。《绿色美学》结合多种领域的概念,造就了全新的公共艺术型态。

Chi Nan university is located in the central of Taiwan, China. Surrounded by mountains, it's a veritable mountain city. It embraces all sorts of natural landscape and assets. There are a series of works and activities using art as its means and focus on returning to natural. Following the path between the shadow, the bus station appeared here and it seems like welcoming the visitors. The Slice/ Illusion of Taiwan artist Chun-Yan Ker stands beside the road like swaying grass, accompany with the people waiting for the buses. Living Willow Pavilion was fixed on the steel structure by the international volunteers, teachers and students , Sanfte Strukturen Art Group. It let the willow to grown at its will and form the best natural patronage. The rainbow of Japanese artist Takahide Mizuuchi goes across the lawn. It's so surprising and uncontrolled that people can only hope for the surprise—shining in the sunlight and at the right time. The Tree of Life, the earth in the water was made into walls with the technique of ramming, which are full of life. The walls will eventually fall, but the seeds inside of it would burst out from the soil and show the new strength.

Echoing the green spirit of university, the public art project Green Aesthetics designed the public art belongs to Chi Nan University based on the local environment. Through various activities, it condensed the nearby residents and school teachers and students, and also creates a everyone shared space of art. It makes people begin to notice this natural environment. Interaction between all kinds of people, human and nature, people and art gradually appeared in the surroundings. The relationship between man and nature has been strengthened. In the process of getting along, we will find that during the coring days, lives disappear and lives begin. Human being is no more then a drop in the bucket in nature. The whole plan not only create works of art but also inspired people's reverence to nature, which will be inherited along with continuous interaction.

The public art program is unique, not only lies in its works of art according to the local conditions, but also because it aroused people's care of natural consciousness, practiced and extended out the thinking. The Green Aesthetics combined concept of a variety of fields, produced a new type of public art.

[解读]

自然的艺术

"天地玄黄，宇宙洪荒，日月盈昃，辰宿列张。"上到浩瀚广漠的天地，下到"野火烧不尽，春风吹又生"的小草，无一不是大自然的杰作。自然，作为最伟大的艺术家，以鬼斧神工的技法，创作出无数令人惊叹的艺术作品。但在现代社会，人们叫嚣着"人定胜天"的口号，试图以现代化的都市进程征服自然，割裂人们与自然的联系。殊不知，自然一直围绕在我们身边，潜移默化地诠释着自己的艺术之道，人们一旦领悟，就会发现在自然面前，我们是多么的渺小可笑，"高山仰止，景行行止"。中国台湾地区暨南大学以自然为灵感，艺术为手段，创作出属于自然的《绿色美学》，真正促进了各种人与人、人与自然、和人与艺术的互动，加强了人和自然之间的关系，"艺术本天成，妙手偶得之"。（闫丽祥）

玉米田时间
Corn Time

艺术家：颜名宏
地点：中国台湾地区高雄市驳二艺术特区
形式和材料：装置艺术
时间：2013 年
推荐人：熊鹏翥

Artist：Ming-Hung Yen
Location：The Boer Art District, Kaohsiung,Taiwan,China
Media/Type：Installation
Date：2013
Researcher：Hsiung Peng-Chu

驳二艺术特区，一幢幢的水泥仓库，中央的广场开辟了一大片玉米田。方正的田地却有着流动的纹理，有如记忆的轨迹及时间的延续。时间在这个计划中扮演了最重要的角色，随着玉米的生长，各个阶段所呈现的画面，都唤起我们不同的回忆。从土地的棕色，玉米生长的绿色，到最后收割的金黄色。孩子们拾起农具，将种子种下，手触摸着土壤，人们再次感受到土地的温度。玉米发芽愈长愈高，田中象征"家"的不锈钢家屋渐渐被隔开，望着它，却碰触不到，就像现代社会中，人们在忙碌工作时，常会产生有家却回不了的无奈。玉米开花前夕，田边竖起稻草人，它们穿着旧衣裳，张开手臂手护着这亩田。收割时，民众能亲自走进田里，协助采收丰满的果实，陪伴着玉米进入它生命的最终章。而田中向天空伸展的梯子就好比回家的途径，也成为之后丝瓜攀爬的依靠。

颜名宏的《玉米田时间——家×记忆的渡口》，不仅是一个环境装置艺术，更利用时间的深度，突破了空间的范围。2013年的暑假，颜名宏跟着朋友到了高雄左营眷村，过去的家屋已变得斑驳不堪，曾经熟悉的环境早不如当初。这种熟悉又陌生的震撼，成为本计划的开端。以"家"为核心的一系列活动，透过玉米的生命周期展开。或许每个人心中对家的记忆不尽相同，但这片位在都市中的玉米田和矗立在田中的不锈钢家屋之间的矛盾犹如我们对家的记忆，暧昧不清。

本计划借助玉米的成长，将抽象的时间化成实体的剧场，随着时间的变化，民众和作物扮演不同角色，逐渐产生与土地的连结，并引导我们回头去思考"家"的认同。

In the Boer Art District are towering cement warehouse and a vast corn field opened up in the center of the square. The square with flow texture seems the track and continuation of memory. Time plays an important role in this plan. From the brown of the earth and the green of the growing corn, to the golden of harvest, the picture of corns in different stages arouse our different memories. The children pick up the tools and plant the seed. Touching the earth, people can feel the temperature of the earth. While the corns grow, they become higher and higher. The stainless steel houses, which characteristics of "home" in the field was separated gradually. People could see them but not touch them. Just like that in the modern city, people being busy for work would be helpless that they can't go home though they have one. Before the corns blossom, a scarecrow would be erected beside the field. Wearing old clothes, they open their arms to protect the field. While harvest, people can go into the field in person, and help to harvest the plump fruit, accompanied with the corns to their end. The ladder stretches into the sky looks like the way home, and becomes the loofah's backing while they climb up later.

Ming-Hung Yen's Corn Time is not only a art of environmental installation, but break through the scope of space using the depth of time. In the summer holiday of 2013, Ming-Hung Yen came to Kaohsiung Zuoying military dependants' Village with his friends. The earlier houses had become mottled and the once familiar environment had changed a lot. This kind of familiar and strange shock was the beginning of the project. Focusing on HOME, this activity starts with the life cycle of corns. Memories of home may is different in everybody's heart, but the contradiction between this corn field in the city center and the stainless steel house stands in the corn field is similar to our memory to the home. Both of them are ambiguous.

Base on the growing of corns, the project turns the abstract time into a physical theatre. Along with the change of time, people and the crop play different roles and their connection with earth generated, which lead us to reflect our thinking of HOME.

[解读]
家的记忆

"少小离家老大回",漂泊在外的游子,重回自己的家乡,蓦然发现,过去的家屋已变得斑驳不堪,曾经的环境变得陌生又熟悉。"家"是每个人内心最柔软的地方,每当回忆起来,人们总是面带微笑,而当真正面对的时候,突然发现现实的家与记忆中的形象相去甚远却又慢慢重叠,这种熟悉又陌生的震撼,成为每个人内心深处的记忆。《玉米田时间——家×记忆的渡口》,通过玉米的生命周期与矗立在田中的不锈钢家屋,将抽象的情感赋予具象的载体,利用时间的深度,突破了空间的范围。那随着玉米发芽慢慢被遮掩的房屋,正如我们拼搏在外与家的隔阂,时刻思念家,却想回而不得的矛盾,犹如我们对家的记忆,暧昧不清。家就在眼前,触手可及却又远隔天涯。(闫丽祥)

海市蜃楼——台湾闲置公共设施摄影计划
Mirage: Disused Public Property "in" Taiwan

艺术家：姚瑞中、失落社会档案室
地点：中国台湾地区
形式和材料：摄影
时间：2010 至今
推荐人：熊鹏翥

Artist：Jui-Chung Yao、
　　　　Lost Society Document
Location：Taiwan，China
Media/Type：Photography
Date：2010-on-going
Researcher：Hsiung Peng-Chu

姚瑞中，艺术家、艺评人及策展人，自 20 世纪 90 年代开始就走访中国台湾地区各地的废墟。直到近年，他将计划扩展到他的学生，组成"失落社会档案室"（Lost Society Document, LSD），把被置之不理的公共建物记录下来，汇集成册，至今已出版了四本书，让台湾当局不得不关切闲置设施对台湾财政的消耗。

该计划被称为《海市蜃楼》，意即在台湾当局提出看似美好的蓝图下，往往本质空洞。表面上编列庞大的金额推动工程，却没有足够的软件经费支持营运。使各种规模的公共设施，小至平房大至工业区，全台湾地区已登录到上百个 "蚊子馆"。这现象反映出台湾当局在规划时，常常思考不周、评估过于乐观、未考虑到使用者需求，最后不仅造成浪费，也压缩其他项目的经费，无疑也剥削了人民福利。

姚瑞中在大学的课堂中，教导学生摄影技巧和冲洗相片手法，并加强他们的社会意识。在调查基本的背景资料后，让学生利用课余的时间，实地调查记录，并带回课堂分享讨论。学生回到家乡时才惊觉，儿时回忆之处，已被冰冷荒废的不当建设给无情取代。摄影作为艺术手法，突显出一幕幕的现实，黑白的对比使画面更加震撼。本计划不同于一般艺术创作，它脱离了美术馆或艺廊的场域，不强调艺术的社会功能，但让观众从废墟的影像，看见体制的腐败。

自第一本书《海市蜃楼——台湾闲置公共设施抽样调查》出版后，台湾当局已表达重视，进而提出解决方针：谨慎监督源头，将当下可能沦为"蚊子馆"的个案与已废置的空间和工程活化或是改变用途，成为青年住宅或展览空间，也有单位选择将设施拆除或绿化。虽然要"蚊子馆"消声匿迹还不太可能，但本艺术计划已让问题浮出水面，更唤起民众的公民意识与参与意愿。

Jui-Chung Yao is an artist, art critic and curator. He began visiting the ruins in Taiwan, China from 1990s. In recent years, he begins to extend his plan to his students, and forms the Lost Society Document(LSD). They record the forgotten building and compile them into books. They have published 4 books till now, which made the Taiwan authorities have to concern about the financial consumption for idle facilities in Taiwan, China.

The project was known as Mirage, meaning that under a good blueprint from the Taiwan authorities is empty. With the surface of huge amount of pushing projects, they don't have enough fund to support the operating. It makes hundreds of different scale facilities in Taiwan, China, from bungalow to industrial area have become useless. This phenomenon reflects that the Taiwan authorities think too hasty and toot optimistic in planning that they didn't think about the requirement of users. At last, they not only waste the public money, but also reduce the fund of other projects, which is exploiting people's welfare.

Jui-Chung Yao taught students photography and film developing technique in the university classroom ,as well as strengthen their social consciousness. After looking for the background information, he let the students do field investigation in their spare time and let them to share in class. After going back hometown, the students found that the memory of childhood had been replaced by the cold waste improper construction. As a means of art, photography highlighted the reality of scenes. The contrast between black and white makes the picture more shocking. Different from the normal art creation, the project had get away from art museum or gallery. It doesn't emphasize the social function of art but also let the audience see the Taiwan authorities from the ruins.

Since the first book Mirage: Disused Public Property in Taiwan, the Taiwan authorities has expressed their concern and then put forward the policy: supervising the source prudently, activate of change the use of the cases and projects which are useless, let them become the youth's house or exhibition space. Some units also choose to remove the facilities or do some afforest on them. Though it's unlikely to removed or change all the useless facilities, this art project has revealed the problems, which also arouse people's citizen consciousness and participation intention.

[解读]
"盗梦"空间

庞大的资金，浩大的工程，中国台湾地区当局雄心勃勃地耗巨资建各种公共建筑，边建设边自矜自己的丰功伟绩，沉浸于自己"海市蜃楼"般的梦想不愿自拔。看似庞大的功绩，细察后会发现，只是表面蒙上一层看似美好的蓝图，耗巨资建成的公共建筑毫无用处，只是成为蚊子安乐的"蚊子馆"。究其原因，只是台湾地区当局的好大喜功，规划时不切实际，如漂浮梦里般荒唐糊涂，最终不仅浪费经费，更是剥夺了人民福利。《海市蜃楼——台湾闲置公共设施摄影计划》就是为了将台湾地区当局从幻想的美梦中叫醒，以强烈的黑白照片记录那些荒废的公共建筑，也让观众从废墟的影像中看见体制的腐败。如果继续任由台湾地区当局美梦下去，我们记忆之处将被冰冷荒废的不当建设给无情取代，所有为唤醒幻想而努力的人，我称之为"盗梦者"。（闫丽祥）

日常民房
Daily Houses

艺术家：梁美萍
地点：中国台湾地区
形式和材料：装置
时间：2009 年
推荐人：里奥·谭

Artist : Leung Mee-Ping
Location : Taiwan, China
Media/Type : Installation Art
Date : 2009
Researcher : Leon Tan

台北"故宫博物院"南部分支机构（NPMSB）是台北"故宫博物院"扩展的计划，已经在中国台湾地区嘉义县的一块 70 公顷土地上发展了许多年。王子大道是一个文化和娱乐所指定的 1 200 米的设计范围，意图向 NPMSB 提供使用权。《王子大道公共艺术项目》是一个受新艺术和文化"区域"委任，制作 6 部永久公共艺术作品的首创项目。该项目选择了四个当地艺术家：陈涂伟、陈政勋、王文志和许川石，还有来自菲律宾的丹·拉拉里奥以及来自中国香港特别行政区的梁美萍。

《日常民房》是梁美萍赠给王子大道公共艺术项目的作品。这部作品是一个用 2 000 多块手工制造的玻璃砖，半建成的传统房屋模型（尺寸为 4m×3m×2.8m），其中在这些砖里面点缀着 10 块从亚洲不同城市收集来的红土砖。这些玻璃砖里面包含着多种多样来自嘉义县当地人日常生活中的物品：玻璃粉、钥匙、鞋子、玩具、一面旗帜、厨房用具、私人所有物，等等。所有的这些物品都是由居民捐献出来的。这所房子没有屋顶和窗户，自然地向天空开放，在视觉上可以跨越的水平线和

NPMSB 所计划的包含空间形成明显的对比。

梁美萍的作品被阐释为记忆的沉思，而且不难看出，是从一个人和家庭那里收集的手工艺品形成的文化记忆，犹如一个时间的锦囊，不到一个预定的时间时不会被看到，作为一个在特定时间、场所中特有的历史，收集其中。《日常民房》邀请观众了解物质文化，激活它所位于的公共空间，所有的物品在公众与它相视的刹那间得到了应有的交流感触。

这些物品平凡的本质，以及这项规模巨大的工程非常引人瞩目，因为它们体现了关于制造文化记忆的一种不同的组织逻辑，它们反对中国台湾地区大多数博物馆，以及历史上委任制作的众多公共艺术作品的收藏实践，因为那些都倾向于等级森严的组织，配备专家管理者以及名人的意愿，以此来决定藏品的内容。后者的逻辑产生了伟大的文化记忆，前者的逻辑产生出所谓的平民记忆。我们也许可以把这些逻辑叫做君主的或者大众的。称赞《日常民房》作为一个社会和艺术的试验，是因为它在公共艺术中优先考虑了"公众"的做法。

The "National Palace Museum" Southern Branch (NPMSB) is a planned expansion of the "National Palace Museum", and has been in development for several years in a 70-hectare plot of land in Chiayi County, Taiwan, China. Prince Boulevard is a 1200m stretch designated for public art, culture and recreation, intended to provide access to the NPMSB. The Prince Boulevard Public Art Program was an initiative to commission 6 permanent public artworks for this new art and cultural "district". Four local artists were selected—Tu Wei-Chen, Chen Cheng-Hsun, Wang Wen-Chih and Hsu Chuen-Shi - together with Dan Raralio from the Philippines and Leung Mee-Ping from Hong Kong, China.

Daily Houses was Leung Mee-Ping's contribution to the Prince Boulevard Public Art Program. The work consisted of a (4 by 3 by 2.8m) model of a traditional house half-built out of over two thousand hand-made glass bricks, among which were interspersed 10 red-earth bricks collected from different cities in Asia. The glass bricks contained a multiplicity of objects from the everyday life of Chiayi County locals, beads, keys, shoes, toys, a flag, kitchen implements, "intimate" belongings, and so on, all of which were donated by the residents. The house was constructed without a roof or windows, opening out physically into the sky and visually across the horizontal plane, in distinct contrast to the contained space of the planned NPMSB proper.

Leung Mee-Ping's work has been interpreted as a meditation on memory, and it is not difficult to see how the accumulation of artifacts from different individuals and families amounts to a form of cultural memory. Not unlike a time capsule, except that it is not stored out of sight until a pre-determined future date, Daily Houses invites audiences to make sense of material culture as a window onto historical forces and collective desires operative in the locale at a given moment in time. It activates the public space in which it stands, the direct relevance of the objects to daily life in Taiwan, China functioning as a magnet for encounters and conversations between visitors.

The quotidian nature of the objects and the participatory dimension of the project are noteworthy for the fact that they incarnate a different organizational logic with regard to cultural memory-making. They stand opposed to the collection practices of most museums as well as to the historical commissioning of much public art in Taiwan, China which have tended towards hierarchical organization, with expert curators and "the will of elites" determining the contents of collections. The latter logic yields histories (cultural memories) of 'great' and wealthy personages whereas the former produces memories of the so-called "common people". We might call these logics monarchical and democratic, commending Daily Houses as a social and artistic experiment that privileges the "public" in public art.

[解读]

平凡人的胜利

"艺术精英论"曾在相当长时间内占据着主要地位,似乎艺术的特权只属于贵族。平凡的食物中不可能蕴含高贵的艺术,来自中国香港特别行政区的艺术家梁美萍,却以日常生活中最常见的事物,做出具有平民记忆的艺术作品。《日常民房》以 2 000 多块手工制造的玻璃砖,以及来自中国台湾地区嘉义县当地人日常生活中的物品,共同打造了一个开放和特定的空间。作品犹如时间的锦囊,将搜集而来的家庭文化记忆,在特定的空间向公众开放,所有的物品在公众与它相视的刹那得到了应有的交流感触。《日常民房》从平民的角度创作艺术作品,没有森严登记的限制,邀请公众了解物质文化,激活所在区域的公共空间,同时优先将"公众"考虑到作品中去,成为一个城市的社会和艺术的试验。(闫丽祥)

浮岛
Floating Island

艺术家：北泽润	Artist : Jun Kitazawa
地点：日本新泻县新泻市信浓川滨水区	Location : Shinanogawa waterfront, Niigata city, Niigata pref, Japan
形式和材料：空间艺术、互动行为	Media/Type : Space Art、Various Interventions
时间：2009 年	Date : 2009
推荐人：北泽润	Researcher : Jun Kitazawa

2009 年，在日本新泻市举办的国际艺术节——"2009 年度水土艺术节"中，作为参加项目之一的《浮岛》，在艺术节期间，举行了为期一个月的项目成果展示。在实施《浮岛》这个项目的新潟市中，有日本最大的河流——信浓川。在这片土地上，陆地被分流的河水包围，形成宛若岛屿一般的形状，因此被人称之为"新泻市"。在这片"新泻市"上诞生"另一个岛屿"，这便是《浮岛》这个艺术项目的内容。

河流的施工所使用的平坦的船只在岸边停留，拥有房屋、灯台还有集会场所，形成岛屿的要素，艺术家将船只作为基础，把船只比喻成"岛"，对波浪摇晃、微风吹动的"浮岛"产生兴趣的当地居民陆续驻足于此，艺术家们和其他访客也共同创造了各种各样岛屿的"日常"。在"浮岛"上栽培蔬菜、制造风车，孩子们手工制作了旗子来装饰岛屿。在一个月展示的最后，开展了"浮岛庆典"，100 多人聚集在"浮岛"，一起为这个艺术项目的成功举办画上完美的句点。最初只是作为基础的船，到最后各种各样的人在此相识，共同制造了作为"另一个日常"的浮岛。

《浮岛》是北泽润"诞生另一种日常生活的艺术项目"的例子。北泽润的一连串的项目共通的概念——"另一个日常"是岛屿、村落、城市、家、客厅、宾馆、博物馆等日常常见的空间，并以此为基础创造另一个空间。在那里，艺术家们只设定作为基础项目的"体制"，并且以居住在当地的人们创造这个项目的过程作为主体，逐渐创造出"另一个日常"。

《浮岛》只是艺术家准备的平坦的船只，并且把这艘船只比作岛屿的一些创作产物。只有一个将船只比作"岛"的制度，然后放任自由，究竟会变成怎样的岛屿或者日常都是偶然。在那里将"创造另一个日常"这句话和"创造一个岛屿"互换，在这个没有绝对答案的问题之下，艺术家们与当地居民平等地交流想法意见。这个结果就是，谁也没有预料的状况被作为"浮岛的日常"逐一创造了出来。通过这个过程，浮岛的外观发生改变是理所当然的，更重要的是，前来拜访浮岛的人们共同参与了岛上的这个活动，形成了一个跨越年龄与地域的社区。

全然不顾现实中的日常，创造"另一个日常"，日常中本身的"问题"会逐一产生，动摇与日常的关联性。从结果上看，只能作为"另一个日常"来实现形成"另一个关联性"。这难道不正是我们应该寻找的，作为社会崭新的公共地区创造的"大众艺术"吗？

Floating Island was one of the projects of the international arts festival held in Niigata-shi, Niigata prefecture, Japan in 2009. The project lasted about one month. The area that Floating Island was conducted on was the Shinano river valley, whose width is the widest in Japan. This area is called [shin Niigata] because its river branched off and made the area like an island. Floating Island is the project to create another island against [shin Niigata].

We let a flat ship for river work (35m×12m) moor. Then made some elements, such as, a house, a lighthouse and an assembly hall to see the ship as an island. Many inhabitants who were interested in Floating Island, the floating, flat ship, participated one after another and they created many of the elements found with artists and other visitors. We cultivated some vegetables and made a windmill on Floating Island, some kids made a flag by themselves and decorated Floating Island. The end of the period, Floating Island Festival was held and over one hundred people came together to celebrate Floating Island. That was the ending of this project.

At first, Floating Island was just a foundation, but a lot of people joined, so it was created as part of "ordinary life".

It is the first example of the arts projects invented "ordinary life" by Zhun Kitazawa. "ordinary life" which Zhun Kitagawa's common concept for a series of his projects means he created something ordinary such as an island, a village, a town, a house, a living room, a school and a museum. To do this, artists set up only a foundation structure of the project, then many kinds of people living in the region complete "daily life" with working on it positively and actively.

With Floating Island, Artists prepared only a flat ship and some stuff to see the ship as an island. It is only the thing that they see it as a ship, and how will "the island/daily life" be is entrusted by a chance. With this project, the words "create another daily life" was switched to the words "create an island", and artists and many residents would each put out some ideas each. As a result, the situation no one expected, came out one after another and "Ukishima's floating island" was created. Through this process, of course Floating Island appearance changed but also it was important that the community beyond ages or regions was formed through people who visited Floating Island sharing the activity on the island.

Although a real life already existed, creating "another daily life" against a preexisting daily life poses serious questions. It shakes the relation of the daily life with "another relation" which only "another daily life" can come true is built up as a result. With that, through "public arts" in terms of arts related the public apace, we could find out the ideal "public arts" is an art that creates a new public area.

[解读]

另类的岛屿

漂泊在水面上的船只，承载着各种人们的经历，记载着人们行为的痕迹，构成"另一个日常"的岛屿，这就是北泽润的《浮岛》项目，它是北泽润"诞生另一种日常生活的艺术项目"之一。将平坦的船只放置水面之上，犹如漂泊在海面上的岛屿，人们在其上进进出出，或施以装饰，或加以改造，为任何限制，自由发展，最后以另一种面貌呈现在人们面前。看着原本熟悉的日常事物，经过各种日常行为，成为另一种常见的日常，这种感觉陌生又熟悉。北泽润十分擅长在各种常见的空间基础上，创造出另一个空间。创作过程并不是由艺术家主导，而是艺术家和当地人平等交流对话，共同创作，诞生出人们无法预料的日常事物。将熟悉的东西陌生化，最后又重归日常，微妙的改变带来了另一种不同。（闫丽祥）

客厅
Living Room

艺术家：北泽润
地点：日本北本市、德岛市、
　　　北秋田市、冲绳县；
　　　尼泊尔加德满都桑库
形式和材料：空间艺术、互动行为
时间：2010年至今
推荐人：北泽润

Artist : Jun Kitazawa
Location : Kitamoto city, okushima City，
　　　　　Kitaakita City，Naha City，Japan；
　　　　　Sankhu，Kathmandu，Nepal
Media/Type : Space Art，Interventions
Date : 2010-on-going
Researcher : Jun Kitazawa

《客厅》是北泽润艺术项目的代表品，在日本国内共有四处，在尼泊尔联邦民主共和国内有一处，并且逐渐向世界各地展开。作品以商店街作为舞台制造出"另一个日常"，目标是将当地房屋内部客厅的景象，作为商街闲置店铺的再创造项目，其体系便是依照这样的趋势形成。

首先，在商店街的闲置店铺内，铺上一层地毯，对周围的住户进行拜访，收集家具及生活用品。收来的东西，将它放在地毯上，犹如家庭的客厅一般布置，不仅将闲置店铺变成客厅，也作为公众的地方再次开业。某种程度上，在收集物品的阶段，是一个可以将"客厅"中所有的东西和家庭里的东西以物换物的场所。空间随着以物换物的展开而逐渐发生变化，人们被这个空间所吸引，展开各种活动，人与人共同编织出一个"客厅"，随着变化的持续，"另一个日常"也逐渐被制造了出来。

初期的铺地毯工作与物品收集配置工作是《客厅》的基础，接下来成为一个任何人可以以物换物的场所。在《客厅》常驻并且实施以物换物和记录的人们，由于原本这是一个商业街的闲置店铺，因此他们被唤作"店员"。此阶段是利用项目开始的几天时间达到的初期阶段，也就是说，几乎所有的项目期间，根据之后以物换物的内容，来访的作为"店员"的人们的活动，持续改变着《客厅》的阶段。更可以说，关于这个初期阶段，是艺术家们预测可能范围的契机。这之后的阶段，是要跨越这个预测范围，艺术家自身也好，与这个项目相关的当地居民也好，要共同作为《客厅》的一份子，探索如何制造"客厅"空间。在此过程中，会形成少许的"另一个日常"。也就是说，《客厅》的状态不仅仅局限于相关人员的以物换物，而是与各种各样的活动息息相关。

与为期一个月开展的《浮岛》项目相对的，《客厅》没有时间上的限定。从 2010 年开始的琦玉县北本市北本地区的《客厅》持续了 5 年。在此期间，以交换钢琴为契机开展了小型演奏会，交换烹饪器具的时候，《客厅》又变成了小餐厅。另外，初期是有意向的当地居民承担《客厅》运行的店员，自主地运营《客厅》这个艺术项目的情景让各地区居民的感性被激发了出来，从而开展进一步的活动，成为日常未来可发展的方向。

《客厅》利用原本当地就有的空间（商店街的闲置店铺）和物品（家具和日常生活用品）的以物换物体系，变成了不可预测的"客厅"面貌的同时，根据在当地生活的人们展开相关活动的特征，将着眼于"场所"的"特定场所归属"这一概念更进一步深化，捕捉将当地居民的"社区"的观点作为根基的"社区·特定场所归属"的概念。

Living Room is a great example of Zhun Kitagawa's art projects. It was conducted four places in japan and in the Federal Democratic Republic of Nepal. Living Room which completes "another daily life" in a shopping district is the art project that tries to re-composition a scene of some living rooms in an empty place such as an empty storefront This process goes as follows:

First, we spread a carpet on the floor of the empty storefront then visit the houses around it and gather furniture or daily utensils. We arrange the furniture on the carpet and make it like a "Living Room" then we open it as a place everyone can be. After gathering certain furniture, we open it as a place that people can trade something in it for something from people's house. The place which is been changed by a barter is going to be a [living room] that active people interweave each other.

The foundation of Living Room is built up by the first carpet work and by the work which is to gather things then arrange them.

Next, we open it in a region as a place that everybody can enter and trade stuff in there. The people who stay in the Living Room and record a barter or work are called shop assistants because it used to be a store in a shopping district.

The stage near there is the first one which arrives in a few days from the beginning. It means almost the whole term is the stage that the Living Room is been changed by the barter and the activity by the store assistants or visitors. Furthermore, we can say the first stage serves as a template for the artists who begin it. Something they will be able to anticipate, but after the stage, it can't be anticipated so the artists will also be the members of the Living Room project as well as the residents and fumble for how to make the [living room]. [Another daily life] is formed gradually in this process. The condition of the Living Room depended on not only the barter between people but also the diverse activities all the time.

Against Floating Island whose term of the project was a month, Living Room doesn't have a term, so the Living Room which was started in Kitagata-danci Kitagata-shi, Saitama prefecture in 2010 kept on for 5 years. In the meantime, there were some various activities. For instance, a concert was held because a piano was traded, the trial that Living Room was temporarily changed a restaurant was conducted because the cooking utensils became its possession, and an activity that the whole shopping district changed a shopping mall with using the stuff which was gathered.

At first "the store assistants", which rested upon the artists, was taken over by the interested people and they managed the Living Room voluntarily. From this project, we could see the future's figure that penetrates daily life, as a function that the art project stimulates residents' sensitivity and creates a new activity in a regional community.

[解读]

屋无常态

屋里的客厅如何装饰，现代简洁风，巴洛克风情，或是富丽堂皇，无论如何都会饰以一种风格。但稍微加以变化，例如移动桌子、放架钢琴，屋子顿时又有另一种感觉。北泽润的《客厅》便是将搜集来的物品放置于闲置的商铺，犹如家庭的客厅一般布置。在搜集的过程中客厅便会随着物品的改变逐渐发生空间的变化，人们可以在客厅中以物易物，屋无定态，当你以交换钢琴为契机，空间便可开展一个小型演奏会，当在其中交换烹饪器具的时候，《客厅》又变成了小餐厅。人们在其中自由交换，自主运营，展开各种活动，共同编织出一个"客厅"，随着变化的持续，"另一个日常"也逐渐被制造了出来。不可预测的"客厅"面貌记录着人们相关活动的特征，成为日常未来发展的另一个方向。（闫丽祥）

我的城市市场
My Town Market

艺术家：北泽润
地点：日本福岛县新地市

形式和材料：空间艺术
时间：2010 年至今
推荐人：北泽润

Artist：Jun Kitazawa
Location：Shinchi town, Soma-gun, Fukushima Pref.
Media/Type：Space Art
Date：2011-on-going
Researcher：Jun Kitazawa

《我的城市市场》是位于日本福岛县沿岸的新地市的一个艺术项目。2011 年 3 月 11 日日本东部发生重大地震，在地震引起的海啸中丧失家园的原住民，从避难所迁移至新地市小川公园，这个公园是以山坡上的球场作为基地建造的应急临时住宅区，于 2011 年 6 月正式成立。

以小川公园应急临时住宅作为舞台而诞生的"另一个日常"——《我的城市市场》位于灾后重建迫在眉睫的新地市，如"市场 =Market"一般，艺术家创造出"我的城市 =My Town"的临时项目。在城市的五分之一被海啸吞没的情况下，城市再建的大复兴行动的过程中有着某种看不见的未来日常，《我的城市市场》犹如市场一样只是临时的，这也是以孩子们作为主角举办的小小复兴。

作品基本上的体系如下：首先利用捆包专用的 pp 绳编制出的 2m 四方形席子，像市场铺地一样将这种席子铺在临时住宅通道上，用一定的标准组合而成。在这之上，配置手工制作的"城市零件"，建立"犹如市场一般的城市"的雏形。在仅有的几小时中，来访的人们与创造着一切的人们融为一体，很快就形成了"另一个日常"。因为作为一个市场会有定期的实行，"另一个日常"的出现会变成习惯性。

震灾的三年内，《我的城市市场》也开展了 11 次，作为当地的例行活动融入了人们的日常。在席子上会创造出怎样的"城市"呢？从方案阶段开始，每回当地居民的相关成员就会利用数月的时间进行准备。每次都能创造出独特的"城市"。在这个过程中，捕捉通过当地居民亲手建立的新的地方"祭典"，成为日本自古以来设立的地方文化与艺术项目模范典型事例。

My Town Market is a project in Shinchi-cho, Hukushima prefecture which is located on coast of the see and was badly damaged by the Great East Japan happened on March 11th, 2011. It was started in Ogawa park emergency temporary housing which was built on the ground of the hill of Shinchi-cho in June 2011 with a situation that people had lived in the area that was floated by the tsunami moved from a place of refuge to the temporary housing.

My Town Market which creates "another daily life" in Ogawa park emergency temporary housing is a project that people create temporary "our town = my town" like "a market = market" in Shinchi-cho which was soon needed a revival from the earthquake. My Town Market is temporary like a market and it can be said "a small revival" conducted by some small children mainly.

The basic process goes follows. First, we make a 2m on all sides mat, "Goza", made of the PP bands for a packing. Then, we lay out the mat on the street of the temporary housing and put them together to make a base. "A town like a market" is created by arranging "the parts for the town" we made on it. Within the limited a few hours, the visitors and the people who made the parts of the town united each other and they let it be "another daily life". Because it is a market, an appearance of [another daily life] is going to be habitual by being carried it out regularly.

It has past three years since the earthquake happened, and My Town Market has been held 11 times. It spread as a customary event in the region. Every time the member of this prepare for it from such a stage that they think about an idea how they create "the town" on the "Goza" spending a few months and they completed the unique "town". The process is in other word a process that the residents create a new [festival] in the area, and it became an example to present resemblance between regional culture have been spun since ancient times in Japan and the art project.

[解读]
日常的重建

2011年3月11日，日本福岛县新地市发生重大地震，地震引发的海啸使许多居民失去家园，城市的五分之一也被淹没，人们只能生活在公园的临时住宅里。万物百废俱兴，重建迫在眉睫，艺术家敏锐地注意到城市的大复兴行动中，蕴含着某种看不见的未来日常。《我的城市市场》以孩子为主角，寓意着城市复兴的希望。艺术家将席子以一定的标准铺在临时住宅通路上，配置手工制作的"城市零件"，建立"犹如市场一般的城市"的雏形，将人们和创作融为一体，形成人们的"另一个日常"。在此后的几年里，《我的城市市场》定期举办，试图将临时常态化，使之成为日常行为，作为例行活动融入人们的日常，成为人们的习惯性行为。（闫丽祥）

太阳自助旅馆
Sun Self Hotel

艺术家：北泽润	Artist：Jun Kitazawa
地点：日本茨城县取手市井野、中国台湾地区台北市	Location：Ino Estate,Toride city, Ibaraki pref；Nanjichang Estate, Taipei city, Taiwan, China
形式和材料：空间艺术	Media/Type：Space Art
时间：2012年至今	Date：2012-on-going
推荐人：北泽润	Researcher：Jun Kitazaw

在茨城县取手市井野地区展开的项目——《太阳自助旅馆》，利用当地限制的房间与照射进当地的阳光，在井野地区利用太阳能开展可提供住宿的"Hotel"。

《太阳自助旅馆》的创造全是以募集的宾馆管理者为中心。面向住宿日的事前准备阶段，管理者要实施定期聚集人员、客房改装、客房服务、"太阳能"制作、更要列出住宿当日的行程表，进行诸如此类一系列的准备工作。另外，在网页上募集有住宿意愿的人，在应募者之中选出特别来宾。

住宿日当天，在上午做完最后准备之后，下午来宾到达。来宾与管理者会面，首先要在井野地区广场上的服务台办理入住手续。在这里，管理者会就酒店、两天一夜的住宿流程和注意事项等——说明。在来宾签署同意书后，管理者会带领来宾进入利用地方的闲置房间改装而成的"客房"。管理者将与来宾共同搭载利用太阳光能发电系统的"太阳能面包车"，一边按下按钮将其蓄电，一边在井野地区兜风。蓄电结束后，来宾体验着管理者为其精心准备的"客房服务"迎来了傍晚。伴随着夕阳，来宾品尝着管理者做的丰盛晚餐。在这之后，服务台的上空会填充氦气体，特制的照明气球，会作为"夜晚的太阳"，浮现在夜空中，这是利用了白天蓄电的电力所发光，在客房内也有着相同的灯光。在享受完"夜晚的太阳"之后，来宾就在客房就寝。伴随着来宾入眠，"夜晚的太阳"也熄灭了光芒。翌日，在品尝完管理者亲手制作的早餐后，来宾办理退宿手续并支付住宿费用。管理者将会在目送来宾离开后，清扫客房，将它变回原本的房间。

这个项目，在以"客厅"与"城市"等人工要素作为要素，成为主题中心的基础之上，将作为人工要素的"宾馆"和作为自然要素的"太阳"相结合，形成专属的特征。《太阳自助旅馆》是根据日本大地震的重灾区新地市的艺术家，通过自身重新认识在"日常"中有着压倒性影响的自然情况，从而发展起来的。以太阳的存在，通过"日常"的再构筑，能不能诞生"另一个日常"？艺术家在这样的联想中，创造了与白天太阳相对而生的"另一个太阳"，地方的房间对应而生的"另一个房间"，同时，太阳能发电的手段也被作为联系"房间"与"太阳"的媒介进行采用。

《太阳自助旅馆》自2012年启动开始，受到了日本各界的广泛关注。从建筑学领域出发，在郊外的土地上完成了宾馆的观点；从节能领域出发，诞生了可利用太阳能发电的观点；从社区规划领域出发，诞生了形成地方性社区的新方法；从戏剧领域出发，当地居民扮演"管理者"这一角色，也类似于一出脱离剧场性的戏剧；2014年2月在东京开展了《太阳自助旅馆》研讨会，从各种各样的专家视角进行了解答。一言以蔽之，可以证明"在另一种日常中诞生的艺术项目"是足以动摇现今社会的既存价值观，并取得轰动的反响。研讨会中，根据"这项艺术可能具有任何领域都无法捕捉既存其专业性观点的力量"的结论，成为近乎可以动摇社会发展新现象诞生契机的先进案例。

2014年，这项活动由日本发展到了中国台湾地区台北市，在南机场地区开展了另一个《太阳自助旅馆》。

Sun Self Hotel started in Ino-danchi Toride-shi Ibaraki is a project that we built as a hotel which people can stay in with solar energy in Ino-danchi by using sunlight which is shining into the empty room of the Danchi and Danchi.

The process of Sun Self Hotel goes as follows. First, we recruit "hotel men" who will be builders. They play a key role in of all stages of Sun Self Hotel building process. They regularly get together and prepare for remodeling of the guest rooms, room service, production of "the sun", and planning for the schedule for coming stay-at day. It is a beforehand stage. They also recruit people who want to stay at the hotel on their website homepage, and choose the guests from all of them. On the designated stay-at day, after the final preparation in the morning, the guests arrive at the hotel in the afternoon. They meet the hotel men and then check into the hotel in the reception desk provided in a public space in Ino-danchi. The guests receive an explanation about Sun Self Hotel, the schedule of staying overnight and matters that require attention then are accompanied to [the guest rooms] ,which are empty rooms in danchi remodeled by the hotel men.

After that, hotel men and the guests walk around in Ino-danchi pushing "a solar wagon" charging the solar power generation system and storing electricity. After finishing charging the generator, the guests wait for it dark, by ordering room service provided by the hotel men. The guests have dinner made by the hotel men when it's dark. After that, the hotel men fill up special balloons with helium gas above "the reception desk" then let them sail the sky as sun of the night. They shine by electricity stored in the daytime and these lights are set up in the guest rooms as well. The guests go to bed after enjoying the "sun" sailing in the night sky. The lights of [the sun of the night] go out with guests' going to bed, they are dismantled. In the next morning, the guests have breakfast made by the hotel men and then check out of the hotel. When they check out, they pay for it. The hotel men which see them off clean the guest rooms and set the guest rooms back to the original empty rooms.

This project is characterized by "the hotel" as an artificial factor and [the sun] as a factor of nature although the theme of it are artificial things such "a living room" or "a town" by then. The background of this development is caused by that the artists realized again nature which affects overwhelmingly "the daily life" through the experience of My Town Market in Shinchi-cho which was damaged badly by the Great East Japan. We designed Sun Self Hotel that creates "another sun" against an ordinary sun and "another room" against the room in the danchi within in the time of two days and stay overnight from an idea that we would create "another room" by re-construction of "the daily life" including a sun. We chose the way of using the sun light as a conductor of "a room" and "a sun".

Sun Self Hotel has received attention attention by a lot of Japanese companies in different fields since the project was started in 2012. The field of the construction pays attention to it in terms of renovating the danchi of the suburb and letting them out as hotels, the field of energy pays attention to it in terms of using a solar power generation, the field of a theater uggests the connection to a play held in a non theater in terms of that the residents played the hotel men.

A symposium whose theme was "what is the Sun Self Hotel" was held in Tokyo in February 2014 and it was analyzed by many experts. "The art project which creates another daily life" became a reverberation to prove that it has a power to look people's existing sense of values of the society over again and became an advanced example of art that approaches the society and look it over again then lets people start to create a new phenomenon by the conclusion that it is "the art has a power that can't be held on to by existing standpoints".

[解读]

夜晚的太阳

在人类历史上，太阳一直是人们顶礼膜拜的对象。中华民族的先民把自己的祖先炎帝尊为太阳神，而在古希腊神话中，太阳神则是宙斯（万神之王）的儿子。无论从情感上还是生活上，太阳一直与我们的生活息息相关。北泽润的《太阳自助旅馆》，重新唤起了我们对自身与太阳关系的思考，以及日常生活的样貌。在《太阳自助旅馆》，白天你可以充分体验太阳的魅力，来宾搭载利用太阳光能发电系统的"太阳能面包车"，边蓄电，边在井野地区兜风，傍晚，特制"夜晚的太阳"会浮现在夜空中，此时的太阳便是白天所蓄太阳能的重新利用。艺术家将作为人工要素的"宾馆"和作为自然要素的"太阳"相结合，无论从艺术手法、建筑领域，还是节能观点，都具有独特的魅力。（闫丽祥）

时间旅行博物馆
Time Travel Museum

艺术家：北泽润
地点：日本京都舞鹤市

形式和材料：空间艺术
时间：2012 年至今
推荐人：北泽润

Artist : Jun Kitazawa
Location : Red-brick distributing reservoir, Maizuru City, Kyoto Pref.
Media/Type : Space Art
Date : 2012-on-going
Researcher : Jun Kitazawa

《时间旅行博物馆》是沉睡在地方文化中的遗迹。将地方的过去、现在与未来串联起来，进行一场宛若时空旅行的再生管理。

京都舞鹤市，根据距今已有 100 年历史的日本海军历史重新建造军港，由之前的渔村风情一举成为簇拥着海军相关设施的海军之城，发生了翻天覆地的变化。

战后很长一段时间，这些设施群无人理会，近年来，部分设施例如红砖仓库作为观光设施实施改建，开始了再利用。在能够俯瞰舞鹤军港的山上，有着从海军时代开始，就被作为饮用水储藏库利用的近代化遗产"红砖配水池"。作为国内重要的文化财产，这个设施内部矗立着可以通过大量水流、高达 5.6 米的巨大红砖墙。

《时间旅行博物馆》被放置了 50 年，在过去曾有大量的水流经过这个地方，如今流淌而过的只有岁月。作为穿越舞鹤的过去、现在、未来的"博物馆"，唤醒了这个沉睡已久的项目。叠立着巨大墙面的这个空间，一边展示着从远古时代的化石、过去的生活、现在的日常、到100年后的未来，一边在地方居民对这场"时空旅行"提出的"时间旅行学员"的联想中开展企划。《时间旅行博物馆》不仅作为舞鹤所有生活史聚集的终点，也是其中任何形成"时空旅行"的出发点，成为编织着这座城市"人"与"时间"之间关系的新视角。

舞鹤的过去、现在、未来，积累了太多的"日常"形态，作为"学员"的当地居民以此为想象依据，对穿越时空诞生的"另一个日常"进行实践。《时间旅行博物馆》在北泽润迄今为止的项目中更具发展前景，经过众人的手创造出"诞生另一个日常的艺术项目"，可以说是现在进行中的影射了社会现象的突破性尝试。

Time Triavel Museum is a project to regenerate the ruins which is still in the area to the museum people travel the daily life of the past, the present, and the future.

In Maizuru-shi Kyoto, a naval base was made by the Japanese navy one hundred years ago. The town changed from fishing village to a navy's town with a lot of different institutions.

The institutions, which stayed for a while after the war, began to change their purpose, such as a red brick warehouse that became a tourist facility. There is a modernization inheritance, "a red brick water supply pond", which was used as a storehouse for drinking water from the era of the navy to 50 years ago on the mountain which can be seen from Maizuru's port. Inside of the institution designated as an important cultural property, the huge walls which are 5.6 m tall and made by red bricks stand in a row to float large amounts of water.

Time Triavel Museum is a project to revive the place that had been left alone for 50 years and used to be flooded with large amount of water, into a museum in which people travel to the past, the present, and the future of Maizuru. We plan a project for the trip of the era with exhibits such as an ancient fossil, an ancient life, a present daily life, and the daily life a hundred years from now, and the future a hundred years later in the place in which the huge walls stand in a row now. It is planned by "the artists of the time trip", consisting of the interested people in the region. Time Triavel Museum is the terminal point that any Maizuru's history of life comes together and it is also a starting point that lets us do [the time travel], so it would be a new a base to spin the relation between "the people" and "the time" in the town again.

With Time Triavel Museum, a lot of the figures of "the daily life" such as the past, the present, and the future are accumulated then the residents become the artists while thinking about it. They practice the project that creates "another daily life" which is actually beyond the time. Time Triavel Museum is the most developed of his projects so far, and it could be said a present progressive trial that reflects the society that "the art project which creates another daily life" is created by the hands of many kinds of people.

[解读]

来一次时光之旅

假如可以掌握时间，你会如何回到过去重温美好的记忆，或是穿越未来，领略不一样的新奇？日本京都舞鹤市的《时间旅行博物馆》，记录了这个地方的过去、现在与未来，人们游览其中，宛若一场时空的旅行。舞鹤市原本是一个平静的渔村，之后根据日本海军历史重建军港，成为一座簇拥着海军设施的海军之城，但战后大多设施却再也无人理会。50 年后，人们发现了被荒废的设施，惊讶地发现上面明显地记录下岁月留过的痕迹，艺术家开始联合众人，将岁月的记忆、现在的痕迹、未来的轨迹，全部呈现在《时间旅行博物馆》。作品不仅成为所有生活史的终点，同时也是"时空旅行"的起点，未来的想象由此出发，成为链接这座城市居民与时间的纽带。（闫丽祥）

白色建筑中的乡村节日
Bonn Phum Nov Bo-Ding

艺术家：莎莎艺术项目组织
地点：柬埔寨金边
形式和材料：空间艺术
时间：2014年
推荐人：凯利·卡迈克尔

Artist：Sa Sa Art Project
Location：Phnom Penh, Cambodia
Media/Type：Space Art
Date：2014-on-going
Researcher：Kelly Carmichael

《白色建筑中的乡村节日》是"2014 我们城市的节日"大事件的一部分。我们城市的节日（简称 OCF）于 2008 年设立，致力于加快柬埔寨的城市化步伐，将艺术家聚集一起，倡导从业者把重点放在城市化及其对当代文化的影响上，展示艺术和建筑为展览的主题。OCF 以活动、表演、放映、教育讲座和参观及研讨会的形式，"考察柬埔寨的现在，回忆过去，并通过创造性的视觉设想未来"。

《白色建筑中的乡村节日》采用多重的、交互平台的展览和活动的形式，贯穿于柬埔寨首都金边附近的白色大楼。它展示了新的媒体作品和表演，以及艺术家在所谓的白色建筑工作的回顾。

白色建筑作为这座城市标志性的站点，有着悠久的历史。最初叫做市政公寓，它是由柬埔寨建筑师 Lu Ban Hap 和俄罗斯建筑师 Vladimir Bodiansky 设计的雄伟的巴萨河前文化融合而成。这栋建筑是 20 世纪五六十年代后期金边巨大变革的一部分，有着新的公共基础设施建设、纪念碑和建筑。

该建筑于 1963 年奠基，含有 400 个现代公寓，并且是第一个试图给中低收入的柬埔寨人提供多层次都市模式。随着红色高棉倒台，人们的返回，这座白色大厦见证了柬埔寨动荡的过去。

现在这座白色建筑结构破旧，居住人口超过 2500 人，并沾染上了贫困、暴力犯罪、危建和卫生条件差的耻辱。它也是金边唯一的艺术家进行艺术实践的经营的间艺术家。

由莎莎艺术项目策划，《白色建筑中的乡村节日》吸引了许多柬埔寨人和一些从未到过白色建筑的外国人。它的关注点是创造一个集体的、可供分享的参与经验，居住其中的人来塑造项目。莎莎艺术项目的负责人 Lyno Vuth 这样叙述："通常两个分开的子社区第一次携手合作一个大项目，该项目是由白色建筑中的两个村村长支持。那种强烈的骄傲就是很好的例证；一个居民这样说：'我们的白色建筑恢复了声誉。'令我们非常鼓舞的是，这个节日有助于增加这种自豪感以及和邻里间的正面形象。"在国际化发展的背景下，Vuth 组织的项目旨在促进社会变革。

作为一个公共节日，《白色建筑中的乡村节日》通过艺术、建筑学的想法启动建设和周边环境。从更深层层次，作为艺术节的一部分，对于艺术在社会上的作用，这个社会参与、供人分享的项目起到了催化剂的作用。

这个节日把金边的人们和具有历史意义的遗址重新连接在一起，也给这座城市一个机会来庆祝白色建筑作为一个积极的、具有创造性的社区或微型城市。

这个节日把社区变成了合作、共享的创造性空间，让那些通常避开白色建筑的人，了解它的历史和它的创造力。

一位参观者随后与白色建筑的孩子们一同进行绘制，由莎莎艺术项目和社区学校提供设施，将学校后面的墙壁变成了一个友好的、丰富多彩的环境，有着寄托儿童希望的特色图像和他们对这个社区的梦想。

Bonn Phum Nov Bo-Ding (Village Festival at the White Building) was a public festival that occurred as part of a larger event, the "Our City Festival 2014". Acknowledging the accelerated pace of urban change in Cambodia's cities, Our City Festival (OCF) was founded in 2008 and brings together artists and creative practitioners to focus on urbanism and its influence on contemporary culture. Presenting art and architecture themed exhibitions, events, performances, screenings, educational talks and tours and workshops, OCF "examines Cambodia's present, remembers its past and envisions its future through a creative vision."

Bonn Phum Nov Bo-Ding took the form of multiple, cross-platform exhibitions and events throughout the White Building neighbourhood in Phnom Penh, Cambodia's capital city. It presented new media works and performances and a retrospective of artists working in what is known as the White Building. An iconic site the in the city, the White Building has a long history. Originally known as the Municipal Apartments it formed part of the ambitious Bassac River Front cultural complex designed by Cambodian architect Lu Ban Hap and Russian architect Vladimir Bodiansky. The building was part of a tremendous transformation of Phnom Penh in the late 1950s and 60s, with an abundance of new public infrastructure, monuments and buildings. Inaugurated in 1963, the building held over 400 modern apartments and was the first attempt to offer multi-story urban lifestyle to lower and middle-class Cambodians. Repopulated in the 1980s as people returned following the fall of the Khmer Rouge, the White Building echoes the country's turbulent past. The now dilapidated structure is home to over 2,500 people and tainted with the stigma of poverty, petty crime, dangerous construction and poor sanitation. It is also home to artists, musicians, dancers and Sa Sa Art Projects, Phnom Penh's only artist-run space dedicated to experimental art practices.

Organised by Sa Sa Art Project Bonn Phum Nov Bo-Ding attracted many Cambodians and foreigners who never visited the White Building before. Its focus was on creating a collective and participatory experience, engaging residents of the building in shaping the events. Lyno Vuth, director of Sa Sa Art Projects, recounts: "Two usually divided sub-communities came to work together for the first time in a large project, supported by the two village chiefs from the White Building. A stronger sense of pride was evident; one resident said, 'Our White Building has reputation again'. We were very encouraged that the festival contributed to the increasing sense of pride and positive image of the neighborhood." With a background in international development, many of the projects Vuth organises aim to facilitate social change.

As a public festival, Bonn Phum Nov Bo-Ding activated the building and surrounding environment through art, architecture and ideas. On a deeper level, and as one part of Our City Festival, this socially engaged, participatory project acted as a catalyst for discourse on the role of the arts in society. The festival reconnected the Phnom Penh's population with a site of historical significance and also gave the city an opportunity to celebrate the White Building as an active and creative community or micro-city. The festival turned the neighbourhood into a large collaborative and shared space of creativity. Demystified, those who typically avoid the White Building learnt about its history and the creativity it houses. One visitor subsequently supported a collaborative drawing workshop with White Building children. Facilitated by Sa Sa Art Project and the community school, a wall behind the school was transformed into a friendly, colourful environment featuring images of the children's hopes and dreams for their neighbourhood.

[解读]
历史的余音

"以史为鉴，可以知得失"，历史无论是文明的，还是丑恶的，都可以成为滋养今日的土壤。伫立在柬埔寨首都金边上的白色建筑，距今已有近 60 年历史，它见证了国家动荡的过去，也成为纪念的标志，到近代，却又成为罪恶的乐园，落后的代表。深厚的历史文化底蕴，不应发酵出丑恶的面貌，莎莎艺术项目联合艺术家和地方社区，开展《白色建筑中的乡村节日》，通过艺术、建筑学的想法，以活动、表演、放映、教育讲座等形式，追溯柬埔寨的过去，打造创造性的未来，将金边的人们和具有历史意义的遗址连接在一起，让那些遗忘、远离白色建筑的人重新了解其历史和创造力，打造合作、共享的创造性空间。善于从历史中汲取积极的营养，将之转化为文明的回音，那么，历史的余音将转化为一首美妙的音响。（闫丽祥）

当代精神的第 31 世纪博物馆
31st Century Museum of Contemporary Spirit

艺术家：卡明·勒猜普拉斯特	Artist : Kamin Lertchaiprasert
地点：泰国清迈和国际卫星区	Location : Chang Mai, Thailand and international satellites
形式和材料：互动行为	Media/Type : Various Interventions
时间：2011 年至今	Date : 2011-on-going
推荐人：凯利·卡迈克尔	Researcher : Kelly Carmichael

卡明·勒猜普拉斯特是一位艺术家和佛教僧侣，这些身份对他的作品产生了极大影响，他的作品探讨目前的文化行动主义，在当代艺术中决定性的主导因素。勒猜普拉斯特的实践秉承协作、公开的精神，并且其核心为社会参与，结合艺术家自己的理念投入到工作。2011 年，对协作艺术的兴趣引导他创建了一个大型互动分享型项目：《当代精神的第 31 世纪博物馆》。这个项目是一个微型乌托邦式的想法，寻求生活和社会交往学习中的平衡。《当代精神的第 31 世纪博物馆》是一个关于全球艺术问题和人类存在的公共艺术，但重视个人的经验和主观性。卡明·勒猜普拉斯特曾经提出："我们的身体就是博物馆，我们的灵魂就是艺术"，并且以此作为创作概念，期望每个人都是艺术家、馆长和博物馆拥有人。31 世纪博物馆侧重于艺术与非艺术是如何实践的想法，它评估了公众的精神层面，质问："我们该如何分享、鼓励人们去正视他们自己的创造力？" 作为社会调查的核心，这个项目问

到:"当代精神究竟意味着什么?"并且当意识到社会中缺乏正能量时,为人们提供确定自己觉醒的机会。

目前,位于清迈的31世纪博物馆有一个共享空间,在全球其他范围也有小型项目。该项目包括多个部分:在清迈,由一个重刷的海运集装箱做成的结构,其轴心做成数字31的样子;在芝加哥,勒猜普拉斯特在芝加哥艺术研究所的学校经营了一个工作室和展览室;在日本,为了新泻艺术节,他创建了一个书籍项目。31世纪博物馆的核心价值就是参与和过程。该项目探索榜样和积极改变之间的关系,以及环境问题的创新和可持续解决方案之间的关系,以及各种诸如此类的问题。

由泰国和国际艺术家在清迈完成的海运集装箱房子的艺术品,提供了讨论空间和每月分享的工作间。集装箱设置在开放场所,没有任何防护和装饰,邀请那些探险者和反材料主义者免费入内饮茶。它不像典型的博物馆那样,参观者可以通过成熟路线穿过展览或建筑物,在这里学习进步;而在这个开放的集装箱里,精心布置的树桩,让游客在纵横交错的区域和空间里,自己去理解和体验这个项目。根据到这里来的艺术家、策展人、艺术史学家、市民,不同的即兴工作坊和讲座经常发生。这里没有任何单一或固定的经验,就像是在生命和灵魂中没有奇特经历一样,每个游客对着摄像机回答几个问题,他们被录下来放到脸书上。问题通常包括"你对这个世界有何看法?"或"如果你有一个让世界更美好的想法,你会分享什么?",一个奇怪的脱稿回答,通常会打乱思维的结果。但是参与和第一个回答问题的人对这个项目来说至关重要,他让参与者来评价我们生活在其中的世界,和那个我们想生活在里面的理想世界。

Kamin Lertchaiprasert's background as an artist and a Buddhist monk strongly inform his work, which seeks to present cultural activism as a critical domain of inquiry within contemporary art. Lertchaiprasert's practice is collaborative, publically located, spiritually based and has social engagement at its core, incorporating the artist's own philosophy into a diverse body of work. In 2011 his interest in collaborative art led to the establishment of a large, interactive and participatory project titled 31st Century Museum of Contemporary Spirit. This project is a micro-utopian idea, seeking to balance learning with life and social engagement.

31st Century Museum of Contemporary Spirit is a public art project about the global questions of art and human existence, but one that values individual experience and the subjective. Kamin Lertchaiprasert has offered "Our Body is our Museum and our Spirit is art" as the conceptual idea of the project, proposing that everyone is an artist, a curator, and an owner of a museum. 31st Century Museum focus on an idea of how art and non-art practices can evaluate public concern about the spiritual dimension and asks 'how can we share and encourage people to see the creative value in themselves?'. As a hub for social research the project asks "What does contemporary spirit mean?" and offers people the chance to identify their own awakening while gaining an awareness of missing positive energy in society.

Currently located in Chiang Mai 31st Century Museum offers a communal space, but has a global reach, incorporating smaller projects in other locations. The project includes multiple strands: in Chiang Mai a structure made of refurbished shipping containers in the shape of the number "31" exists as the hub, in Chicago Lertchaiprasert ran a workshop and exhibition at the School of the Art Institute of Chicago and in Japan he created a book project for the Niigata Art Festival. At the heart of 31st Century Museum are the values of participation and process. The project explores issues such as the relationship between role models and positive change, and innovative and sustainable solutions for environmental problems.

In Chiang Mai the shipping containers house artworks by Thai and international artists and provide spaces for discussions and monthly creative sharing workshops. Set in an open space with no fence or property demarcation, the containers invite exploration and anti-materialism. Entry is free and so are cups of tea. Unlike a typical museum, where visitors progress via a considered route through exhibition or building, here open containers and thoughtfully placed tree stumps allow visitors to criss-cross zones and spaces, creating their own understanding and experience of the project. Impromptu workshops and lectures often occur, depending on the artists, curators, art historians and members of the public who come by. There is no singular or fixed experience, just as there is no singular experience of life or spirit.

After visiting guests are invited to answer a few questions on camera which are then uploaded to the Museum's facebook page. Questions often include: "What is your opinion about this world?" and "If you had a wish to make this world better, what would you like to share?" An odd jumble of unscripted and sometimes confusing thoughts results, but participation and first person response is central to this relational project which invites participants to respond to and reconsider the world we currently live in, and the world we want to live in.

[解读]

开放的当代精神

印象中的博物馆总是充满说教意味，告诉你什么是对，什么是错，将人们困定于条条框框之中，一板一眼地教育人们。但是每个人都是一个独特的个体，都有着自己的看法和见解，有着自己对于美好世界的理解，鼓励每个人表达自己的看法，无拘无束，开放自由，这才是当代应有的精神。卡明·勒猜普拉斯特不仅是一位艺术家，同时也是一位佛教僧侣，他注重于每个人的经验和主观性，认为每个人的身体就是博物馆，灵魂就是艺术。《当代精神的第 31 世纪博物馆》便是构建一个开放的空间，没有任何防护和装饰，让人们自己去理解和体验这个项目，将人们的参与和过程作为艺术的核心，挖掘每个人内心深处的潜力，为人们提供确定自己觉醒的机会。(闫丽祥)

群岛电影院
Archipelago Cinema

艺术家：奥雷·舍人	Artist：Ole Scheeren
地点：泰国攀牙湾瑶诺岛	Location：Yao Nvi Island, Phang Nga Bay, Thailand
形式和材料：空间艺术	Media/Type：Space Art
时间：2012 年	Date：2012
推荐人：里奥·谭	Researcher：Leon Tan

泰国攀牙湾瑶诺岛上的电影是由纳特·萨拉萨拉，宗文·维拉瓦拉，以及 2010 年金棕榈奖获得者阿彼察邦·韦拉斯哈古所创立。

在 2012 年 3 月 9—13 日举办的首届年度艺术电影节上，德国建筑师奥雷·舍人被委任建造一个室外放映的漂浮观众席。舍人研究了当地的基地环境，重点调研了漂浮的龙虾养殖场的建造工程，即当地的渔民在木筏上面养殖龙虾。舍人利用木制的结构，将橡皮带系在包裹的泡沫板上，然后设计一个能够装配到一块浮动平台上的模块组件体系，这些组成成了日后的《群岛电影院》，这个建筑实质上是一个用可循环材料建成的超大尺寸的木筏。

在节日期间，被邀的宾客在晚间乘船来到漂浮的观众席。背景幕与超群的岩石和大海形成视觉平衡，群岛电影院给人一种自然和移动影像契合的壮观体验。当描述自己的工程时，艺术家写道，"在大海上空盘旋，在这个难以置信的环礁湖空间中央的某处，聚焦于跨越水中的移动影像，一副快乐结合在一起的风景画。一种世事短暂性和随意性的感觉，几乎就像漂流木那样"。在某种意义上，群岛电影院开启了电影院接近自然环境的传统。而电影的放映次数由潮汐的节奏决定，电影的故事和风以及水声音一起进行。

艺术电影节的管理者，蒂尔达·斯文顿和阿彼察邦·韦拉斯哈古负责为群岛电影院选择电影题目。伴随着电影的放映，他们也会举行一些活动或者交流，作为加深观众融入电影或者自然环境的方式。此外，斯文顿和韦拉斯哈古建立了一个"瓶中信"的活动，在这里客人们可以带来一个以电影、书籍、雕塑，或者分享画形式的"瓶中信"。这些信息最后到一个专用的"当地图书馆"，将它们留给未来到这个岛上的参观者。

当这个电影节闭幕的时候，群岛电影院会被拆除，重新还给建造它的那个村庄。舍人认为群岛电影院是属于他们的东西，仅仅是被借来而已。如同舍人另一个受到赞扬的工程：马尔法免车载电影院（得克萨斯州沙漠的一个免下车电影院），群岛电影院是一个有目的的、简单的室外电影体验，它有限的寿命周期，表达了舍人对短暂不间断环境的兴趣。对他而言，"短期的空间占领"是全球流动性增加，和以当代为特征的经济快速增长的征兆。作为一个议题，群岛电影院表明建筑师可能会有利可图的聚焦于设计临时性的服务，以及周旋公共冲突来捕捉日益流动（短暂）的社会团体的注意力。将一个电影院（或者剧院或者游乐场）看作一个模块化系列，然后被装配或者根据需要被掩盖在一座木筏上，也正是舍人对"紧急文化变迁"最好的表达方式。

Film on the Rocks Yao Noi was founded by Nat Sarasas, Chomwan Weeraworawit and 2010 Palme d'Or winner Apichatpong Weerasethakul as an annual festival for art and film on Koh Yao Noi (Yao Noi Island) in Phang Nga Bay, Thailand.

For the inaugural edition (9-13 March, 2012), the German architect Ole Scheeren was commissioned to create a floating auditorium for outdoor screenings. Scheeren studied local spatial practices, specifically, the construction of floating lobster farms – "Local fishermen farm lobsters on rafts. Wooden frames are tied by rubber straps to foam blocks wrapped in mosquito nets." - and devised a system of modular components that could be assembled into a floating platform. Subsequently given the name Archipelago Cinema, the structure was essentially a super-sized raft built out of recycled materials.

During the festival, invited guests were taken by boat at nighttime to the floating auditorium. Set against a dramatic backdrop of towering rocks and sea, Archipelago Cinema afforded spectacular experiences of the convergence of nature and moving image. Describing his

project, the architect wrote, "Hovering above the sea, somewhere in the middle of this incredible space of the lagoon, focused on the moving images across the water. A landscape of pieces playfully joined together. A sense of temporality, randomness. Almost like drift wood."
In a sense, Archipelago Cinema opened the tradition of cinema up to the natural environment; the screening times were dictated by tidal rhythms while the cinematic narratives had to work alongside the sound of wind and lapping water.

The festival curators, Tilda Swinton and Apichatpong Weerasethakul, were responsible for the selection of film titles for Archipelago Cinema. Alongside the screenings, they also facilitated activities and conversations as a means of deepening audience engagement with the films and the natural environment. Additionally, Swinton and Weerasethakul instigated the "Message in a Bottle" project, where guests were asked to bring a "message in a bottle" in the form of a film, book, sculpture or painting to share. These messages ended up in a dedicated "local library" presumably for future visitors to the island.

At the conclusion of the festival, Archipelago Cinema was dismantled and gifted to the village where it was built. Scheeren thought of it as "something that belongs to them, that was merely borrowed." Like the architect's other acclaimed project, the Marfa Drive-In (a drive in cinema in the Texas desert),Archipelago Cinema was an intentionally brief outdoor cinematic experience, its limited life cycle expressing Scheeren's ongoing interest in the condition of transience. For him, "the short-term occupation of space" is symptomatic of the increase in global mobility and intensification of economic precarity characterizing our contemporary moment. As a proposition, Archipelago Cinema suggests that architects might profitably focus on the design of temporary services and mediated public encounters to capture the attention of increasingly mobile (and transient) communities. A cinema (or theatre or playground) conceived as a modular series of rafts assembled and dissembled as needed is the perfect expression for what Scheeren has described as an "emergent culture of change."

[解读]

废墟上的重生

当你划船穿过大海,经过泰国攀牙湾瑶诺岛时,会看到一束光柱,投影在银幕上,此时声、光、电在自然的背景下完美展开,这就是德国建筑师奥雷·舍人的《群岛电影院》,一座坐落在大海上,群礁环绕、独一无二的电影院,群岛电影院给人一种自然和移动影像契合的壮观体验。位于漂浮观众席上的宾客,在银幕与超群的岩石和大海间,"游目骋怀,足以极视听之娱"。奥雷·舍人以临时影院的艺术形式,表达了对短暂不间断环境的兴趣。对他而言,这也正是当代经济快速增长的一个征兆。人们的关注度会随着时间的延长而逐渐消失,同时拉长的战线又会消耗巨大的人力财力,如何平衡设计与公众短暂注意力之间的关系,做好应对"紧急文化变迁"的方式,这正是艺术家带给我们的启示。(闫丽祥)

案例研究
北美

CASE STUDY
Northern America

Mapping Skin Deep	皮肤深处的绘图
Untitled Project for New Museum Triennial 2015	2015 年新博物馆三年展的未命名项目
Red Shoes	红鞋子
Between the Door and the Street	门与街之间
Connect the Dots: Mapping the High Water Hazards and History of Boulder Creekt	连接到点：勘察高水的危害和博尔德溪的历史
Create: The Community Meal	创作：社区膳食
Sandy Storyline	珊迪故事情节
The Music Box: A Shantytown Sound Laboratory	音乐盒：棚户区声音实验室
All City Canvas	全城画布
Metamorphosoup	年度万圣节室外木偶表演
Dorchester Projects	多尔切斯特项目
Waiting for Godot in New Orleans	在新奥尔良等待戈多
Gramsci Monument	葛兰西纪念碑

皮肤深处的绘图
Mapping Skin Deep

艺术家：克劳蒂（亚·埃斯皮诺萨·拉莫斯）	Artist：Cerrucha（Claudia Espinosa Ramos）
地点：加拿大蒙特利尔	Location：Montreal, Canada
形式和材料：视听艺术	Media/Type：Audio-visual Art
时间：2014 年	Date：2014
推荐人：杰西卡·费亚拉	Researcher：Jessica Fiala

《皮肤深处的绘图》是从 2014 年 3 月 1—16 日在 Place d'Armes 地铁站公开发起的视听项目，它是蒙特利尔地下艺术节的一部分。蒙特利尔地下艺术节起始于 2009 年，每年邀请艺术家在蒙特利尔地下城进行创作，地下城是连接和延伸到市区的室内公共场所和充满生气的地下通道。

在《皮肤深处的绘图》项目中，艺术家克劳蒂展示了来自墨西哥、伊朗、埃塞俄比亚、哥伦比亚、海地、喀麦隆和萨尔瓦多的难民和移民者的画像和故事。每幅经典黑白画像在后期制作时都在画像主角的身体上加上一道伤疤，追踪他或她一路来到加拿大的历程。伤疤的不同严重程度反应了不同旅程和经历对每个人的不同影响。画像还伴有采访录音，记录了记忆、期望、挑战、恐惧和洞察力。

《皮肤深处的绘图》是克劳蒂近期探索的延续，包括数字化照片和对话。In-Visible Part Ⅱ (2012) 系列包括修改背景各异的个人照片，添加其各自国家的歧视性词语的纹身，来吸引人们对此类语言产生的影响的关注。Take Time / Prends le Temps (2013) 在塔隆市场搭建了一个摊位，邀请路人与艺术家和其他陌生人进行对话，希望讨论时间是如何流逝及繁忙世界中关系的展开。通过《皮肤深处的绘图》项目，这些策略互相融合，形成处理的照片和延续的谈话，以引出以前不曾见过或听过的奇特个人经历。

全国有近 164 000 名难民，每年有成千的难民移居加拿大，个人的故事和经历在人口数据及全球冲突的报道面前轻易被掩盖。《皮肤深处的绘图》不仅提高了对被边缘化人群的关注，而且将难民经历变成鲜活的故事。通过采访和画像，该项目得以分享独特的视角和故事，强化每个人经历的独特历程。

该项目还提供了有关加拿大文化景观对话的机会，以及更大社区中个人的划分。一名受采访者提出这样的疑问："何为身份？为什么我的身份是伊朗人？……我出生在伊朗，我在别处学习，但这不意味着我的身份是伊朗人，身份问题不是国家、不是土地、不是语言……身份是我保留的知识、经历以及能够与他人分享我的这一切。"以这种方式，该项目还反驳了移民带来压力的论断，不仅强调了移民者的经济贡献，还有每个人为更大社区带来的独特洞察力和态度。

以前只有在地下艺术节公开展示的《皮肤深处的绘图》，现在这些图像可以在克劳蒂的网站上查看，以及随附的采访音频，由云录制播放。

Mapping Skin Deep is an audio-visual project publicly installed from March 1-16, 2014, in the metro station Place d'Armes as part of Montreal's Art Souterrain Festival. Founded in 2009, the festival annually invites artists to install projects in Montreal's Underground City, a network of indoor public spaces and lively underground passageways that connect and extend the city's downtown.

In Mapping Skin Deep, CERRUCHA (Claudia Espinosa Ramos) exhibited portraits and stories of refugees and immigrants from Mexico, Iran, Ethiopia, Colombia, Haiti, Cameroon, and El Salvador. Each classic black and white portrait was modified in post-production to add a scar to the body of the subject, tracing her or his mapped path to Canada. The severity of the scars is varied to reflect the impact that the distinct journeys and experiences had on the life of each person. The portraits are accompanied by audio recordings of interviews describing memories, expectations, challenges, legal fears, and insights.

Mapping Skin Deep is a continuation of recent explorations by CERRUCHA involving digitally manipulated photographs and dialogues. The series In-Visible Part II (2012) involved altering photographs of individuals from disparate backgrounds, adding tattoos of misogynistic phrases from their countries of origin as a means of drawing attention to the impact of such language. Take Time / Prends le Temps (2013) established a booth at the Jean Talon market to invite passersby to enter into a conversation with the artist or with other strangers, with the goal of bringing into question the nature of how time is spent and how relationships unfold in a busy world. With Mapping Skin Deep, these strategies feed into one another, combining manipulated photographs and extended conversations to bring out the unique experiences of individuals often unseen and unheard.

With nearly 164,000 refugees in country and thousands of refugees immigrating to Canada each year, individual stories and experiences can be easily lost in data about populations and reporting on global conflict. Mapping Skin Deep not only increases visibility generally for a frequently marginalized population, it also humanizes the refugee experience. Through interviews and portraits, the project shares distinct perspectives and stories, underscoring the unique journey taken by each individual.

The project also presented the opportunity to enter into a larger dialogue about the cultural landscape of Canada and the categorization of individuals within the larger community. As one interviewee questioned: "What is identity? Is my identity Iranian?⋯ I was born in Iran, I learned something there, but it doesn't mean my identity is⋯Iranian⋯the issue of identity is not a country, not a land, not a language⋯the identity is preserving my knowledge, my experiences, and being able to share it with others." In this manner, the project also counters assertions that immigration represents a burden, emphasizing not only the economic contributions of immigrants, but also the distinct insights and attributes brought to the larger community by each person.

While only publicly on view during the Art Souterrain Festival, the images that comprise Mapping Skin Deep are now available on CERRUCHA's website, with the accompanying audio interviews embedded for streaming via SoundCloud.

[解读]

伤痕记忆

再深的伤口，也会随着时间的流逝慢慢愈合，但是那存在脑海的伤痕记忆，却永远挥之不去。漂泊流离的生活、冷漠歧视的白眼，犹如锋利的刀刃划过皮肤，在移民的内心留下了深深的伤痕。艺术家克劳蒂在《皮肤深处的绘图》项目中，展示了来自墨西哥、伊朗、埃塞俄比亚等地难民和移民者的画像和故事。以黑白画像上的伤疤，记录旅程中的情感；以采访录音，记录了移民的期望和恐惧。艺术家将内心的伤痕表现于皮肤之上，使人们可以直面移民者的伤痛，唤起人们对这些处于边缘化，并且被伤害的人群的关注。人们无法如摩西般有莫大的法力，面对红海的阻拦，渡海如履平地，只能用脚步一步一步丈量苦难，并任由其一次次狠狠地划过自己的伤口。只有所有人的正式和关注，这些藏于内心的伤痕才会逐渐愈合。（闫丽祥）

2015年新博物馆三年展的未命名项目
Untitled Project for New Museum Triennial 2015

艺术家：卢克·威利斯·汤普森
地点：美国纽约
形式和材料：雕刻的混凝土和表演

时间：2015年
推荐人：彼得·尚德

Artist : Luke Willis Thompspn
Location : New York , USA
Media/Type : Carved concrete and perromance
Date : 2015
Researcher: Peter Shand

卢克·威利斯·汤普森是第一位受邀参与新美术馆三年展的新西兰艺术家，也是第一位受邀的太平洋斐济岛人后裔。这次展览给他提供了一个机会，让他展开一个精确缜密又开放的艺术实践。这个实践是基于公共空间和公开性两者细微差别的校准，以及两者关系对于艺术创作和艺术作品的影响。如此一来，他就能更大范围地收集公众对于艺术以及艺术创作来源的猜想与疑问，也能更大程度地为公众关于艺术价值、艺术目的的讨论增添活力，这样更能令他重新思考艺术对于公众的潜在意义。

汤普森的全部作品可以说是以其巧妙地适应社会地理的细微差别为特点。他的项目总是来源于自传中的经历，但他却能够把这些经历迅速发散开来并考虑到这些经历所涉及的方方面面，还将更广的社会政治重要性及影响纳入考虑范围之内。他能够清晰而且批判地抓住当代艺术创作的要义，能够灵巧地展开调查，他引领观众，或者可以说，他为观众提供了一个自我发现的机会，或者他用这个机会来解释艺术成果和艺术实现方法，而不是仅仅对公众宣布艺术结果，也不是用艺术或者政治性的东西来震慑公众。他作品中的含义是多方面的，而且是能让观众慢慢体会的。他把经历所带来的影响做了仔细的考虑，同时他也考虑到如何将这些影响潜移默化地带给他的观众。

汤普森在他的 2015 年项目中已经研究出一种微妙的、带有挑战性和富有启发性的行为艺术。而这项工作是询问市民在公共领域的经历是怎样的，问他们有没有意识到他们依赖于各种不同的监控设备、防护控制设备，及居住设备又或者他们与这些设备有着怎样的联系。这项行为艺术在它举行的现场会产生直接而且迅速的影响，它在历史意义方面、社会生活方面、日常所经历的东西描述为不稳定的、意料之外的、新的东西，让我们对熟系的东西产生一种难以消除的陌生感，或至少是一种能力，让人把他认为是熟悉的东西想象为一种完全不同或者是陌生的物体。在这个特殊的例子中，影响发生了转移，在观众的心里具体化，这个作品是一个精确的模拟事件，让观众们感受其他市民的不安经历、不稳定甚至是带有暴力性质的事件。

这个项目里的行为艺术是由一些非白种人的年轻人来完成的。这些表演者会带着他们的小组人员走过新当代艺术博物馆周围的大街小巷，他们是没有预先规划好线路的。这些"巡回表演"或"游行"将反映表演者经历过的历史事件和社会事件，如果说在过去的 25 年曼哈顿发生了很大程度变化的话（如果我对小组领导者的年龄理解正确），那么只要走过这些路线，这些参与者的心情就会发生微妙的变化，因为在 20 世纪 80 年代晚期和 90 年代早期，这些参与者是不可能走过字母城的波威里街，或者下东区的其他地方的。这也暗示着这些游行本质上是在追溯以小组领导人为代表的一代人在这些地区的经历。

在这个项目中与这一代人种族特征相关的是该委托作品反映了一次联邦集体诉讼（弗洛伊德等人控告纽约市）。该案件控告纽约市警察局有种族形象定性行为。多亏了这个案件，"拦截搜身"这个政策才得以停止，他们指出这个政策带有种族歧视及政府强行实行色彩。与此同时，发生在纽约的这种类似案件似乎转移到了其他地方。据说，在合法的执行范围内，或者以维护社会秩序为由，发生了多起由政府准许的针对年轻非裔美国人的暴力行为（以特雷文·马丁[①]、迈克尔·布朗[②]、约翰·克劳福得[③]为例）。

然而汤普森这个项目里最有深刻见解、切中要害的地方就在于这些小组领导人本质上是邀请别人来监督他们的。他们会在新博物馆外面与他们的小组碰面，但他们不会面对面见对方，而是扭头离开，开始这次游行，

[①] 特雷文·马丁，17岁，非裔美国人，2012年2月16日，在佛罗里达州被28岁的乔治·齐默曼枪击致死。
[②] 迈克尔·布朗，18岁，非裔美国人，2014年8月9日，在密苏里州被白人警察达伦·威尔森枪击致死。
[③] 约翰·克劳福得，22岁，非裔美国人，2014年8月5日，在俄亥俄州被两名警察枪击致死。

其他组员就会跟着他们走。在游行结束时（是否结束不是由时间和路线来定义的），他们会转身面向带领小组，并宣布表演已经结束。除此之外，在穿行有社会历史意义的大街小巷中，还有跟着他们小组组长行走的过程中，这些小组组员会无意地受到暗示——他们是弗洛伊德等人中的一员，跟着别人走会令他们觉得不安，在其他情况下对方很有可能会误会他所不认识的这位年轻人想对他做什么（任意一种情况，或是这两种情况都会带来执法人员的关注）。

这个项目也完全颠覆了以下这些艺术传统（维托·阿肯锡[①]、苏菲·卡尔[②]），因为不安的感觉深深地存在于这些参与者的真实生活中。因为在暴力与控制这个体系里，有着一种无法逃避的暗含之意，而且只要一个眼神的凝视，就能把监督与表演结合在一起，这也把精神分析弄得很复杂，所以这很有可能颠覆一些政策。

我认为这个项目带有一种微妙又慎重的政治意图，而且它还反映了在公共领域聚居的复杂而且有意义的方式，也反映了我们作为公民的行为，所以这个作品应该纳入公共艺术奖得奖作品的考虑范围中。

Luke Willis Thompson is the first artist of New Zealand or Pacific (in his case Fijian) descent to be invited to participate in the New Gallery Triennial. It represents an opportunity for him to extend a precise, rigorous and open practice that is predicated on a nuanced recalibration of public space and publicness and their relationship to art-making and the artwork. As such, it extends his capacity to invite public questioning of the assumptions of art and the arenas of its experience, to invigorate a public debate around the value and purpose of art and to enable searching reconsideration of art's potential meanings for citizens.

Thompson's oeuvre may rightly be characterised as acutely attuned to subtle nuances of social geography. His projects invariably derive from autobiographical experience but are extended quickly and with complexity to considerations of broader socio-political importance or significance. He achieves this with an astute and critical grasp of contemporary artistic practice and intellectual inquiry and leads or presents opportunities for audience self-discovery or achievement of elucidation or realization rather than from overt declaration or the risk of artistic or political hectoring. The implications of his work, then, are multiple and slow-burning, with a fine regard for the affect of experience and how this will shift unexpectedly for public audiences of his practice.

---

① 维托·阿肯锡，1940年出生于美国纽约，是一名建筑师、装置艺术家，同时，他也是观念主义艺术家、诗人和表演家。
② 苏菲·卡尔，1953年出生于巴黎，被称为一个打乱秩序和难以归类的艺术家。

For the 2015 project, Thompson has developed a subtle, challenging and illuminating performative work that interrogates how citizens experience the public realm and how they are (or are not) aware of, reliant on or in different relationship to different apparatuses of surveillance, protection, control or habitation. It is a project that responds directly and immediately to site but does so in a manner that is layered historically, socially, politically and psychologically. It seeks to present the known or daily experienced as unstable, unexpected, new —an abiding strangeness of the familiar or at least a capacity for that which one thinks of as familiar to be rendered different or strange. In this particular instance, that shifting of affect embodies in the audience of the work a precise analogue for the disturbing, destabilizing or violent experiences of other citizens.

The performative component of the work is to be led by non-Caucasian young men. These performers will lead groups through the streets surrounding the New Museum on undeclared routes. The "tours" or "walks" will reflect the particular histories and social experiences of those men, Given the extraordinary degree of change in this area of Manhattan over the past two and half decades (if I take this to be a fair understanding of the age of the group leaders) the routes will subtly unravel the immediate experience of participants who may have been less likely to walk in the Bowery, Alphabet City or other lower Eastside locations in the late 1980s or early 1990s. What will also be implied here is that the walks are essentially retracing the experiences of a generation represented by the group leaders.

Of pertinence to the generational specificity of the work is that the commissioned piece responds to a Federal Class Action Lawsuit (Floyd et al v City of New York) that charges the NYPD with having engaged in racial profiling. The case is credited with having led to the cessation of "stop and frisk" policies and the politics of racism and state authority they manifested. At the same time as there may have been a shift in approach in New York, persistent incidents of state-sanctioned or approved violence against young African-American men allegedly within the frame of legal enforcement or the maintenance of civil order (one thinks of Trayvon Martin, Michael Brown and John Crawford III, for instance).

What is insightful and incisive about Thompson's project, however, is that the group leaders essentially invite their own surveillance. They will meet groups outside the New Museum but instead of facing the group will walk off on the tour, with the group following. At the end of the walk (defined neither in time or route) they will turn to face the group they have lead and declare the performance complete. In addition, then, to the experience of the social history of the locations traversed and of the men leading them, the groups will themselves be implicated in attitudes that were embedded in Floyd et al – the discomfort of following someone and the potential that one might understand differently one's relationships with young men not of one's acquaintance in other situations (either or both of which led to political justifications for law-enforcement officials' activities).

The work also turns on its head an artistic tradition of following works (Vito Acconci, Sophie Calle) insofar as the seat of discomfort is firmly within the lived experience of the audience participant. This is potentially more politically upending because of the inescapable sense of implication in systems of violence and control and more psychoanalytically complex in the way by which it draws together surveillance and spectacle within the operation of the gaze.

It is as a work of subtle and very deliberate political intent and one that reflects in complex and meaningful ways on the habitation of public space and our behaviour as citizens that this work ought to be considered for a public art award.

[解读]

不安的追随

艺术家卢克·威利斯·汤普森善于引导出我们精微的内心感受，当我们追随着这些非白种人穿过那些曾经发生过一些暴力事件的街道，当我们以他们的视角去感受这种不安、不稳定和暴力的气息，而这些都是他们切实经历过的。非裔美国人遭遇的一系列暴力对待，是否代表种族的区分和偏见如今依然存在，而参与者去观察、去跟随着他们穿过那些标志性的街道的时候，内心的变化和感受是否在这样的引导下而变得不同。也许乌托邦或者桃花源只是人们的一个美好想象，一个社会和国家中的矛盾与问题会永远存在，但是在发现和解决的过程中，国家的法律和体制都会愈加完善，艺术家是这个社会敏感的医生，他们发现了问题，让我们去感受那些不为人注意的社会与人性的角落。自由平等与博爱的世界也许一直在前方，等待我们去实现。（李田）

红鞋子
Red Shoes

艺术家：艾琳娜·萧维　　Artist：Elina Chauvet
地点：墨西哥华雷斯　　Location：Ciudad Juárez/initially México
时间：2009 年　　Date：2009
推荐人：皮特·莫拉莱斯　　Researcher：Peter Morales

2009 年 8 月，墨西哥视觉艺术家艾琳娜从华雷斯城（墨西哥的一个城市）收集了 33 双女式鞋子。鞋子或本来是红色的，或被染成红色的。因此萧维在墨西哥的众多购物中心开始摆放红鞋子，进而这个风潮传播到了全球。萧维认为这些艺术形式不只是属于他自己一个人的创作，而是集体创作。因为其涉及的是对一个事件的集体回忆，这个事件表面上看来非常外交化，只是因为其已被艺术家艾琳娜公诸于众，事实上它却具有极度的个体色彩。

就像萧维所描述的那样，红鞋子的出现使人想起那些孩子们，母亲们以及妻子们由于消失、遭受强奸、煎熬以及谋杀之后所带来的消失感以及空虚感。这项设计的一部分灵感是来自亲人：艾琳娜的姐姐被其丈夫谋杀。当艾琳娜在寻找一种聊以安慰的方式时，开始调查在华雷斯城死亡的女人们，然后他发现鞋子是一个共同的主题。这些消失的女人很多都在鞋厂上班，或者准备去鞋厂工作，又或者正在买鞋子。他们出事后是通过鞋子来辨别身份的。在墨西哥华雷斯城有很多鞋子加工厂，鞋子是这个行业的标签。并且萧维和他姐姐会换穿鞋。所以她发现对于大多数犯罪来说鞋子是一个强有利的隐喻。

女性灭绝指的是仅因为是女人就谋杀。这个术语随着 20 年前华雷斯城事件的曝光和社会大众对其呼吁获得了越来越多的关注。现今工作变得更国际化，女人工作变得更普及化，甚至在某些地方已婚女人和未婚女人大量消失是不曾见到的。因此萧维的工作引发了一个不容回避的话题，即暴力对待女人。

尽管一些人可能争论在华雷斯城消失的和被谋杀的女人的实际数量到底代表了什么。就百分比来说，华雷斯城女性夭折的数量不如世界上任何一个地区的多，哪怕跟美国比。但是这个问题还应得到重视。许多要求政府给予回应并且采取行动的女权主义激进分子由于将这个问题摆到显著位置而受到威胁，遭受袭击，有的甚至被暗杀。

红鞋子活动在世界各地重复上演着。此活动邀请市民捐赠红鞋子，或者在活动上演当天在当地染红鞋子。此项活动承诺留出足够时间通知媒体和大众以保证民众的出席——带着自己的红鞋子在指定时间（通常会设在特定时间）到指定地点（通常是一个公共广场）。活动结束后，排定的讨论事项会如期进行。

In august of 2009 Mexican artist, Elina Chauvet, collected 33 pairs of shoes from women in Ciudad Juárez, Mexico. The shoes were either red to begin with or were painted red and thus Chauvet began a series of installations in plazas throughout Mexico that later expanded internationally. Chauvet considers these installations a collective work that does not belong to her, since tit deals with a collective memory of something that is terribly private and seemingly foreign, only that now it is made public by the artist. The work of art forces us to confront the issue of feminicide, something for which Ciudad Juárez came to be know starting tin the 1990's.

As Chauvet describes it, the presence of the red shoes calls up the absence and emptiness left by the disappearance, rape, torture and murder of daughters, sisters, mothers and wives. The inspiration for the work is in part intimate; Chauvet's sister was murdered by her husband. As she sought to find solace in artistic expression she undertook to investigate the deaths of women in Juárez and she found that shoes were a common theme. Many of the missing women of Juárez worked in shoe shops, were on their way to find work in a shoe shop, or were shopping for shoes. Many were later identified by their shoes. There are many maquilas in Juárez and shoes are a staple of that industry. Chauvet shared shoes with her sister. The artist discovered that shoes were a powerful metaphor for these crimes.

Feminicide is defined as the murder of women for the simple fact of being women; the term has garnered more attention since the publicity and outcry over the events of the last two decades in Ciudad

Juárez. As the work travels internationally, even in places where the disappearance of women and girls in large numbers is not experienced, Chauvet's work provokes a conversation on a topic that cannot remain hidden —that of violence against women.

Although some may dispute what the actual numbers of the disappeared and murdered women in Juárez actually represent— as a percentage, say, in that the proportion of fiminicides in Juárez compared to total homicides is actually no greater than it is in other parts of the world including the United States—the issue nonetheless is worthy of attention. Many women activists demanding answers and action from their government have themselves received threats, suffered attacks and have even been killed for bringing this issue to the fore.

Red Shoes is replicated in other locations by inviting citizens to donate women's shoes that are red of have been painted red, some are invited to paint shoes on site on the day of the installation. The process of installation is undertaken to allow for enough time to notify the media and the public of the possibility of participation—by bringing a pair of red shoes to a predetermined location, usually a public square, on a specific date. Care is also taken to allow for the possibility of scheduled discussions afterward.

[解读]

消逝的红

墨西哥《宇宙报》在 2010 年曾发表过一篇题为《华雷斯，全球最暴力的城市》的文章，墨西哥的华雷斯城以最高的暴力死亡比例成为全球之首。而针对女性的暴力事件更是层出不穷，艺术家萧维将红鞋子作为女性身份的标志，将不同的红鞋子展示出来，在一片灰暗的色彩中，无数的红鞋子显得尤为触目惊心，每一双红鞋子就代表着一位女性，而女性在暴力中大量的死亡，只有她们的鞋子留了下来，成为识别她们身份的标志，而红鞋子本来是女性之美的表现，是美好，是热情，是新娘的嫁衣，是天边的晚霞，如今却成为哀伤的红色，它们孤零零地被展示在那里，人们看到这些各种各样的红鞋子，是否会想到他们的主人——女性这个群体到底面临着怎样的危险境地。而那些伴随着女性一起消失的红鞋子，也让女性的身份的认同感随着红鞋子一点点消减，而艺术家提出了这个问题，如何解决它仍是路漫漫其修远。（李田）

门与街之间
Between the Door and the Street

艺术家：苏珊·拉齐
地点：美国纽约布鲁克林
形式和材料：社会参与公共艺术

时间：2013 年
推荐人：梅根·古尔伯

Artist：Suzanne Lacy
Location： New York, the USA
Media/Type：Socially engaged public art

Date：2013
Researcher：Megan Guerber

苏珊·拉齐被认为是社会参与公共艺术的先驱。她曾与自 20 世纪 70 年代设立的社区一起工作。她的许多项目重点打造播出边缘化的声音的平台。她先前的作品，包括《火焰之上的屋顶》（1993—1994）。该项目集合了 220 个奥克兰 CA 公立高中学生参加政治对话，该对话普遍排斥年轻人，包括性、毒品和家庭。对话在停放的车之间进行，100 个奥克兰市民被邀请来不间断地收听。

《门与街之间》的功能与《火焰之上的屋顶》非常相像，很大程度上依赖于拉齐丰富的社区组织经验。和纽约的伙伴一起创建，以公共艺术组织为基础，在布鲁克林博物馆的女性主义艺术"创意时代和伊丽莎白"A. 赛克勒中心，该项目的展出仅持续一天，却是好几个月的对话、宣传和研究的成果。

经过 5 个月的课程学习，苏珊·拉齐遇到了一群 19 岁的女性艺术家。这些学生帮助她筹划关于当今性别政治的重要问题。她们同样也在寻找参与这个游廊对话的参与者上做出了很多的贡献。总共有78个组被邀请，每组都表现出安全的参与。为了尽可能从不同角度重现，在寻找各种不同的组上花费了很大的精力。

布鲁克林博物馆的台阶布满了和这个项目相关的男女平等的问题。问题包括："谁在照看孩子？"以及"比赛是不是男女平等问题？"这些问题曾作为前奏，在即将到来表演进行讨论的重要议题。

在2013年10月19日，《门与街之间》在60门阶上持续约一个半小时。389名参与者出现。每个游廊都有一个"保卫"，拉齐通过改变讨论中的信号给每个人进行暗示，虽然几组选择跟随线索，让谈话自然发展。交流小组包含有4—7人。

公众被邀请来聚集在每个门阶周围，倾听正在发生的对话，然而，对话却对外围的人是"关闭的"，而且参观的人被邀请进来只是倾听。

参与者身穿黄色羊绒以便参观者辨别。当对话结束的时候，小组转向观众问他们是否有什么问题；或者想不想要对话继续。桌子放在街道的中心，并提供甜甜圈和热苹果汁。音乐家Arooj Aftah进行表演，让这个项目在温馨的节日气氛中结束。

《门与街之间》在谈论男女平等问题上，聚集大量不同的女性谈话组上是非常成功的。估计有2500名参观者倾听了他们的对话。该项目引起不少讨论和争议，点燃了公共艺术和行动主义的复杂性。

因为它严重依赖于志愿者，经济差异被这个项目的组织者所忽略。一些为照料小孩、创造时间的参与者被召集在男女平等问题上，不要考虑作为母亲的需求，而这些对这次表现产生了消极性，或许这也是这种公开辩论的亮点，并且许多女权主义者的对话也在全国范围内激发讨论，这是该项目最大的成功之处。

Suzanne Lacy is considered a pioneer of socially engaged public art. She has worked collaboratively with communities since the 1970s. Many of her projects focus on creating platforms to broadcast marginalized voices. Prior work includes The Roof is on Fire (1993—1994). The project assembled 220 Oakland, CA public high school students to participate in political conversations that generally exclude youths, including sex, drugs and family. The conversations were held within parked cars and 100 Oakland citizens were invited to listen in without interrupting.

Between the Door and the Street functioned much like The Roof is on Fire and relied heavily on Lacy's extensive experience with community organization. Created in partnership with New York—based public art organization Creative Time and the Elizabeth A. Sackler Center for Feminist Art at the Brooklyn Museum, the project's performance only lasted one day but was the result of months of conversations, outreach and research.

Over the course of five months, Suzanne Lacy met with a group of nineteen female activists. These advisees assisted her in mapping out important issues related to gender politics today. These women also played a large role in finding activist groups to participate in the stoop conversations. In all, 78 groups were invited, each of which secured performance participants. Great effort was put into finding diverse groups of women in order to represent as many perspectives as possible.

The steps of the Brooklyn Museum were covered with feminist questions in correlation with the project. Questions included "Who watches the kids?" and "Is race a feminist issue?" These questions served as a prelude to important topics discussed during the upcoming performance.

On October 19, 2013 Between the Doors and the Street took place across 60 stoops for an hour and a half. 389 participants were present. Each stoop had a "guard" who was given cues by Lacy to signal a change in discussion, although few groups chose to follow the cues and instead let the conversations develop naturally. Conversation groups consisted of 4—7 people.

The public was invited to gather around each stoop and listen in to the conversations taking place, however the conversations were "closed" to outsiders and visitors were only invited to participate by listening in.

Participants wore yellow pashminas to help visitors identify them. At the close of the conversation, groups then turned to the audience to see if anyone had any questions or wanted to continue the conversations. Tables were brought to the center of the street where mini doughnuts and hot cider were provided. Musician Arooj Aftah performed, giving the close of the project a warm and festive air.

Between the Door and the Street was successful in bringing together a large and diverse group of women to talk about feminist issues. An estimated 2,500 visitors listened in to their conversations. The project sparked much debate and controversy, bringing to light the complexities of both public art and activism.

Because it relied heavily on volunteers, economic disparities were overlooked in the project's organization. Some participants struggled to find childcare and Creative Time was called out for not considering the needs of mothers during a feminist project.

While this created some negativity prior to the performance, it was perhaps the spark of this controversy, and the many feminist conversations it inspired nationwide, that was the project's greatest success.

[解读]
第二性

当我们谈到女性权利的时候，我们在谈什么，是波伏娃的《第二性》里所提到的相对于男性存在的他者，是《创世纪》里脱胎于男性代表亚当身体的夏娃，天主创造了她，是为了让男性不感觉到孤独。女性是大莉拉和尤滴，阿斯帕西亚和路克雷蒂亚，潘多拉和雅典娜，同时也是圣母玛利亚。她是一个偶像、一个女仆，是生命的源泉、黑暗的势力。几千年来，女性被赋予了各种各样的地位和形象，如今的"她"要摆脱他者的身份，而一群女性艺术家组织了这场讨论，观众在其中作为倾听者，也是活动的参与者，男性和女性都在反思两性应当如何平等，如何摆脱过去的桎梏。语言是交流的媒介，而门与街之间给了我们一个愉悦的平台，女权主义不如说是女性主义，达成绝对的平等不如相对的平等，而这场活动所引起的讨论不光是对参与者来说的，还波及了更广泛的区域。（李田）

连接到点:勘察高水的危害和博尔德溪的历史
Connect the Dots: Mapping the High Water Hazards and History of Boulder Creekt

艺术家:玛丽·密斯
地点:美国科罗拉多州
形式:当代装置艺术
时间:2007 年
推荐人:梅根·古尔伯

Artist:Mary Miss
Location:Boulder, Colorado, USA
Media/Type:Contemporary Installation
Date:2007
Researcher:Megan Guerber

玛丽·密斯被训练成为一名雕刻家,并且在 20 世纪 60 年代开始她作为一名园地艺术家的职业生涯。自从那时起,它的练习就逐渐从探索一个人是如何通过景观来迁徙到探讨当今社会的重大政治和经济问题。当代主要项目包括:在双塔倒塌的纽约世贸遗址周围建造一个临时纪念碑,并且建设一个新德里 17 世纪展馆的临时药用植物园,作为重新激活空间的手段。

按照她的主页所说,密斯的作品"打造一种情况,强调网站的历史,它的生态,或被忽略的环境方面"。她频繁地与科学家、生态学家、工程师、建筑师以及公共管理者一起来创作。

密斯当前的创作致力于用视觉来展现大众的抽象问题,大多数具体的环境问题,以帮助他们理解为什么这些问题很重要。把这些问题可视化,巩固了他们作为具体的、可识别的关注,并且帮助市民加强并采取行动。

《连接到点：勘察高水的危害和博尔德溪的历史》是 2007 年的一个项目，试图培养博尔德，CO 社区潜在的约 500 年洪水威胁。由于 500 年一遇的洪水似乎离我们的日常关注非常遥远，这个地区没有任何装备来应对理潜在的损害。

解释 "500 年一遇的洪水"这样的定义非常重要，每年发生的概率有 0.2%。虽然这样的概率很低，飓风卡特里娜的发生概率是 0.25%。因为只有这么一点的准备，飓风卡特里娜摧毁了这个地方，导致许多人丧命。因为全球变暖，气候变化，密斯挑战我们的防备自然灾害的能力，力求以有意义的方式教育公众。

科罗拉多州博尔德位于博尔德峡谷口，使得它非常容易遭遇山洪暴发。百年一遇的洪水（发生概率只有 1%）在 1894 年发生，导致明显的大面积破坏。自从那时起，博尔德经历了多次小洪水，分别在，1914 年、1919 年、1921 年、1938 年和 1969 年。

对于连接到点，玛丽密斯和一位水文学家和地质学家一起研究博尔德的河漫滩，并且绘制了 3D 地图，安装在市中心。该装置显示了潜在的 500 年一遇洪水的水位。总共 300 个蓝点被放置在建筑物，树木，灯柱和石块上。当人们走在城市中心，他们就会关注这个灾难性的洪水。

相较于收到抽象的数字和图形信息，市民会更清楚的知道水位比自己的脑袋高多少。连接到点目前作为露西莱帕德的天气预报的一部分：艺术和天气变化，一个与博尔德博物馆的当代艺术和生态艺术放在一起的大型展览，一个基于巨石的艺术组织调查气候变化，并通过性能，展览，讲座，创造一个可持续发展的未来。总计有 51 个项目展示在 10 个室外和 5 个室内场所。其他参展艺术家包括 Eve AndréeLaramée、Patricia Johanson、Subhankar Banerjee and Future Farmers.

连接到点的成功在于它能够把抽象的信息变得简单化。密斯使用了她作为合作者的技巧与科学家合作，并让公众了解他们的环境。通过把一个抽象的问题具体话，密斯帮博尔德社区彻底明白气候变化是如何影响他们的未来。

Mary Miss was trained as a sculpture and began her career as a land artist in the 1960s. At the time, her works were largely architectural. Since then her practice has grown from exploring how one moves through a landscape to addressing important political and ecological issues of our time. Major contemporary projects include building a temporary memorial around Ground Zero, NY following the fall of the Twin Towers and installing a temporary medicinal garden in a New

Delhi 17th-century pavilion as a means to reactivate the space.

According to her website, Miss' work "creates situations emphasizing a site's history, its ecology, or aspects of the environment that have gone unnoticed". She frequently collaborates with scientists, ecologists, engineers, architects and public administrators to create her work.

Miss's current practice focuses on bringing visual representation of abstract problems, most specifically environmental concerns, to the masses in order to help them understand why such issues are important. The visualization of these issues solidifies them as concrete, identifiable concerns and helps citizens to step up and take action.

Connect the Dots: Mapping the High water Hazards and History of Boulder Creek is a 2007 project that sought to educate the Boulder, CO community about the potential threat of a 500-year flood. Because a 500-year flood seems so distant from our everyday concerns, the area is unequipped to handle the potential damage.

It is important to note that the definition of a "500-year flood" is a flood with a 0.2% chance of occurring each year. Although this is an extremely low percentage, Hurricane Katrina had only a 0.25% chance of occurring. With so little preparation in place, the impact of Katrina devastated the area and killed many. With changing weather patterns due to global warming, Miss challenges our preparedness for natural disasters and seeks to educate the public in meaningful ways.

Boulder, CO is located at the mouth of Boulder Canyon, making it highly susceptible to flash flooding. A 100-year flood (a flood with 1% change of occurring) last occurred in 1894 and caused significant widespread damage. Since that time, Boulder has experienced smaller scale flooding in 1914, 1919, 1921, 1938 and 1969.

For Connect the Dots, Mary Miss partnered with a hydrologist and a geologist to learn about the Boulder floodplain and create a "3D map" installation throughout its downtown. The installation demonstrated potential flood levels of a 500-year flood. In all, 300 blue dots were placed on buildings, trees, light posts and boulders. As people walked through the city center, they could physically place themselves within the context of a catastrophic flood.

Rather than receiving the information via abstract numbers and graphs, citizens could readily understand how high above their heads water levels would be. Connect the Dots was commissioned as a part of Lucy Leopard's Weather Report: Art and Climate Change, a major exhibition put together in collaboration with The Boulder Museum of Contemporary Art and Eco Arts, a Boulder-based arts organization investigating climate change and the creation of a sustainable future via performance, exhibitions, talks and more. In all, 51 projects were displayed in 10 outdoor and 5 indoor locations. Other participating artists include Eve Andrée Laramée, Patricia Johanson, Subhankar Banerjee and Future Farmers.

The success of Connect the Dots lied in its ability to make abstract information easily accessible to the masses. Miss used her skills as a collaborator to work with scientists and inform the public about their environment. By making an abstract issue concrete, Miss helped the Boulder community think though how climate change may affect their futures.

[解读]
危险的讯号

蓝色的圆点遍布了整个城市，它代表着艺术家玛丽·密斯调查出来的大洪水发生时的平均水位。它们连接在一起便是艺术家对于未来的一个警示，可能发生的洪水会让我们的城市变成什么样子，我们在被建筑物、树木、灯柱上的蓝色圆点包围的时候，我们是否有一丝惶惑，我们是否可以想象自然的暴怒会将这个城市淹没到何种程度，而城市的景观又会如何变化。当艺术与生态联系在一起，艺术家以一个上帝的视角告诉我们大洪水也许会来临，而我们的"诺亚方舟"可否迎接这一切，应对这一切。艺术家有时候是美的创作者、有时候是科学家、有时候是救世主、有时候也是受自然操控的普通生命，他以简单的明了的方式说明了气候和自然会如何得影响我们的生活以及一切。（李田）

创作：社区膳食
Create: The Community Meal

艺术家：赛图·琼斯
地点：美国明尼苏达州
形式和材料：社区食物和
　教育、临时装置艺术
时间：2014 年
推荐人：梅根·古尔伯

Artist：Seitu Jones
Location：Saint Paul，Minnesota, USA
Media/Type：Temporary
　participatory installation
Date：2014
Researcher：Megan Guerber

许多都市团体陷入了"食物荒漠"类型。美国的农业部把"食物荒漠"定义为"城市地区以及没有新鲜健康充足事物的乡镇"。通常情况下，这些社区对事物的选择是有限的，都市便利店和快餐店，在那里很难找到新鲜和健康的饭菜。食物是一类问题。在低收入的都市社区，他们没有选择可言。对于廉价的快速和方便食品的依赖，会加剧这些社区一些健康并发症如，肥胖症、糖尿病和心脏病，因为这些食物含有更高的脂肪和糖类。对此，越来越多的食品正义运动已经席卷美国。社区成员们聚集在一起，增加他们对健康食品的获取渠道和知识。

社区关注公共，艺术家赛图·琼斯在他的实践中探索食物和土壤中的问题已经不是新鲜事了。他的《创作：社区膳食》项目，为了易于种植，带着可降解的名片传播甘蓝种子。类似地，从堆肥压实，粘土和种子制作而制成的手榴弹形状的种子"炸弹"，被他称为了食物传播而不是破坏的"和平武器"。

正如琼斯告诉 TPT 的 MN 原创节目——一个每周播放的节目，特色是明尼苏达州的艺术家，"你应该让你的社区比你发现它更美丽"。这样的信念指导着它的行动，并且在他近期的作品中更加突出。例如《创作：社区膳食》，2 000名居民围聚在一张桌子旁，延伸半公里长，来聚会、吃饭，谈论、判断他们的社区内的食物。琼斯住在青蛙镇，具有最多不同邻居社区，明尼苏达州的最多样化的街区之一。透过前窗，他可观察人们走到附近的便利店，看他们买食品。这样的观察刺激了他的欲望，与他的邻居一起拓展找的健康食物的方法。

从美国农业部的获得的政府补助允许琼斯做食品的评估，在代顿的布拉夫，青蛙镇以及高层会议——大学社区。这些评估表明，成本、运输和缺乏教育是让人们远离健康饮食的最大元凶。2013年琼斯荣获了乔伊斯奖，并与公共艺术圣保罗合作组织了"创作"活动。他花费1年时间从他所在的青蛙镇收集关于食物的故事，并把利用故事来做菜单。考虑包括，文化和家庭传统以及附近餐馆的特色菜。

同时，该项目时的重点是开始谈论食品，但性能、诗歌和视觉艺术元素也被包含了进去。诗人 G. E. 帕特森在吃饭前会读格雷斯。Ananya Chatterjea 教授给志愿服务者一组表演动作，当上菜的时候来表演。艺术家玛丽徐利用从生物垃圾手工制成的纸与邻居合作制作了2000个餐垫，一群由詹姆斯·贝克领导的当地的主厨制作了一顿由鸡肉和蔬菜的饭食。大部分食品是来源是50英里以内，由农民种出的。这些农民就在当场，帮助讲解健康食品，在哪里可以找到它，以及如何准备呢。

这个项目成功在于，在他的社区聚集了2 000名邻居，来讨论、判断食物。这是对公民采取的形势和机构作出重要改变的第一步。这也连接青蛙镇附近的当地大的农业社区，也增加了周围所有人的文化的理解。随着对如何采购和烹调健康食物的教育，公共健康也会得到改善。

Many urban communities fall into the category of "food deserts". The US Department of Agriculture defines food deserts as "urban neighborhoods and rural towns without ready access to fresh, healthy, and affordable food". Often the accessible food options in these communities are limited to convenience stores and fast food restaurants, where it is difficult to find fresh and healthy meals. Food is a class issue. It is lower—income urban communities that do not have access healthy options. A reliance on inexpensive fast and instant foods, which are higher in fat and sugar, leads to a dramatic increase in such health complications as obesity, diabetes and heart disease in these communities. In response, a growing food justice movement has been sweeping across the United States. Communities are coming together to increase their access to and knowledge about healthy food.

Community-focused public artist Seitu Jones is not new to exploring issues of food and soil in his practice. His Collard Field project disseminated collard seeds within biodegradable business cards to allow for easy planting. Similarly, his grenade shaped seed bombs, made from compacted compost, clay and seeds, created what he referred to as "weapons of peace" that spread food rather than destruction.

As Jones told TPT's MNOriginal program, a weekly program featuring Minnesotan artists, "You should leave your community more beautiful than you found it." This principal guides his practice and is prominent in his latest work, CREATE: The Community Meal, a gathering of 2,000 residents around a table stretching a half mile long in order to gather, eat and speak about food justice within their community. Jones resides in Frog town, one of the most diverse neighborhoods of Saint Paul, Minnesota. From his front windows he observed many people walking to the nearby convenience store to do their grocery shopping. This observation incited his desire to work with his neighbors on expanding access to affordable healthy food.

A grant from the USDA allowed Jones to make food assessments in the Dayton's Bluff, Frogtown and Summit–University neighborhoods. These assessments showed that cost, transportation and lack of education were the biggest barriers keeping people from eating healthy food. In 2013 Jones received the Joyce Award and partnered with Public Art Saint Paul to organize the CREATE event. He spent a year collecting "food stories" from his Frogtown community and used them to create the meal's menu. Considerations included cultural and family traditions as well as items featured at neighborhood restaurants.

While the focus of the project was on starting conversations about food, elements of performance, poetry and visual art were also included. Grace was read prior to the meal by poet G.E. Patterson. Choreographer Ananya Chatterjea taught the volunteer servers a set of movements to perform while serving the food. Artist Mary Hark collaborated with neighbors to create 2,000 placemats from handmade paper using bio waste from the neighborhood. A group of local chefs were led by James Baker to create the meal consisting of chicken and vegetables. Most of the food was sourced within a 50-mile radius and raised by farmers of color. These farmers were present to help teach about healthy food, where to find it and how to prepare it.

The projects successes include gathering 2,000 neighbors to talk about food justice within their community. It was the first step toward citizens taking agency of the situation and making important changes. This also connected the Frogtown neighborhood to the larger local agricultural community and increased cultural understanding all around. With education improving on how to source and cook healthy food, public health will improve.

[解读]

美味与艺术

食物中的营养构成我们的血肉,而我们所食用的,会让我们变成不同状态和外形的人,肉体是精神的神殿,无论里面供奉着什么,都应该更精致、更清洁。当我们度过了曾经对食物不加选择,只求生存下去的状态,我们对方方面面包括食物也有了更加健康的要求。如今食品安全问题成为城市居民十分困扰的问题,艺术家琼斯用粘土和种子制作而制成的手榴弹形状的种子"炸弹",自己种菜来达到健康和环保的要求,而这样的炸弹不会带来战争和破坏,而是带来绿色健康与和平。他还将艺术文化将饮食结合起来,故事、诗歌和视觉元素都融入在食物里,此时的食物不仅成长为我们的血肉,更潜移默化地渗入进我们的精神。在社区的人们一起欢乐宴饮的时候,美味与艺术巧妙地结合在一起,让肉体与精神都更加健康。(李田)

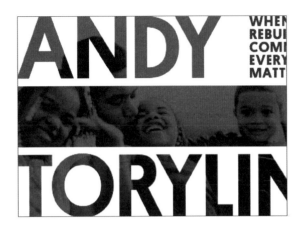

珊迪故事情节
Sandy Storyline

艺术家：瑞秋·法尔科内、迈克尔·普莱莫	Artist：Rachel Falcone Michael Premo
地点：美国纽约和新泽西	Location：N/A, USA
形式和材料：参与型纪录片	Media/Type：Participatory Documentary
时间：2012年至今	Date：2012-on-going
推荐人：梅根·古尔伯	Researcher：Megan Guerber

《珊迪故事情节》是一个多媒体的参与纪录片，记载那些经历了飓风珊迪的人们的经历以及随后的救援工作。许多志愿者前来帮助这个项目，并且尽可能多的记录故事。这个项目的目标是捕捉记录自然灾害，允许那些真正经历过的人创建叙事（而不是外人、新闻媒体和一个小的专业团队），并在他们的社区的重建过程中，给予社区成员提供一些机构。

"飓风珊迪"应该为182人的死亡负责，并且估计导致650亿美元的损失，这让它成为美国历史上第二大海岸飓风。风暴潮在2012年10月29日袭击整个纽约，给新泽西州和沿东海岸造成了严重的破坏。直至今天，救援工作仍在进行中。

《珊迪故事情节》项目在三个部分存在：第一种是大量的网站，其提供书面记录的故事（包括音频，视频和照片）。所有的故事都在第一时间告诉给人。第二，珊迪故事情节已经在纽约组织了活动，来分享故事和信息，以及进一步促进桑迪幸存者的聚会。第三，为青少年和成人进行教育项目（包括研讨会和媒体培训），这些已经协助与包容被主流媒体边缘化的声音，尤其是年轻人的声音。雷切尔·法尔科内和迈克尔普莱默成立项目，法尔克的经历十年口述历史文件、无线电／多媒体广播制作的产品，普莱默的经验是作为一个摄影师、记者、纪录片讲故事的人，并且社区组织者帮助塑造了这个工程。

《珊迪故事情节》是一个开放的互动节目。任何人都可以通过拨打免费电话或 SMS 或 MMS 来提交自己关于珊迪的故事。手机图片和音频／视频记录也受到了欢迎。对于那些没有获得技术支持的人，通过记者或多媒体制作，可以接受记者采访。由于在风暴后许多地区停电很长一段时间，志愿记者和生产者的宣传工作使我们能够分享许多人的故事，并记录并处理他们的经验。为了这一天，故事仍然继续收集。该项目完全由参与者塑造，并且已经放大了社区的声音，以及建立一个叙事文，优先考虑他们自己的经验，需求和欲望。它帮助这些在风暴中损失最严重的社区，促进的包容和民主重建。

此外，它帮助解决珊迪的破坏留下的根深蒂固的问题，包括气候变化、可持续发展和经济不平等。综上所述，对于公共艺术来说，珊迪故事情节是一个革命性的方法，因为它通过使用技术和故事分享，创造了公民对话。这个项目的成功，源于它捕捉自然灾害的社会叙事，并给予这些社区在关于他们自己的未来的发言权。易于接受的在线形式的访问允许许多人来贡献，以及可以很容易的信息传播。主动的教育，允许那些平时没有政治发言权的人（包括青年和那些没有途径获得技术者）学习多媒体技能，贡献自己的思想，记忆和观点。最终，这个项目唯一地捕捉到了人类在面对飓风时候的一面，一个容易被历史遗忘的一面。它把人们引导在一起，并继续通过活动，展览，研讨会和讲座这样做。

Sandy Storyline is a multimedia participatory documentary chronicling the experiences of those who lived through Hurricane Sandy and the relief efforts that followed. Many volunteers have come together to assist with the project and document as many stories as possible. The goal of the project is to capture the narrative of a natural disaster, to allow those that actually experienced it to create that narrative (rather than outsiders, news outlets and a small team of professionals) and to give community members some agency in the rebuilding of their neighborhoods. Hurricane Sandy was responsible for the deaths of 182 people and is estimated to have caused $65 billion in damages, making it the second–costliest hurricane in US history. The storm surge hit New York City on October 29, 2012 and caused severe destruction throughout New York, New Jersey and along the Eastern seaboard. Relief efforts are still underway today.

The Sandy Storyline project exists in three parts. The first is the extensive website, which provides written and recorded stories (including audio, video and photography). All stories are told in the first—person. Second, Sandy Storyline has organized events and exhibitions throughout New York to share stories and information as well as to facilitate further gatherings for Sandy survivors. Third, educational programs (including workshops and media training) for both youths and adults have assisted with the inclusion of voices often marginalized by the mainstream media, particularly those of young adults. Rachel Falcone and Michael Premo founded the project. Falcone's experience with oral history documentation and radio/multimedia broadcast production as well as Premo's experience as a photographer, journalist, documentary storyteller and community organizer helped to shape the project.

Sandy Storyline is an open and interactive program. Anyone is welcome to submit their own Sandy story by calling the toll-free number or sending a SMS or MMS message. Cell phone pictures and audio/video recordings are also welcomed. For those without access to technology, an interview may be scheduled through the project with a journalist or multimedia producer. Since many areas were without power for an extended period of time following the storm, the outreach efforts of volunteer journalists and producers made it possible for many to share, record and process their experiences. To this day, stories continue to be collected. This project is shaped entirely by its participants and has helped to amplify community voices as well as to construct a narrative that prioritizes their own experiences, needs and wants. It has assisted in facilitating an inclusive and democratic reconstruction of those communities hit worst by the storm.

In addition, it has helped to address problems entrenched within Sandy's destruction, including climate change, sustainability and economic inequality. In summary, Sandy Storyline is a revolutionary approach to public art due to its creation of civic dialogue via the use of technology and story sharing. The project is successful because it captures the community narrative of a natural disaster and has given these communities a say in their own future. The accessibility of the online format has allowed for many to contribute as well as for an easy dissemination of information. The education initiatives have allowed for those normally without a political voice (including youths and those without access to technology) to learn multimedia skills and to contribute their thoughts, memories and viewpoints. Finally, the project has uniquely captured the human side of Hurricane Sandy, a side easily forgotten by history. It has brought people together and continues to do so through events, exhibitions, seminars and talks.

[解读]

天地不仁，以万物为刍狗

当个人独自面对巨大的自然灾难，我们被上帝之手操控，而只能默默承受瞬间便来临的摧毁性力量。我们的命运犹如风中柳絮，水中浮萍。而在灾难之后，我们如何去面对他，面对这种可怕的力量，面对死亡以及面对我们的创伤。人之所以为人，是因为人可以彼此交流分享，人是社会性的。《珊迪故事情节》便是人们面对灾难的一种自救，它给了公众一个平台，让缄默的个体发声，让彼此能够听到，能够理解，能够懂得。灾难对人类来说是再可怕不过的事情，但是灾难中人们的团结，以及灾难之后人们彼此分享的温暖，也让我们可以在某种程度上超越灾难本身。人们的故事会留下，而灾难本身会逐渐远去，活着的人们会继续生活下去，而伤痛会被存留在一个可以被容纳的地方。（李田）

音乐盒：棚户区声音实验室
The Music Box: A Shantytown Sound Laboratory

艺术家：多位艺术家

Artist：Multiple, Project led by New Orleans Airlift

地点：美国路易斯安那州新奥尔良

Location：New Orlean / LA, USA

时间：2011—2012 年

Date：2011—2012

推荐人：梅根·古尔伯

Researcher：Megan Guerber

路易斯安那州新奥尔良市是音乐、艺术和文化混合的城市。当它在 2005 年遭受飓风卡特里娜时，整个城市的生命和房屋全部被摧毁。虽然重建工作帮助解决了许多破旧物业，但是多年之后，仍然可以很容易找到飓风卡特里娜破坏的遗迹。

2008 年，飓风卡特里娜过去之后，新奥尔良空运倡议成立，用来支持艺术家奋力重建自己。由艺术家管理员和音乐家杰伊·彭宁顿，安装多媒体艺术家马丁·德莱尼成立，该项目旨在支持地下的生活民俗和新奥尔良的街头艺术文化。根据他们网站描述，通过促进本地项目的国际旅游以及当地的。民族的艺术节和国界艺术家之间的合作，新奥尔良空运寻求"为了共同的强大目标，学习和团结完全不同的群体"。

2011 年，马丁和边宁顿购买了一处房产，带有一个折叠的 18 世纪的平房，位于第 9 区水岸部分，建设者发现了巨大的潜能，材料再利用，生活的焕然一新,以及玩弄于股掌的房产。从音乐和新奥尔良社区的强有力的创造性来汲取灵感,《音乐盒:棚户区声音实验室》就这样诞生。

该项目包含,用从他们新买的房产附近打捞上来的材料建成的临时安装的"音乐架构"。超过25个艺术家协助创作这个音乐剧的结构,此外,在性能阶段(2011年10月22日—2012年12月10日),还有70多个音乐家被邀请来表演。

这类演唱会是售门票的,因为空间有限并且还有一些大牌来演唱,否则《音乐盒:棚户区声音实验室》会向公众开放,以探索和玩乐和互动。此外,一周有4次的免费游览机会,直到2012年1月底。在短暂的安装阶段,超过15000参观者驻足,包括一些当地居民、艺术爱好者。游客和500多名学生包括《心跳屋》在内的奇形怪状的音乐建筑,一个窝棚——像一个在屋顶装有扬声器,另一端接到一个听诊器上的结构。通过把听诊器拿到心脏附近,参观者可以听到自己的心跳旋律。《回音壁》是把弹簧和钢薄片附着于铜水暖管件所构造出来。"噪音地板"特邀嘉宾用脚创作音乐,《瞭望塔》结合了一个带由从教堂管风琴上找到的管的螺旋楼梯。

这个点的计划还继续。是一个大项目的一部分,《音乐盒:棚户区声音实验室》作为Dithyrambalina的试运行,而Dithyrambalina则是一个永久性的音乐结构。此外,新奥尔良空运的总部将设在由布鲁克林设计的大厦里,基于街头艺术家swoon,对先前的居住地,他贡献了小麦糊艺术。目前,《音乐盒:棚户区声音实验室》结构以"流浪的村庄"形式存在,可以穿越整个新奥尔良和美国,分享音乐盒和新奥尔良艺术社区的俏皮灵感。

总之,《音乐盒:棚户区声音实验室》是一个非常成功的项目,给新奥尔良沿着水岸的地区和艺术团体都带来了希望。通过再利用回收的材料和识别建筑史,重要的新奥尔良文化和历史被保存在毫无生气的,财产快速发展的新时代。除此之外,在该项目中,街头和民间等地下艺术团体被赋予了新生活和合法性。该项目也提供给了改革、希望和玩乐的空间,告诉当地居民,飓风卡特里娜并没有带走所有的东西,越是大的社区,重新利用的可能性越大。

New Orleans, Louisiana is a city of music, art and blended cultures. When it was hit by Hurricane Katrina in 2005, lives and homes were destroyed throughout the city. While rebuilding efforts helped to fix up many blighted properties, years later the remains of Katrina's destruction could still readily be found.

In 2008 the New Orleans Airlift initiative was founded to support artists struggling to re-establish themselves after Hurricane Katrina. Founded by artist manager and musician, Jay Pennington, and installation and multi-media artist, Delaney Martin, the program seeks to support the underground, living folk and street art cultures of New Orleans. According to their website, by facilitating international travel for local projects as well as collaborations between local, national and foreign artists, New Orleans Airlift seeks to "empower learning and unite disparate groups in common and powerful goals".

In 2011 Martin and Pennington purchased a property on which sat a collapsed 18th-century cottage. Situated in the Bywater section of the 9th Ward, the founders saw great potential for material reuse, life renewal and play within the property. Drawing inspiration from the music and potent creative energies of the New Orleans community, The Music Box: The Shantytown Sound Laboratory was born. The project consisted of a temporary installation of "musical architecture" built using salvaged materials from the fallen cottage on their recently purchased property. Over 25 artists assisted in the creation of the musical structures and another 70+ musicians were invited to perform within the space during a performance series (October 22, 2011—December 10, 2012).

While tickets were sold for the concerts due to limited space and the inclusion of some big names, the Music Box space was otherwise open to the public for exploration, play and interaction. In addition, free tours were conducted four times a week through the end of January 2012. During its temporary installation, over 15,000 visitors stopped by, including local residents, art lovers, tourists and 500+ students. The whimsical musical buildings included "The Heartbeat House", a shack--like structure with a speaker perched on its roof and attached at the other end to a stethoscope. By holding the stethoscope to their heats, visitors could project the rhythm of their own heartbeats. "Echo Wall" was constructed out of springs and steel sheeting and attached to a copper plumbing pipe. "Noise Floor" invited guests to create music with their feet and "Lookout Tower" combined a salvaged spiral staircase with pipes from a found church organ.

Plans continue for the site. Part of a larger project, The Music Box served as a trial run for Dithyrambalina, a permanent settlement of musical structures. In addition, New Orleans Airlift's headquarters will be housed in a building designed by Brooklyn-based street artist Swoon who contributed wheat paste art to the previous settlement. Currently The Music Box's structures are living on as a "roving village" that will travel throughout New Orleans and the United States to share the playful inspiration of the Music Box and New Orleans art community.

Overall, The Music Box was a very successful project that brought a lot of hope to both the Bywater neighborhood and art community of New Orleans. By repurposing salvaged materials and recognizable pieces of architecture, important New Orleans culture and history was preserved in an era of lifeless, fast new property developments. In addition, the street, folk and underground art communities have been given new life and legitimacy within the project. The project also provided a space for innovation, hope and play, showing the local community that Hurricane Katrina did not take everything away and the larger community the potential of creative reuse.

[解读]

音乐乌托邦

每个城市也许都有一个伤痛，飓风袭击下的新奥尔良即使在重建之后仍隐约存在着一些破损的痕迹，而《音乐盒：棚户区声音实验室》项目让城市的伤口里开出一朵花来。"音乐盒"这个名字让我们想起精致的工艺品里旋转的芭蕾女孩，而异曲同工之妙的是，艺术家通过《音乐盒：棚户区声音实验室》构建了一个又一个独特的乌托邦——独立的音乐建筑，在里面人们可以聆听音乐、感受音乐、创造音乐，形成一个整体的独特的属于新奥尔良自由俏皮的氛围。《音乐盒：棚户区声音实验室》从新奥尔良慢慢走向其他的区域，街头和地下音乐人参与其中，他们建立了一个"流浪的村庄"，去歌唱、传达飓风所没有摧毁的新奥尔良的希望和美好。艺术有时候产生于苦难，当然也可以将人们拯救于苦难，欢乐和反思在这里聚集，也在这里消散，进入到每个参与者的心中。（李田）

全城画布
All City Canvas

艺术家：罗伯特·清水、
　　　　贡萨洛·阿尔弗斯、
　　　　维克多·雨果·塞拉亚
地点：墨西哥城
形式和材料：壁画
时间：2012 年
推荐人：杰西卡·费亚拉

Artist：Robert Shimizu
　　　　Gonzalo Alverez
　　　　Victor Hugo Celaya
Location：Mexico City
Media/Type：Mural (Painting)
Date：2012
Researcher：Jessica Fiala

2012 年 4 月 30 日至 5 月 5 日，墨西哥城举办了《全城画布》（ACC）活动，一项包含墙壁、文字和作品三类元素的街头艺术节。活动吸引了艺术家、学者、街头艺术拥护者，以及公众的支持，通过会议讨论、壁画、画廊展示的形式，该艺术节成为都市艺术艺术市场的一部分。

壁画起自 Paseo de la Reforma，一条横穿墨西哥城中心的大动脉，它是本活动最显眼、最恒久的影响因素。8 位享誉国际的街头艺术家受邀创作，使用临时支架和臂式云梯在高大建筑的墙壁上创作多层艺术作品。涂鸦/街头艺术通常在废弃和衰败的地点，而本活动为其提供在精致房产地段进行专题创作，从游客和旅人经常光顾的奢华 W 酒店，到墨西哥最久远报纸的摩天大楼中心，再到坐落于拉古尼利亚的巴拉圭大楼，此处自前西班牙时代便有人居住。地点经过精心挑选，既覆盖全城，又将壁画置于重要历史文化或人流多的地点。

活动时间为一周，据估计将有 22,000 名狂热爱好者。支持者及将前来观看由 Interesni Kazki（乌克兰）、El Mac（美国）、Saner（墨西哥）、Sego（墨西哥）、Roa（比利时）、Herakut（德国）、Vhils（葡萄牙）和 Escif（西班牙）创作的壁画。然而，组织者的目的并不单单是吸引众多观看者。活动选址的部分目标和策略是将街头艺术展示给从未接触它们的观众。活动组织者之一的罗伯特·清水在接受《布鲁克林街头艺术》的采访时，对这些被视为破坏性行为（包括在奢靡的 W 酒店进行街头艺术）作出了评论，对清水而言，这项活动反对定性，那种"任何人都将无法想象……都市艺术家在他们经常吃中饭或参加会议的地方创作超大壁画"的想法。对他而言，这项活动期望拓宽通道、拓展观众，以"向世界宣布，都市艺术不仅仅是年轻人居住或靠近的地方——它是人人的都市"。

地点是本活动的一个至关重要的因素——将通常在边缘地带的行为带到城市中心，并且通过扩大规模和选址，为以往被摒弃或视作犯罪的艺术行为展示、凸显和支持。从这一层面上讲，两者关系是相互的，一方面墨西哥城宏大的建筑以其巨大的影响力支持街头艺术，另一方面艺术和艺术家为活动地点乃至整个城市带来活力、个性和视觉力量。

此次将街头艺术拓展到全新领域，招致了一系列争论。一些艺术家称赞在墨西哥城作壁画，鉴于其以往强大的壁画历史。然而另一些人却与涂鸦历史和政府支持的壁画作品划清界限。随着街头艺术越来越普遍、引人注意，一股潜在的紧张关系也随之出现，且面对着不可计数的文化行为。历史传承的次文化或反文化的立场与行为与市场化风格之间的界限为何？市场化行为在拓展其影响范围，与它的起源已渐行渐远。在墨西哥城，《全城画布》为这些话题提供了空间和舞台。

2012 年活动之后，视频短片被公布出来，它记录了艺术节和 ACC 组织的一次全球系列活动，世界各个范围内的个人艺术家都表达了对壁画的支持。ACC 现在正开展调查、组织活动、建立视频博客和社交媒体帖子，报道国际的街头艺术场景。

From April 30—May 5, 2012, Mexico City hosted All City Canvas (ACC), a major street art festival organized in 3 components—Walls, Words, and Works. Events gathered artists, academics, street art supporters, and the public around conferences and discussions (Words), major large-scale murals (Walls), and a subsequent a gallery exhibit to foster art markets for urban art (Works).

The murals, oriented around Paseo de la Reforma, a major artery that cuts through the center of Mexico City, are certainly the most visible and lasting impact of the events. Eight internationally renowned street artists were invited to create the pieces, using scaffolding and boom lifts to paint multi-story artworks across the walls of major buildings. Often relegated to abandoned or decrepit sites, here graffiti / street art was given a featured home on coveted real estate. Buildings ranged from the luxurious W Hotel frequented by tourists and business travelers to the skyscraper hub of Mexico's oldest newspaper to the Paraguay Building located in the La Lagunilla neighborhood, inhabited since pre-Hispanic eras. Locations were chosen with care, to both cover an expanse of the city and to locate the murals in historic, culturally significant, or highly frequented sites.

Over the course of the week, it is estimated that 22,000 aficionados, supporters, and passersby gathered to watch murals being created by Interesni Kazki (Ukraine), El Mac (USA), Saner (Mexico), Sego (Mexico), Roa (Belgium), Herakut (Germany), Vhils (Portugal) and Escif (Spain). Organizers, however, did not seek merely to reach a large number of viewers. Part of the goal and strategy for site selection involved bringing street art to new audiences. Robert Shimizu, one of the organizers of All City Canvas along with Gonzalo Alverez and Victor Hugo Celaya, commented on the disruptive act of including street art at the "posh" W Hotel in an interview with Brooklyn Street Art. For Shimizu, the project counters assumptions, the idea that "nobody could have imagined…urban artists painting a huge mural on the same terrace where they usually eat their lunch or have their business meetings." For him, the project sought to expand access and audiences, to "make a statement to the world, that urban art is not only for young people that live in and around big cities—it's for everybody—doctors, politicians, business people, Moms, merchants…"

Location is a significant component of the project—bringing practices often relegated to the margins into the center of the city and, through scale and site, giving a presence, visibility, and sanction to art practices that have historically been dismissed or associated with criminality. The relationship is in this regard two-way, Mexico City's significant buildings lend some of their power to support street art while the art and artists provide a dynamic new character and visual strength to neighborhoods across the city.

This expansion of street art into new territory brings out an array of debates. Some artists spoke of the honor of painting murals in Mexico City, given the strong history of muralism in Mexico. Others, however, drew a sharp distinction between the resistant history of graffiti and government supported mural works. As street art becomes more widespread, and more visible, a potential tension arises that has faced innumerable cultural practices. What is the boundary between a historically grounded subcultural or counter-cultural stance or practice and a marketable style that, in expanding its reach, becomes disconnected from the impetus that spawned it? In Mexico City, All

City Canvas has created a space and a stage for these conversations to unfold.

Following the events in 2012, a short film was released documenting the festival and ACC organized a Global Series, supporting murals by individual artists in cities around the world. ACC now conducts research, organizes events, and creates video blogs and social media posts covering the international street art scene.

[解读]
"下里巴人"的逆袭

街头涂鸦壁画一直作为"下里巴人"似的存在，处于艺术的边缘区域，壁画也大多绘在一些偏僻之处。与美术馆、博物馆中精美的艺术品不同，壁画处于露天区域，甚至街头闹市之中。而此次的全城画布，艺术家将街头艺术带入城市中心，甚至高档酒店、历史悠久的建筑中，尝试将街头艺术拓展到全新领域，为以往被摒弃或视作犯罪的艺术行为寻找新的展示空间和支持。随着街头艺术越来越普遍，历史传承的次文化与市场化风格之间的界限越来越模糊，市场化行为与它的起源已渐行渐远。

　　"下里巴人"的街头涂鸦艺术，成功地拓宽了艺术的通道，拉近了艺术与大众的距离。《全城画布》为艺术和大众提供了对话的空间和舞台，同时为整个城市带来了新的视觉力量。（闫丽祥）

Metamorphosoup:
年度万圣节室外木偶表演

艺术家：Bare Bones 电影制作公司
地点：美国明尼苏达州明尼阿波里斯
形式和材料：表演
时间：1993 年至今
推荐人：杰西卡·费亚拉

Artist：BareBones Productions
Location：Minneapolis, MN, USA
Media/Type： Performing Arts
Date：1993-Present
Researcher：Jessica Fiala

美国万圣节已演变成一项重大商业活动，这是一个关于化妆、糖果、恐怖和鬼魂的节日。但是 10 月底也标志着时光流逝、季节变迁，从生机勃勃的收获季到冬天的死亡。正是在这最后的时光里，位于明尼阿波里斯的 BareBones 电影制作公司推出了《年度万圣节室外木偶表演》，它既给社区带来盛大表演，同时追忆和祭奠亡魂。

每年，活动准备工作都要在社区集思广益。选择一个主题之后，社区成员受邀制作木偶，并排练参加表演。参与形式各不相同，每个作品需 250 多个参与者，其中包括音乐人、专业演员、社区参与者、建筑师、技术工、志愿者及组织者。为符合节日和季节，主题围绕死亡和蜕变，但是也倾向于家庭和温暖。演出，是幽默、恐怖与灵魂的结合，正如共同导演马克萨福德所说，尽管是表现死亡的主题，主旨仍然是积极向上的，"在死亡密集的秋季庆祝生命的轮回"。

所有观众请到大型公园后，表演小插曲、艺术设施和演员们开启晚会活动，并开通通往表演场地的多条路线。压轴表演有大型木偶戏、喷火仪器、踩高跷、音乐、空中舞蹈和社区木偶戏。而 2014 年的 Metamorphosoup，其最初灵感来自一个有关石头汤的民间传说，石头汤强调了合作、分享和社区活动。故事中，饥饿的旅行者到达一个村镇寻找食物。当地居民不愿意分享他们的食物，旅行者们在河边搭建营地，在锅中装满水和石头。村里人经过时询问他们在干什么，他们得到的回答是他们在做"石头汤"，此汤还需一些东西才可以做好。每个村民都愿意献出一点食材，最后，凑成了一锅营养美味的汤，每个人都可以分享。对于他们的作品，BareBones 将故事与原生汤结合，它是宇宙初生时生命构成的基础元素。就这样，每年一个新主题带来了想象和表演的机会，有诡异的、超俗的和社区庆祝的。尽管灵感每年都换，一个年年重复的核心活动是请观众大声说出去年死去的挚爱之人的名字。通过这一公开又私密的互动艺术环节，参加者有机会追忆和悼念逝者。

表演结束后，将邀请观众欣赏现场音乐、装置艺术和由当地一家免费有机食物发放中心柯莱特姐妹提供的免费食物。

自 1993 年第一次之后，一年一度的活动已由 100 名观众、5 名演员的表演发展到每年吸引 6000 名参加者。表演已然成为当地社区的一项常年而独特的主要活动。BareBones 仍然是非盈利组织，《年度万圣节室外木偶表演》成为商业气息浓重的万圣节期间的另一更受欢迎的免费之选。过去 20 年里，表演培养了新秀，为社区意见提供平台，并直面通常避而不提的死亡话题，而重点却是重生、感恩和蜕变。除了秋季表演之外，组织在一年之中推出了其他活动，包括室外表演、冬季游行和使用回收利用材料制作木偶大赛等。

Halloween in the U.S. has grown into a major commercial event, a festival of costumes, candy, horror, and ghouls. But the end of October also marks a time of passing, a seasonal transition from lively harvest to the symbolic death marked by winter. It is in this latter vein that Minneapolis-based BareBones Productions developed The Annual Halloween Outdoor Puppet Extravaganza, both a community-created spectacle and a time for reflection and honoring the dead.

Each year, preparations begin with a community brainstorm. After a theme is chosen, community members are invited to work crafting puppets and practicing to participate in the performance. Involvement varies, with every production gathering over 250 participants ranging from musicians and professional performers to community participants, builders, technicians, volunteers, and organizers. In-line with the season, themes center around death and transition, yet there is also an orientation towards families and accessibility. The performances, described by Co-Director Mark Safford, are a combination of elements ranging from humorous to spooky and spiritual. Although traversing topics of death, the tenor remains positive, "honor[ing] the circle of life by celebrating its seasonal arc of death in the Fall." Welcoming audiences to a large public park, performative vignettes, art installations, and actors launch the evening's events and line a series of paths to the performance grounds. The main spectacle weaves together large-scale puppetry, flaming apparatuses, stilt performance, music, aerial dance, and community-based puppetry. For the 2014 iteration, Metamorphosoup, the base inspiration came from a folktale about stone soup that emphasizes cooperation, sharing, and community-oriented action. In the story, hungry travelers arrive in a village seeking food. When residents are unwilling to share their provisions, the travelers set up camp near a river and fill their cooking pot with water and a stone. As villagers pass and inquire about the traveler's activities, they are each told that they are preparing "stone soup," which just requires a bit more of something before it is done. Each villager is willing to part with some small ingredient and, in the end, there is a nourishing soup to be enjoyed by everyone. For their production, BareBones interwove this story with the idea of the primordial soup, the rudimentary elements that combined to form life at the dawn of the universe. In this manner, a new theme each year provides opportunities to explore imagery and scenes that vary from eerie and otherworldly to celebrations of community. While inspirations change annually, a core recurring moment is an invitation to audience members to announce the names of loved ones who have died in the past year. Through this public recognition, and private interactive artistic installations, visitors are offered an opportunity for personal reflection and remembrance.

Following the performance itself, audiences are invited to enjoy live music, art installations, and free food provided by Sisters' Camelot, a locally based free organic food distribution collective.

Since beginning in 1993, the annual event has grown from one performance for a crowd of 100 to five performances, drawing over 6,000 attendees each year. The performance has become a reliable and unique staple in the local community. BareBones remains a non-profit organization and the Annual Halloween Outdoor Puppet Extravaganza offers a welcoming free alternative to the heavily commercial activities surrounding the Halloween holiday. Over the past two decades, the production has cultivated emerging talent, provided a forum for community voices, and offered an opportunity to engage directly with the often-avoided topic of death, with a focus on renewal, appreciation, and transition. In addition to their fall performance, the organization creates other projects throughout the year, ranging from outdoor performances to Winter Pageants and puppetry competitions based in reused and repurposed materials.

[解读]

生存还是毁灭

"生存还是毁灭,这是一个值得考虑的问题"。当数百年前的哈姆雷特在生存和死亡面前思考徘徊时,已经完成了灵魂的蜕变,破茧成蝶,将生与死上升到了艺术的高度。一年一度的万圣节除了关于化妆、糖果、恐怖和鬼魂外,更是寓意着从生机勃勃的收获季到冬天的死亡,节日的目的不是为了庆祝死亡,而是完成死亡后的蜕变、灵魂的重生。由 Bare Bones 电影制作公司策划的《年度万圣节室外木偶表演》,围绕死亡的主题,完成灵魂的蜕变,"在死亡密集的秋季庆祝生命的轮回"。每次活动的核心便是请观众大声说出去年死去的挚爱之人的名字,追忆和悼念逝者。死亡是恐怖的,但是每一次死亡都伴随着另一种重生,正如秋天枯萎飘落的树叶,来年却更加碧绿生机盎然的出现在枝头,"落红不是无情物,化作春泥更护花"。(闫丽祥)

多尔切斯特项目
Dorchester Projects

艺术家：西斯特·盖茨
地点：美国芝加哥伊利诺斯州
形式和材料：空间艺术
时间：2006年至今
推荐人：杰西卡·费亚拉

Artist：Theaster Gates
Location：Chicago, Illinois, USA
Media/Type： Space Art
Date：2006-present
Researcher：Jessica Fiala

对于一个社区而言，要阻止其经济的慢慢衰退是件难事。一旦生意倒闭，家园空置，废弃的地产将很快陷入绝望，并成为犯罪的目标和场所。当艺术家西斯特·盖茨观察到在芝加哥南区的多尔切斯特社区发生的这一切时，他采取了积极的措施对抗这一被视作不可挽救的衰败势头。

首先，盖茨于2006年搬入这一社区，他当时的想法是通过他自己的存在和房屋维修建立一个可行的"剩余效应"。然而全国范围内，房产市场崩盘带来的经济困境和影响正在加剧，且渐渐地社区的住户越来越少。为了阻止大批人的离开，盖茨设想通过将空置的房产变成艺术和社区场所来重振社区。他购买并重建自己房屋周围的房屋，利用回收和废弃的材料，包括北区保龄球馆的木头。然后，他从最近倒闭的当地书店里买回14,000册艺术和建筑书籍，以及由芝加哥大学艺术史学院捐赠的60,000册存档的玻璃幻灯片。

在这些最初的努力成功之后，在附近房屋进行的额外项目继续展开，他额外注意到社区中大量非洲裔美国人的存在。盖茨建立黑人影院，向他们提供筛选、电影课及讨论。开放精神食粮、提供晚餐一起吃饭、学习非裔美国人历史、欣赏表演并谈论艺术、社区和文化的机会。一家当地音乐店倒闭后，他将自己房屋的一部分变成音乐小屋，里面放有成千上万张唱片。

该项目目前已覆盖了6处房产，且还在继续扩大。活动和工作受到津贴和捐助的支持，但是盖茨也发展另外的方式来资助和维持进一步的发展，并支撑当地经济发展。他最近购买了一家废弃的银行，它的大理石将进行再利用，变成艺术品进行出售，以帮助该建筑的修整。这是目前最大的项目，银行用石板铺盖，里面存放了John H. Johnson——*Ebony-Jet*杂志的创办者的永久图书馆，以及一家餐厅、社区房间和表演、展示场地。一家筹建中的当地砖厂将为一系列多尔切斯特项目提供材料，当地居民被雇佣来进行工作。进一步的扩张计划是要对36个单元的房屋建筑群进行重修，为当地居民和新兴艺术家建成混合收入房屋，配有一个供表演、展示和其他活动的社区艺术中心。

2010年，盖茨成立重建基金来支持多尔切斯特社区的项目，并进一步扩大到圣路易斯和奥马哈的类似项目。

除了特别的建筑项目和相关项目外，设想中还有一种独特的社区重振方式，盖茨问道："除了建造40000平方英尺的文化中心或基督教青年会或全新的大型图书馆外，我们难道不能利用现有的房屋而做出更好、更深思熟虑的决策吗？"这样的思维方式比将机构加盖到已衰败的城市中或修缮已有建筑寻求的利益更多。对盖茨而言，考虑一栋建筑能发挥的作用和社区意义是十分必要的。在这方面，他工作的重心是重修和项目、创造形体和功能兼具的空间。空间和社区因此互相关联，空间建造和社会艺术行为相互因而彼此加强、共享和支持。

It can be difficult for a neighborhood to resist the slow degradation of an economic downturn. Once businesses fail and homes are emptied, abandoned properties can soon fall into disrepair and become objects and havens for crime. When artist Theaster Gates saw this progression beginning in Chicago's South Side Dorchester community, he sought to take an active stance against what might have been considered an inevitable downslide.

Gates began by moving himself into the neighborhood in 2006 with the idea in mind of creating a possible "residual effect" through his own presence and property maintenance. Nation-wide, however, economic hardship and the impact of the housing market crash intensified, and one by one, neighborhood properties were vacated. To counter this exodus, Gates envisioned revitalizing the neighborhood by turning empty properties into arts and community spaces. He purchased and restored the home adjacent to his own, using recycled and salvaged materials when possible, including wood from a North Side bowling alley. He then installed 14,000 art and architecture books from a recently closed local bookstore and a collection of 60,000 archived glass lantern slides donated by the University of Chicago's Art History Department.

Following the success of this initial endeavor, additional projects in nearby homes ensued, with an added focus on celebrating the large African-American presence in the neighborhood. Gates created a Black Cinema House to offer screenings, film classes, and discussions. Open Soul Food dinners became an opportunity to share a meal, learn about African-American history, enjoy performances, and engage in conversations about art, community, and culture. After a local music store went out of business, he turned part of his own home into a Listening House stocked with thousands of records.

Now covering half a dozen properties, the project continues to expand. Events and work have been supported by grants and donations, but Gates is also developing alternative means to finance and sustain continued growth, and to bolster the local economy. He recently purchased an abandoned bank whose marble will be repurposed, turned into artworks to be sold to help fund the building's renovation. The largest project to date, the bank is slated to house the permanent library of John H. Johnson, founder of Ebony and Jet magazines, as well as a restaurant, community rooms, and performance and exhibition spaces. A planned local brick factory will furnish building supplies for a variety of Dorchester Projects and local residents have been hired to carry out the work. A further expansion is schedule to renovate a 36-unit housing complex to create mixed-income housing for both local residents and emerging artists, replete with a community arts center for performances, exhibitions, and other programming.

In 2010, Gates established the Rebuild Foundation to support efforts in the Dorchester neighborhood and has since expanded its scope to include similar work in St. Louis and Omaha.

Beyond specific building initiatives and related projects, what is envisioned is a different way of approaching neighborhood revitalization. Gates asks: "Instead of building 40,000-square-foot cultural centers or large YMCAs or big new libraries, couldn't we use some of the existing housing stock and make better, more thoughtful incremental decisions?" This mode of thinking seeks to do more than stamp institutions onto blighted portions of city maps, or refurbish existing structures. For Gates, it is necessary to consider the roles and community purposes that a building can serve. In this vein, his work is concerned with renovation and programming, creating spaces of possibility both physically and functionally. Place and community are herein interconnected, with placemaking and social practice art each serving to enhance, inform, and support the other.

[解读]
重塑希望

"哀莫大于心死。"对于一个人而言,一旦丧失希望、心灰意冷,那便很快陷入绝望,无药可救。对于一个社区同样如此,一旦生意倒闭、家园空置、废弃的社区将很快陷入绝望,并成为犯罪的目标和场所。为了挽救这股衰败的势头,艺术家西斯特·盖茨购买并重建空置的房屋,建造精神寄托的书店,建立黑人影,开放精神食粮,重新利用废弃的社区,使之焕发出新的生机和活力。对于盖茨而言,如何发挥一栋建筑的作用和社区意义成为他关心的重点,为此他做出诸多多努力,建立空间和社区的联系,同时空间建造和社会艺术行为相辅相成,带给人们新的希望。"从善如登,从恶如崩",将濒临破灭边缘的社区重新纳入正途,遏止不可挽救的颓废势头,如登山般艰难困苦,但值得庆幸的是,我们有着愿意为之努力奋斗的人,这也正是希望所在。(闫丽祥)

在新奥尔良等待戈多
Waiting for Godot in New Orleans

艺术家：陈保罗
地点：美国新奥尔良
形式和材料：表演艺术
时间：2007 年
推荐人：杰西卡·费亚拉

Artist：Paul Chan
Location：New Orleans, USA
Media/Type：Performance Art
Date：2007
Researcher：Jessica Fiala

2005 年卡特里娜飓风的破坏之后，新奥尔良的居民面临着一种更漫长的灾难：政府管理不当、残留物、做作的重建、沮丧以及等待……再等待……还是等待……艺术家陈保罗在 2006 年第一次到达这里，回忆那段经历时说："那里的街道模样依旧，就好像时间把房子也卷走了一般我的朋友说这座城市看起来就像科幻电影中荒凉的背景……我意识到它不像电影场景，而像一部我看过多次的电影舞台。没错。那空旷的街道，光秃的树木……沉寂。"这位艺术家随后想出在这座荒废的城市演出塞缪尔贝克特有关无为和萧条的荒诞剧——《等待戈多》。

与制作商创意时光和哈莱姆古典剧院合作，该项目发展出更多的演出。陈保罗在城市里待了 9 个月，与艺术家、活动分子和组织者一起工作，与社区联系发展演出。该项目受到社区对话和倾听会、中等圆桌会议和百乐餐的巩固和启发，以哈莱姆古典剧院为社区人群和高中学生提供表演工作坊。

更深程度的参与是该项目成功的重要因素。在卡特里娜灾难之后,新奥尔良居民了解到了未完成的梦想并参观艺术项目,这些项目并未损毁社区表面一丝一毫,且对地面毫无影响。当地居民为在新奥尔良等待戈多提供支持,并组建网络参与进来。Ninth Ward 居民 Robert Lynn Green 讲述了这部戏与他自身以及邻居面临的境况和经历的相关性。他们不仅要共同等待——等待联邦应急管理局的援助或红十字会迟来的约定——还要面对戏剧主角面临的困境,在知道援助可能不会到来后是否还要继续等待或是放弃,心里明白如果一旦离开,将失去一切。因此,该项目在主题和合作伙伴将自己置身到社区两者之间找到了共鸣,寻找投入和洞察力,并推出一系列免费活动调动当地居民。

当地特有的表演在两个社区中展开——Ninth Ward 的交叉路口以及 Gentilly 的一所废弃房屋前院。四场免费演出开始前都有秋葵汤晚餐和乐队。陈保罗本人主要关注幕后组织工作,与哈莱姆古典剧院紧密合作,戏剧由该剧院推出和表演。说到该项目的内在联系,他不仅强调创作一部戏剧与人分享,而且强调了产生一种环境和公众。在此情况下,作品为那些与贝克特的戏剧有共鸣之人所表演。他们是投入观众,陈保罗称他们为"一群分裂、疲惫然而依旧等待事情到来的公众"。

这些活动本身之外,陈保罗努力向社区提供直接支持。与创意时光合作,他建立了"影子基金"来为当地社区收集捐款,捐款金额近 50 000 美元,然后分成每份 1 000 到 5 000 美元给当地组织进行重修,并在 2010 年,出版了一本附有图片、筹划文件和论文的创意时光项目书籍。

Following the immediate destruction of Hurricane Katrina in 2005, residents of New Orleans were confronted with a prolonged, slow disaster, characterized by governmental mismanagement, lingering debris, stilted rebuilding, frustration, and waiting...waiting...waiting. Artist Paul Chan first visited the city in 2006 and recalled the experience: "The streets were still, as if time had been swept away along with houses...Friends said the city now looks like the backdrop for a bleak science fiction movie...I realized it didn't look like a movie set, but the stage for a play I have seen many times. It was unmistakable. The empty road. The bare tree...The silence." It was then that the artist developed the concept of staging Samuel Beckett's absurd play of inaction and stagnation, Waiting for Godot, in the devastated city.

Partnering with the producer Creative Time and The Classical Theatre of Harlem, the project developed into much more than a series of performances. Chan spent 9 months in the city working with artists, activists, and organizers to both connect with communities and develop the performance. The project was enhanced and informed by community dialogues and listening sessions, moderated roundtables, and potluck dinners. In conjunction with the main project, Chan taught public contemporary art seminars and The Classical Theatre of Harlem offered acting workshops for community groups and high school students.

This deeper level of involvement was integral to the project's success. In the aftermath of Katrina, New Orleans residents became familiar with unfulfilled promises and visiting art projects that barely scratched the surface of the community and had little impact on the ground. Local residents provided vital support to Waiting for Godot in New Orleans and rallied their networks to get involved. Ninth Ward resident Robert Lynn Green, Sr., spoke to the relevant nature of the play for the conditions and experiences faced by himself and his neighbors. They dealt not only with the common experience of waiting—waiting for assistance from the Federal Emergency Management Agency or for delayed appointments with the Red Cross—but also the conundrum encountered by the play's protagonists, whether to continue waiting knowing that help may not come, or give up, knowing that one could lose out completely if one leaves. The project therefore struck a chord in both subject matter and in the collaborating partners' commitment to embedding themselves in the community, reaching out for input and insights, and developing a variety of free activities to engage local residents.

The site-specific performances were staged in two neighborhoods—the middle of an intersection in the Ninth Ward and the front yard of an abandoned house in Gentilly. Each of the four free performances began with a dinner of gumbo and a band to march the audience to the performance site. Chan himself focused primarily on the behind-the-scenes organizing work, partnering closely with The Classical Theatre of Harlem who staged and performed the play. Speaking of the cohesive nature of the project, he emphasized the importance not just of creating a performance to share, but of producing an environment and a public. In this context, the work is performed for individuals whose experiences are echoed in Beckett's spectacle. They are an invested audience, as Chan describes, "a public that is incredibly divided, and tired, and waiting, still, for things to come."

Beyond the events themselves, Chan sought to provide direct support to the community. Working with Creative Time, he established a "Shadow Fund" to gather donations for the host communities, raising nearly $50,000 that was allocated as $1,000-$5,000 donations to local organizations working to rebuild. In 2010, a book was released about the project by Creative Time, with photographs, planning documents, and essays.

[解读]

戈多到来

1952年,塞缪尔·贝克特的《等待戈多》以戏剧化的荒诞手法揭示了世界的荒谬丑恶、混乱无序的现实,写出了在这样一个可怕的生存环境中人生的痛苦与不幸。剧中代表人类生存在一个凄凉而恐怖的背景,处于孤立无援、恐惧幻灭、生死不能、痛苦绝望的境地。在今天,这个现实世界,卡特里娜飓风过后的新奥尔良的居民,同样面临着这种荒诞的境况:空旷的街道,光秃的树木……沉寂,政府管理不当、沮丧以及等待……再等待……还是等待……人们明知援助可能不会到来,但还是继续等待,因为他们心里明白,如果一旦离开,将失去一切。值得庆幸的是,他们等待的戈多,却以另一种方式慢慢到来。艺术家陈保罗与哈莱姆古典剧院紧密合作,推出《在新奥尔良等待戈多》剧场表演,成功引起人们的共鸣,唤起政府的注意,慢慢迎来了久违的"戈多"。(闫丽祥)

葛兰西纪念碑
Gramsci Monument

艺术家：托马斯·赫塞豪恩
地点：美国纽约
形式和材料：装置艺术
时间：2013年
推荐人：里奥·谭

Artist：Thomas Hirschhorn
Location：New York，American
Media/Type：Installation Art
Date：2013
Researcher：Leon Tan

葛兰西纪念碑是瑞士艺术家托马斯·赫塞豪恩纪念碑系列中的第四个，也是最后一个工程。正如名称所表明的那样，它是奉献给意大利马克思主义理论家——安东尼奥·葛兰西。就像其他三个用同名的人命名的纪念碑那样，斯宾诺莎（阿姆斯特丹，1999），德勒兹（阿维尼翁，2000）和巴达伊（卡塞尔，2002），葛兰西也许是以一个"次要的"（使用德勒兹名义非主流的）思想者为特征的作品。葛兰西在意大利共产党中起着重要作用，于1926年被法西斯政府逮捕，直到1937年死于脑出血之前一直被囚禁在艰难的环境中。他因发现了了文化霸权的概念而被人们怀念，这个概念解释了对不同社会群体盛行的权利分配，表面上"自发的"赞成。

赫西霍恩的公共作品是受迪亚艺术基金会委任，作为一个在"森林房屋"的临时装置。迪亚艺术基金会是一个在纽约布朗克斯的房屋委员会发展机构。按照艺术家的思想，它的目的是"建立一个纪念碑的定义，驱使遭遇战，创造一个活动，思考今天的葛兰西"。这个纪念碑由一系列的室外展览馆组成，包括一个来自于葛兰西基金会研究所的照片展览，一个由葛兰西创作和关于葛兰西的藏书，一个网络空间，一个休息室和一个酒吧，大部分由当地的居民建造和经营。葛兰西纪念碑也包含了讲座和活动，从艺术讲习班到哲学和诗歌讲座，再到自由言论会议以及葛兰西剧院的戏剧。赫西霍恩同时为这个工程创造了一个网站，提供来自于公共项目的记录材料以及关于艺术品的说明。

赫西霍恩的作品不同于典型的纪念碑，他经常使用不同于平常的材料建造。"我在我周围得到的东西都是一些包装材料；厨房里面有一些铝的金属薄片，楼下的街上的一些纸箱和木板。对我有意义的是：我使用的材料，没有精神与灵魂的力量。它们就是普通的材料，是些这个世界上每个人都熟悉的材料。" 赫西霍恩这样解释自己的审美选择。

因此，葛兰西纪念碑有一个简易的外表，我们也许会把这称为自己动手做的审美观。作为一个定义，它表明了把一个纪念碑看成是一个制造回忆的过程，或者通过抓住葛兰西的生活思想，对社会共同体的一个激活方式，而不是仅仅作为主要代表思想家生活的盲目循环。它批判了那种作为一个文化典型产物，意图胁迫观众盲目接受一个胜利者叙述的传统纪念碑，在某种程度上，它却等同于那种具有葛兰西思想的文化霸权的传统纪念碑。

今天想起葛兰西意味着去发现他在集体生活中所标志的概念的相关性。它意味着对进行集体主义想法和行动的条件进行试验。至关重要的是，这种尝试必须要在公民中进行，因为对葛兰西来说公民社会是一个建立和争夺文化霸权的王国。葛兰西纪念碑提醒我们对争夺文化霸权批判性思维的重要性，以及非法西斯主义生活的重要性。与此同时，它给我们留下了这一思想活动在根本上是不确定的议题。它或许会被倡导，但永远不会如传统纪念碑所暗示的那样被永恒的捕捉。

Gramsci Monument (2013) is the fourth and last project in the Swiss artist Thomas Hirschhorn's Monuments series. As the title suggests, it is dedicated to the Italian Marxist theorist and dissident, Antonio Gramsci. Like the namesakes of the other three monuments, Spinoza (Amsterdam, 1999), Deleuze (Avignon, 2000) and Bataille (Kassel, 2002), Gramsci might be characterized as a "minor" (non-mainstream, to use Deleuze's term) thinker. He played an important role in the Italian communist party, was arrested in 1926 by the fascist government, and imprisoned in harsh conditions for a decade until his death in 1937 from a brain hemorrhage. He is remembered for developing the concept of cultural hegemony to explain the apparently "spontaneous" consent of different social groups to prevailing power arrangements.

Hirschhorn's public work was commissioned by the DIA Art Foundation as a temporary installation at "Forest Houses", a Housing Authority development in The Bronx, New York. According to the artist, its purpose was "to establish a definition of monument, to provoke encounters, to create an event, and to think Gramsci today". The monument consisted of a series of outdoor pavilions, including an exhibition of photographs from Fondazione Instituto Gramsci, a library of books by and about Gramsci, an Internet space, a lounge and a bar, all built and largely run by local residents. GramsciMonument also comprised a public program of lectures and activities, from art workshops, to philosophy and poetry lectures, to open-mic sessions and a play called Gramsci Theatre. A website was created for the project, providing documentary materials from the public program as well as notes on the artwork. (It is now archived on the Dia Foundation site—http://www.diaart.org/gramscimonument/index.php).

Unlike typical monuments, Hirschhorn's work is often constructed out of everyday materials. Explaining his aesthetic choices, the artist explains, What I've got around me is some packing material; there's some aluminum foil in the kitchen and there are cardboard boxes and wood panels downstairs on the street. That makes sense to me: I use the materials aroundme. These materials have no energetic or spiritual power. They're materials that everyone in the world is familiarwith; they're ordinary materials.

Consequently, Gramsci Monument had a makeshift appearance; we might call this a DIY aesthetics. As a definition, it suggests that a monument might be better conceived as an memory-making process, an activation of the community through engagement with Gramsci's living thought, rather than as a mindless recycling of the major representations of a thinker's life. It critiques the traditional monument as a cultural artifact intended to intimidate audiences into blind acceptance of a victor's narrative; in a way, it equates the traditional monument with the cultural hegemony that occupied a great portion of Gramsci's thought and life.

Thinking Gramsci today means to discover the relevance of the concepts he signed in collective (communal) life. It means to experiment with the conditions for collective thought and action. Crucially, such experimentation must take place in civil society, since for Gramsci civil society was the realm in which cultural hegemony was both established and contested. Gramsci Monument reminds us of the importance of critical thought to the contestation of cultural hegemony, and thus to non-fascist life. At the same time, it leaves us with the proposition that the event of thought is fundamentally precarious. It may be nurtured and encouraged, but can never be captured for eternity as traditional monuments suggest.

[解读]

献给大众的纪念碑

不同于以往庄严肃穆的英雄式纪念碑，艺术家托马斯·赫塞豪创作的《葛兰西纪念碑》仅有一个简易的外表，这件向意大利共产主义思想家安东尼奥·葛兰西的献礼，以室外场馆的形式，由艺术家和15人的本地居民小组共同构建，自成一体，犹如一个机构。艺术家抓住葛兰西的生活思想，通过自己动手做的审美观，使纪念碑成为一个制造回忆的过程。赫赛豪恩在讲解自己的作品时说，"纪念碑并不是为艺术爱好者或专业人士而定制"，而是为了"让住在周围的邻居们来一同欣赏"。艺术家让每一名参与这个项目的人都将自己当成作者，为他们的成果负全责。通过艺术创造出对话和理解的可能，同时激励所有人产生巨大的能量，这也正是艺术家的初衷。（闫丽祥）

案例研究
非洲

CASE STUDY
Africa

Looking into the Future	展望未来
Invasion	侵略
An Installation of Animal Forms from Disused Bicycles	由废弃自行车组成的动物装置
Breeze across Borders	穿越国界的清风
Nakivubo Channel	通道
El Mattam El Mish Masry	马塔姆米什马斯丽餐厅
Bakaboza Campaign	Bakaboza 竞选活动
Anthea Moys VS The City of Grahamstown	安西娅·莫瓦斯 vs 格雷厄姆斯敦
Dlala Indima	发挥你的作用
Burning Museum	燃烧的博物馆
Sermon on the Train	火车上的布道

展望未来
Looking into the Future

艺术家：亚当·麦德比
地点：津巴布韦
形式和材料：雕塑
时间：1985 年
推荐人：拉斐尔·奇古瓦

Artist : Adam Madebe
Location : Republic of Zimbabwe
Media/Type : Sculpture
Date : 1985
Researcher : Raphael Chikukwa

《展望未来》是一座有名的雕塑品，它创作于 20 世纪 80 年代早期，并于 1985 年完成。C.H. Naak (Pvt) Ltd R. 和 G. Gwelo 将其购买到布拉瓦约城，并安置在布拉瓦约城的议会大厦。这尊高四米的裸身男性雕像展示的是一个身无分文的年轻人对未来的展望。在 1985 年的早期，这个作品曾在布拉瓦约城引起过争论，因为这尊雕像一开始是展示在市政机构外面的，因其赢得了某项竞赛，然而之后这尊雕像又被挪到津巴布韦庭院的国家美术馆。但在此之后，其 2010 年又被挪动，现在放置在津巴布韦储藏所的国家美术馆。而如今津巴布韦美术馆正与当局合作想把这尊雕像重新挪回庭院，如其他雕像一样作为美术馆的中心装饰物。

麦德比生于 1954 年津巴布韦市，1999 年他搬到南非，并在这里生活工作，如今他供职于南非娱乐集团。他是津巴布韦最富盛名的金属雕刻家，并以和事物大小相符或者更大的艺术作品赢得了很多奖项。《坠落的勇士》是一尊引人注目的钢铁雕塑品，如今坐落于津巴布韦。它制造于 1989 年，在津巴布韦市的 Pachipamwe 工作室，并由 Triangle Arts Trust 组织。在 Pachipamwe 工作室，麦德比与英国艺术家 Sokari Douglas Camp 和 Chris Ofili 共同工作过。

麦德比是在 Mzilikazi 艺术中心和布拉瓦约艺术学校接受的培训教育。就是在 Mzilikazi 的艺术课程中，他第一次对雕塑产生了浓厚的兴趣，最开始他学习的是模型雕像和动物黏土工艺。对于创造比黏土更大型作品的愿望促使寻找新的工艺材质媒介。在焊接车间里，他灵光一闪，意识到可以通过金属材料焊接，并通过弯曲和捶打使其成为自己想要的形状来做艺术品。

当在创作新雕像比如《坠落的勇士》时，麦德比先制造一个黏土模型，接着再用大约 1.6—2mm 厚的金属板将其包裹，并将钢板融化焊接为这个形状。当作品完成后，他再将遗留在里面的无用粘土清理出来。并且这些雕塑从不涂防护层，Madebe 认为腐蚀生锈的过程也是艺术的一部分。

Madebe 在多个国家展出过他的作品，包括英国、博茨瓦纳、荷兰、挪威、瑞典、美国和南非等国家。

Looking into the Future is a famous sculpture created in early 80s and completed in 1985. C.H. Naak (Pvt) Ltd R. Gentile and G. Gwelo purchased it for the City of Bulawayo; and it was installed at Bulawayo City Council Tower Block. This four-meter-high statue of a nude male represents a young man looking to his future with no possessions. Early 1985 the piece caused controversy in Bulawayo, where it was initially displayed outside the municipal offices after winning a competition but later moved to the National Gallery of Zimbabwe courtyard. While in the National Gallery of Zimbabwe courtyard it was then removed again in 2010 and it now lying down in the National Gallery of Zimbabwe storage. The National Gallery of Zimbabwe is working with the authorities trying to get it back in the gallery courtyard where it became a centerpiece of the Gallery like many other sculptures.

Profile: Born in 1954 in Zimbabwe and currently living and working in South Africa where he moved in 1999 and today he works under the Recreation Africa group. He is Zimbabwe's most famous sculptor in metal and has won many awards for his works, which are often life-size or larger. Fallen Warrior is a striking steel sculpture, which appeared within the Ground Force Africa Garden. It was made in 1989 at the Pachipamwe workshop in Zimbabwe, organised by the Triangle Arts Trust. During the Pachipamwe Workshop Madebe worked along side British artist that includes, Sokari Douglas Camp, and Chris Ofili.

Madebe trained at Mzilikazi Art Centre and Bulawayo School of Art. It was during his art course at Mzilikazi that he first became interested in sculpture, initially learning to model figurines and animals out of clay. A desire to create larger works than clay would allow inspired his search for a new medium. While attending a workshop in welding, he struck upon the idea of joining metal off-cuts and bending or hammering them into shape.

When creating a new sculpture such as "The Fallen" Warrior, Madebe first creates a clay model. Next he fits metal sheets about 1.6 to 2 mm in thickness around it and melts and welds them into shape. When the piece is finished he removes the clay leaving the sculpture hollow inside. The sculptures are never painted: Madebe sees the rusting process as a part of the art.

Madebe has exhibited his work in many different countries, that include, Britain, Botswana, Holland, Norway, Sweden, America, and South Africa to mention but a few.

[解读]

属于未来的少年

这个体量惊人的雕塑让我们想起文艺复兴时的雕塑巨匠——米开朗琪罗，他的代表雕塑《大卫》远比真人更为高大，让人感受到一种高度所带来的宏伟与震撼。Madebe 的《展望未来》雕塑身高 4 米，如果米开朗琪罗的《大卫》是捕捉到少年极为紧张，同时身体在动静交错一瞬间的爆发力，那么《展望未来》则更为轻松，他仰着头，一手遮挡住灿烂的阳光，宛如新生。这个少年似乎刚刚诞生或者清醒过来，带着无限的迷茫和欢欣，像被上帝刚刚创造出来的亚当一般。不加处理的金属表面留下了时间流逝的痕迹，似乎我们通过对金属质感变化的观察就能看到他的年龄和历史。正如雕塑的名字一般，《展望未来》其实是对这个国家未来的一种期待，就如这的少年一般，沐浴着阳光，从晨光中慢慢醒来。（李田）

侵略
Invasion

艺术家:瑟奇·奥力维耶·弗科瓦
地点:喀麦隆
形式和材料:装置艺术
时间:2008年
推荐人:拉斐尔·奇古瓦

Artist : Serge Olivier Fokoua
Location : Cameroon
Media/Type : Installation
Date : 2008
Researcher : Raphael Chikukwa

通过这个《侵略》项目,奥力维耶论述的是一个在当时备受喀麦隆关注的议题。据奥力维耶所说,喀麦隆受到侵略;它需要的是最高级别的国家防护安全。奥力维耶认为,喀麦隆这个在以往被称为"安全岛"的天堂,正处于受到外界威胁的境地。恐怖组织和黑帮派时不时地闯入国家三角区进行野蛮破坏活动。这个侵略项目不仅仅干涉到喀麦隆国家的安全问题,其在更大范围里也涉及 AU。

瑟奇·奥力维耶·弗科瓦1976年出生于杜阿拉市。如今生活并工作于喀麦隆的雅温得。他在由驻喀麦隆西班牙大使馆组织的工作室里形成了他的视觉艺术理念,并于德国的汉堡完成了文化管理的进阶培训。其主要艺术作品形式是装置和行为,参与了很多喀麦隆、尼日利亚、南非、法国、德国、日本、加拿大、芬兰等国家的会展和项目。他最近还获得了美国佛蒙特州工作中心(Vermont Studio Center)的居留许可权。作为 Palettes of Kamer 集团的创办成员,他自2008年担任了雅温得 Rencontres d'Arts Visuels(称作 RAVY)的董事,并于2006年成为国家互联网节日组织——IC ZONE 的成员。

The Invasion project Olivier addresses an issue that concerns Cameroon at the moment. According to him, Cameroon has been invaded; it needs highest level of state security. Olivier believes Cameroon, which has always been considered a safe haven, to be within the scope of external threat today. Terrorist organizations and gangs regularly burst into the national triangle to commit barbaric acts. Invasion project is an intervention to question not only Cameroonian state security but also AU at large.

Artist profile: Serge Olivier Fokoua was born in 1976 in Douala. He lives and works in Yaoundé (Cameroon). Formed in visual art at workshops organized by the Spanish Embassy in Cameroon, he has also completed advanced training in cultural management in Hamburg, Germany. Artist working mainly on installation and performance, Fokoua participated in many exhibitions and projects in Cameroon, Nigeria, South Africa, France, Germany, Japan, Canada and Finland. He has recently obtained a grant of residence for the Vermont Studio Center in the USA. Co-founding member of the group Palettes of Kamer, he is since 2008 the Director of Rencontres d'Arts Visuels in Yaoundé called RAVY. He is a member of the international network of festivals, IC ZONE since 2006.

[解读]

位卑不敢忘忧国

艺术作品有些表达美好与希冀，有些揭露哀痛和黑暗，有些是艺术家个人的表达，有些则忧国忧民。艺术家创作作品时，不论是独善其身或者兼济天下，当他的作品有了观众的时候，他的想法和气息便有了接收者。Fokoua 的装置作品《侵略》，便表达了他对喀麦隆国家安全问题的担忧，是一系列的条形组合而成的三角形的房子，房子顶面牵满了红色的线状物，整个装置看似坚固，但是可以从外面透过缝隙隐约看到里面，而一排点状物从这个房子里横穿过去，一直延展到展厅的墙壁上。而这个简易的装置代表着喀麦隆，它看似完整而坚固，却受到隐约的威胁，红色的屋顶好像是一种危险的预警，同时也让人们想到了喀麦隆国徽中央的红色金字塔，是一种团结的象征。艺术家对祖国的忧虑和热爱让我们想起喀麦隆的国歌，你在野蛮时代度过童年，但现代你就像太阳在升起，一步一步脱离蛮荒永远向前……你是我们的欢乐和生命，光荣和爱非你莫属。（李田）

由废弃自行车组成的动物装置
An Installation of Animal Forms from Disused Bicycles

艺术家：迪洛普瑞莱克
地点：德国杜塞尔多夫
形式和材料：装置
时间：2005 年
推荐人：拉斐尔·奇古瓦

Artist：Dil Humphrey-Umezulike aka Dilomprizulike
Location：Dusseldorf Germany
Media/Type：Installation
Date：2005
Researcher：Raphael Chikukwa

迪洛普瑞莱克1960年生于尼日利亚埃努古，是一名致力于雕塑、行为和绘画的当代艺术家，并有"来自非洲的拾荒者"的绰号。他取得了位于尼日利亚Nsukka尼日利亚大学的美术学学士学位，并取得了英国顿提大学的美术学硕士学位。在他的艺术作品中，他会回收和使用一些旧衣物和在街上捡到的碎片零件来创造雕塑类装置，以此来反映许多非洲人们被剥夺公民权利后的生存现状。他在拉各斯创办了"丑陋笨拙物品的废旧品博物馆"，这是由拾捡到的废品构建而成的雕塑环境，并于威尔士的兰迪德诺为Mostyn美术馆也设置了这样的展出。

Dilomprizulike依照以下几点对他的装置类艺术品——Wear and Tear做解释：Wear and Tear是一个试着将被忽视或者低估的非洲城市公共生活元素展示出的理念。他身处的非洲隔离社会变得很悲哀。一种在非洲实施文明的意识始终在他的内心做抗争。他既不能回归过去拾捡起父辈褴褛的文化碎片,也不能跟得上白人世界的文化步伐。

在 2005 年,他的作品在伦敦的海奥德美术馆中展出。他也在煤气厂画廊做展出,同时在为维多利亚和阿尔伯特博物馆准备作品。在 2010 年,他为以色列的赫兹里亚当代美术馆创造了《闹市》这一作品,它由当地的垃圾做成,反映的是以色列的消费社会状况。

Dil Humphrey-Umezulike aka Dilomprizulike (born 1960 in Enugu, Nigeria) is a contemporary artist who works in sculpture, performance and painting and has adopted the moniker "The Junkman From Afrika". He got a BA. (Hons) in Fine Arts from the University of Nigeria, Nsukka, Nigeria and has an MFA from the University of Dundee, Scotland. In his work he recycles and transforms heaps of old clothing and other debris found on city streets, creating sculptural installations and performances that reflect the disenfranchised situation of many African people. He created the "Junkyard Museum of Awkward Things", a sculptural environment built up from found objects, in Lagos and has also made a version of it for the Oriel Mostyn Gallery in Llandudno, Wales.

Dilomprizulike explains his installation Wear and Tear in these terms: Wear and Tear as a concept attempts to expose the often overlooked and underrated elements of the African-Urban communal life which largely influence it. The alienated situation of the African in his own society becomes tragic. There is a struggle inside him, a consciousness of living with the complications of an imposed civilization. He can no longer go back to pick up the fragments of his father's shattered culture; neither is he equipped enough to keep pace with the white-man's world.

In 2005 his work was exhibited in Africa Remix at London's Hayward Gallery. He was also a resident at the Gasworks Gallery while he prepared a piece for the Victoria and Albert Museum. In 2010 he created Busy Street for the Herzliya Museum of Contemporary Art in Israel. The piece, which was made from local rubbish, was a reflection on Israel's consumer society.

[解读]

夹缝里的艺术

废品在最终被丢掉之前首先是产品，是消费品，然后变成人们生活的一部分，最终离开人们的生活变成废弃物。废品曾经参与和影响过人们的生活，生活垃圾其实是社会消费和文化的一个侧面反映，废品一定程度可以表现出非洲人们的生活现状和挣扎。当我们看到改装成动物的废弃自行车，他们像一群站立的兽，在荒草丛生的原野上，他们属于城市？不，他们早已被抛弃。他们属于自然？不，他们从被生产出来的一刻就远离了他。艺术家迪洛普瑞莱克将被忽视的非洲生活元素展示出来，他其实展示了一种在南非长期存在的种族认同缺失，有这样一群人们游离在被强行打碎的父辈文化与占统治地位的白人世界文化之间，他们的生活现状令人堪忧。他的作品不是优雅的，甚至是丑陋的，由垃圾组成装置作品将这个普遍存在但是并非主流的群体表现出来。（李田）

穿越国界的清风
Breeze across Borders

艺术家：布雷兹·约克	Artist : Breeze Yoko
地点：多个国家	Location : Multi
形式和材料：壁画	Media/Type : Mural
时间：2014 年	Date : 2014
推荐人：拉斐尔·奇古瓦	Researcher : Raphael Chikukwa

生于南非的布雷兹是一个电影制作人和视觉艺术家，在几年里他调查并记录了南非的青年文化包括街头涂鸦。2014 年他在津布巴韦，Njelele 艺术站的 Afropolicity 开办展出。他也对非洲的其他地区很有兴趣，并且他的作品在博茨瓦纳、莫桑比克、德国、塞内加尔、肯尼亚、马里、瑞士以及津布巴韦等国家参与展出。在上面提到这些国家的展出，布雷兹参与的是街头艺术项目。他称自己为多媒体艺术家，就是可以用最好的方式来做自己想做的任何事情。布雷兹的作品是基于幻想主义和现实主义，富有力量的一类。用他自己的话就是："我的作品是我们所处环境的反映，也是人道和阴暗方面的体现。对我而言，艺术可以代表很多事情，其中有一些是颇具质疑色彩的以吸引其他人的注意，当然也会给予人希望、信念、梦想的机会。"布雷兹从不热衷于美术馆里面的创作，大街上的墙壁就是他的美术馆，并用公共空间将艺术带给大家。他的主要作品之一——《比科的孩子们》为他摘得了许多国际大奖。

布雷兹在参加"看不见的国境线横跨大陆"公路旅行开始于津布巴韦。带着染料，布雷兹从尼日利亚开始了自己的旅行并于波斯尼亚结束。布雷兹的大陆公路旅行使他得以在超过 9 个非洲国家和欧洲国家使用公告空间创作街头艺术。排除万难他们跨过 5 个民族分界并遇到多次来自入境关员的挑战。在这个旅行中，他们与几位当地的的艺术家合作，以便在使用公共空间问题上做下调停，互相交流并了解当地的文化。他们游历过的国家包括：尼日利亚、贝宁、多哥、加纳、象牙海岸、塞内加尔、毛里塔尼亚、西撒哈拉、摩洛哥、西班牙、法国、比利时、荷兰、德国、捷克共和国、奥地利、匈牙利、克罗地亚。

Born in South Africa, Breeze is a filmmaker, and a visual artist, and over the years he has 2014 he exhibited at the Afropolicity at the Njelele Art Station in Zimbabwe. He has also taken interest in other parts of Africa, and his work has been exhibited in Botswana, Mozambique, Germany, Senegal, Kenya, Mali, Switzerland and Zimbabwe to mention but a few. In above-mentioned countries Breeze has participated in street art projects. He calls himself a multimedia artist using the approach that works best for whatever he wants to do. Breeze's work is powerful but based on fantasy, and realism. In his own words he says, "My work is a projection of where we are and would like to be in terms of humanity and blackness. For me art is a lot of things and part of that is questioning things and bringing them to other people's attention but also providing hope, faith and a chance to dream". Breeze never keen on galleries and street walls are his galleries, taking art to the people using public spaces. One of his major project Biko's Children won him international awards.

ATTACHMENT A: Artwork Description—Breeze Across Borders Breeze Yoko's year started in Zimbabwe before taking part in the Invisible Borders Trans—Continental road trip. Armed with his paint, the journey kicked off in Nigeria and ended in Bosnia October 2014. Breeze Across Borders allowed him do some of these interventions in public spaces in more than nine African Countries and in Europe. Against all odds they crossed five borders confronted several challenges with immigration officers. During this tour they worked with several local artists in order to intervene in public spaces, exchange and to get an understanding of local cultures. The countries they have covered include, Nigeria, Benin, Togo, Ghana, Ivory Coast, Mali, Senegal, Mauritania, Western Sahara, Morocco, Spain, France, Belgium, Netherlands, Germany, Czech Republic, Austria, Hungary, Croatia.

[解读]

流浪的墙壁

有些艺术品保存在美术馆和博物馆里；有些艺术品融入公众，成为城市的一部分；还有一些艺术家的作品不仅影响当地，而且是流浪着的。Breeze 的《Breeze across Borders》就穿越了超过 9 个非洲和欧洲国家，他的作品所到之处，展现了一种文化融合的街头艺术。他的作品像春风一样拂过诸多不同的民族和地区，而这些不同地区的墙壁上则像被春风呼唤的生灵一样在他的画笔下呈现出不同的特色与精神。有些艺术是大众的，它可能就存在在一个不经意的转角，带给当地的人们不同于自己文化的气息，而这种穿越边境的艺术活动也带给了艺术家本人更多的灵感和与自己国家和民族完全不同的思维方式。艺术是没有国界的，它是所有民族的人们共通的属于世界的语言，它是自由的，流浪着的。（李田）

通道
Nakivubo Channel

艺术家：萨姆森
地点：乌干达坎帕拉
时间：2014 年
推荐人：拉斐尔·奇古瓦

Artist : Samson Ssenkaaba aka Xenson
Location : Kampala, Uganda
Date : 2014
Researcher : Raphael Chikukwa

萨姆森于 1999 年毕业于玛格丽特·特劳尔学校工业和美术专业，和 Makerere 大学的平面设计和绘画专业。从那以后他的作品就在乌干达和国外的多次展会和时尚表演会上展出。最近，他在坎帕拉的 AfriArt 美术馆和埃明帕夏酒店以及毛里求斯的 Maritime 酒店举办了个人展会。萨姆森曾参与了 2012 年和 2014 年在 MishMash 美术馆举办的 KLA 艺术展会，坎帕拉当代艺术展会。

萨姆森的 KLA ART 012 项目是一个对倾泻现象多个层面的视觉和概念上的分析。在一个层面上来说，如标题所说的，这个项目论述的是发生在坎帕拉的倾泻过程，在这个城市城市的大部分污水排放到 Nakivubo 通道，这个通道已成为污染维多利亚湖的多个水道之一。在社区开始承担他们对环境的责任时，他向布干达反映了这一情况，作为 bulungi bwansi 社区工作。

如今乱丢垃圾已成为坎帕拉的一个普遍而又不健康的生活习惯，而这逐渐显现的不良影响很少引起人们的注意。通过引起大家对这些问题的关注,这一项目使人们认识到倾泻垃圾的危险并强调社会废物管理的重要性。"如今我们可以发现在非洲国家的许多城市都有大规模二手产品的倾泻现象。乍一看,这似乎给人以慈善和支付能力的假象。

"然而,这似乎是将他们不需要的物品从'西方'经济国家倾泻到'第三世界'经济体的周密计划。这些物品的使用期限都很短暂,他们会很快的成为了废物"。大多数二手产品都是通过海运集装箱运来的,而KLA ART 012 也将使用这些海运集装箱作为展览空间。将两个海运集装箱叠放在一起,萨姆森在这里堆满了塑料和不可降解废物,像海洋一样,从两个箱子里倾泻而下,堆到地面上。

萨姆森的项目是在全球市场范围内对霸权关系的有力参考。这引起了对慈善潜在双重性的关注和对当代消费主义文化的质疑。

He graduated in 1999 from the Margaret Trowell School of Industrial and Fine Arts, Makerere University majoring in graphic design and painting. Since then, his works have been shown in numerous exhibitions and fashion shows in Uganda and abroad. Most recently, he held solo exhibitions at the Emin Pasha Hotel and AfriArt Gallery in Kampala and at the Maritime Hotel in Mauritius. Xenson has taken part in KLA ART 2012 AND 2014. In 2010 Xenson exhibited at MishMash Gallery, Kampala Contemporary Art 2012 and 2014

Xenson's KLA ART 012 project is a visual and conceptual analysis of the dumping phenomenon on a variety of levels. On one level, as indicated by the title, the project looks at dumping processes that take place in Kampala where the majority of the city's waste ends up in the Nakivubo channel, one of the waterways polluting Lake Victoria. He reflects on a time in the Buganda Kingdom when the community assumed responsibility for their environment, acted out in bulungi bwansi community works.

Today littering has become a common and unhealthy habit in Kampala and the pending gross repercussions are seldom taken into consideration. By drawing attention to this problem, the project creates awareness of the dangers of dumping and addresses the importance of society's waste management. "Today we are seeing a huge dumping phenomenon of second hand products in many African cities. At first sight this gives the false impression of charity and affordability.

However, there seems to be a deliberate initiative from 'western' economies to dump what is no longer needed into 'third world' economies. The life span of these products is very short and they soon end up as waste." Most of these second hand goods come by sea in the shipping containers that are used as exhibition spaces for KLA ART 012. Making use of two of these containers, one on top of the other, Xenson creates an overwhelming sea of plastics and non-biodegradable waste that pours from the double story height onto the ground below.

Xenson's project is a formidable reference to hegemonic power relations within the global market. It draws attention to the potential double sided character of charity and challenges the contemporary culture of consumerism.

[解读]

"海洋"

我们应当如何保护我们的环境，才能不被生活垃圾污染；我们应当如何保护我们的河流，才能不被生活废物淹没；我们应当如何保护我们的国家，才能不会成为发达强权国家阴谋的受害者；我们应当如何保护我们的人民，才能不会从此远离清洁、从此必须与垃圾和过剩的廉价物品为伍。艺术家 Xenson 的装置作品展示了这种廉价的倾销，和对第三世界国家命运的担心。也许弱肉强食是从远古时期就传达在基因里必然，西方在大航海和工业革命之后的迅速崛起，让他们早早就可以掌握世界的命脉并且倾轧第三世界国家，而如今文明发展到如此程度，人们都更期待建立一个真正自由、平等、博爱的世界，并且为了填补彼此巨大的差距不懈追寻，那个众生平等的乌托邦到底存在吗？这也许正是人们苦苦寻找努力的原因吧。（李田）

马塔姆米什马斯丽餐厅
El Mattam El Mish Masry

艺术家：Asunćion Molinos Gordos
地点：埃及
形式和材料：空间、表演、装置

时间：2014 年
推荐人：纳赫拉·阿尔·塔瓦

Artist : Asunćion Molinos Gordos
Location : Egypt
Media/Type : Space, Performative, Installation

Date : 2014
Researcher : Nahla Al Tabbaa

《马塔姆米什马斯丽餐厅》是家快餐店，位于众多非正式居民区的其中一个，它得名于一段特殊时期，在那时埃及人民无休止地相互询问对所谓国家事业的贡献。随着开罗市民的增加这些非正式的居民区，亦可称作"贫民区"在此涌现。人口急剧增长、国家经济危机以及环境卫生设施的欠缺促使这些居民区在没有规划许可的情况下驻扎在农用耕地上，这里不再长出健康的食物，取而代之的是生活垃圾以及污水。

艺术家也在努力地向人们强调这样的事实——99% 的埃及人得不到健康的食物供给，因为当地生产的优质食物都会立即出口向欧洲，仅留下微乎其微的部分加工食用。

这家餐馆特以每周一换的频率推出四类不同菜系，这基于埃及社会不同的花销预算等级。第一周烹制的佳肴是面向一小部分的顶级社会埃及人民的，那时一位在开罗最高档酒店供职的米其林明星大厨会亲自掌勺，全部用的是最优质进口食材。随着星期的逐个推移,饭菜会越来越廉价,并聘用各种级别厨师来烹制这些他们熟悉的饭菜。最后就不得不选用"当地产出的农作物和资源",这道菜反应的则是被占用的耕地土壤里所"生长"出的任何东西。

艺术家在不再耕作的土地上寻找蔬菜等农作物来烹制饭菜,却除了烟头、垃圾、被污染的土壤以及污染物外什么都没有找到。然而这些东西被开玩笑似地端上了餐桌并提供给了大众。

Gordos邀请了社区的很多厨师来帮她做这些菜,而餐馆本身则被装饰为典型的流行餐馆,这可以使人们认识到可贵的一点,那就是这块土地在失去它的农用潜能。并且她惊喜地发现她的听众不仅局限于"艺术社区"的人们,她还吸引了居住在这个居民区里的整个社区人民。

她还有意给人们提供一个平台或者出口,在这里人们可以自由讨论关于莫尔西执政时期埃及政治状况的"现实生活"话题,不管是大街小巷,还是茶余饭后。在莫尔西政府以新"道德家"理念为方向专注于埃及生活时,这个艺术品(餐馆)将以往被掩盖的,最基本的日常实际事务反映在所有埃及人面前。

然而,对于促进以往大众对话改变的方式以增强意识,Gordos感到很失望。当局政府开始以关闭这家餐馆作为威胁手段,然而由于预算经费问题,这家餐馆不能更多地设立在这座城市的其他位置,Gordos曾设想这座城市可以萌发出更多的自由言论。

虽然如此,这一意义重大的艺术作品,在很大程度上是伪装成了一个社会平台,力图利用这个媒介使用其传达的信息有力地解放广大听众。它很醒目大胆地驻扎在待解决问题的中心地带,而不是局限于博物馆里。它也不必在身上标贴什么艺术话语,因为这会使埃及公民难以理解领会。它有大规模的参与人员,并具有恰如其分的说明性和直观性,尤其是当最后一道菜端给顾客的时候。

El Mattam El Mish Masry ("The restaurant that is not Egyptian") was a pop up restaurant situated in one of Cairo's many unofficial settlements, named during a time when Egyptians were constantly questioning each other's dedication to a so called national cause. These unofficial settlements, or "slums", have been growing exponentially as Cairo's population has increased. This rapid increase, together with the country's economic crisis, and the lack of established sanitation practices, had led to these settlements being built with no planning permissions on agricultural lands, supplanting healthy food produce with residential rubbish and sewage.

The artist also sought to highlight the fact that healthy produce has been made unavailable to 99% of the Egyptian population, as most of the premium locally produced food is immediately exported to Europe, leaving behind very little to work with.

The restaurant featured 4 categories of different meals served on a weekly basis based on the budgets of different classes of Egyptian society. The first week featured meals that a tiny portion of upper class Egyptians could access, and a Michelin star chef working in one of Cairo's most luxurious hotels was commissioned to create them, using only the finest imported ingredients. As the weeks go by, the meals gradually become cheaper and cheaper to produce, recruiting different cooks familiar with preparing these meals. Finally, having to rely on 'local resources and produce', the menu started to reflect whatever was available from the local soil on the appropriated farmland.

As the artist explores these no longer cultivated lands in search of vegetable produce for the meals, she finds nothing but cigarette butts, rubbish, contaminated soil and pollution. Playfully, these found items are prepared as meals in the kitchen and served to the masses.

Gordos had invited many cooks in the community to help her make these meals, the restaurant itself was decorated to represent a typical popular eatery, and it helped raise some valuable awareness towards the area's diminishing agricultural potential. She was pleasantly surprised to find that her audience was not limited to any "artistic community" as such, but rather appealed to the entire community living in that settlement.

She also intended to provide an outlet where "real life" topics about the political situation of Egypt during the Morsi presidency could be discussed more openly in a "real" context—on the streets, over food. Whilst the Morsi government were focusing Egyptian life towards a new 'moralist' direction, the artwork served to represent the very basic day-to-day practical issues facing all Egyptians that were being glossed over.

However, in terms of stimulating a form of change past people's conversations though raising awareness, Gordos felt dissatisfied. The authorities began to issue threats that they would shut down the restaurant, and due to budgeting issues, the restaurant couldn't grow into more pop ups around the city, where more conversations could sprout as she had intended.

Nevertheless, this is a seminal work of art, due in no small part to it having been disguised as a social platform, and one that actively sought to enfranchise a very wide audience through its medium into its message. It was boldly located in the heart of the problem it addressed, and not confined to a museum space. It did not necessarily carry with it an artistic discourse that would have been challenging for the Egyptian citizen to comprehend. Its participation was overwhelming in numbers, and it was accurately demonstrative and straightforward, especially when the final meals were served.

[解读]

土地的死亡

人类是自然的产物，我们产生于辽阔的海洋，成长于森林覆盖的陆地，我们在成为社会性的人类之前，其实首先是自然的人，土地造就了一切我们赖以生存的一切，我们曾经是多么依恋他，不论是农耕民族亦或是游牧民族。可是在埃及，原本耕作的土地被占领，土地已经不能生长出农作物，取而代之的只有垃圾。艺术家 Gordos 的餐厅用十分直白的方法将这个问题提出来，从第一周新鲜高档的美食到最后食材无处寻找，最后将遍布于土地上的垃圾放在盘子里端上餐桌，想必观者是惊愕的，当我们的土地上一寸食材也长不出来的时候，是不是宣告土地已然死亡。Gordos 的餐厅不仅带给人们体验，同时也是人们尽情讨论政治看法的平台，莫尔西政权是否没有将重心放在经济与人们生活上，而是推动埃及走向伊斯兰主义。（李田）

Bakaboza 竞选活动
Bakaboza Campaign

艺术家：妮妮·阿雅琪	Artist：Nini Ayach
地点：埃及开罗	Location：Cairo，Egypt
形式和材料：空间、表演、装置	Media/Type：Space，performative，installation
时间：2011 年	Date：2011
推荐人：纳赫拉·阿尔·塔瓦	Researcher：Nahla Al Tabbaa

在 2011 年第一届总统选举期间，妮妮·阿雅琪基于 Ard el Lewa（开罗市的一个非正式的居民区）埃及公民心中的愿望，创造了这一个虚构人偶——Bakaboza。《Bakaboza 是一种乌托邦式的象征，来替想象中的新型埃及社会发声。

妮妮·阿雅琪勇敢地用这种媒介形式来执意表达政治乌托邦以及实际情况下不可避免的失望交织而成的那苦乐参半的本质，这一媒介形式必然是深思熟虑的结果；一般来说木偶会被世界各地的人民用作某个政治人物的讽刺形象。在埃及，木偶更是作为当局政客司空见惯的形象标志，然而值得注意的是这一人偶来自 Ard el Lewa 的埃及人民，他们居住于开罗市"非正式"发展市区的其中一个居民区内。

Bakaboza 许诺会在城市的每个街道种植芒果树并配备自动唱机，承诺会有更高的薪水，终结腐败和文盲问题，甚至设立由 Steven Seagal

传授的合气道公共课程。通过直接的社会交往，这个项目有效地使其参与者得到反思、希望、挑战以及勾画出哪些构想深埋在他们对自己祖国的想象里。同时人偶行为举动以及人物肖像的影响力可以提供一种可信的美感，使孩子以及成人都可以投入到这个艺术里，同时知道它是乌托邦式的，不切实际并超于现实。

Bakaboza 在孩童和成人的簇拥下在街上游行，和他们聊天和拥抱，并慢慢拥有了大批追随者，但是再次重申这是个基于大家对更美好埃及愿望的虚构人物。Bakaboza 鼓舞了人们发出内心的声音，并使其追随者有了主人翁意识和归属感，并凸显了真希望闪烁的那一时刻。

Artellewa 艺术空间也举办了全体选举活动项目，这里有海报、演讲、合唱还有一个大型的人偶，这在整体上体现了本艺术作品的形式。由于选举活动的消息给人以嘲讽的意味，这也被当做十分具有政治色彩的温和，机敏手段，通过这种方式来突出政治动荡不可避免地给埃及人民要带来的可能性和不可能性。

作为一种艺术，妮妮·阿雅琪所传递的信息是经过深思熟虑的，也许这个艺术品本身是用简单的材料构成，但是其高敏感度的方式可以激起公共反响。很多受到"Arab Spring"激发的艺术家会采取单方面的公共艺术形式，比如壁画或者涂鸦或者口号标语。而 Ayach 在受当下政治激发的同时，试着创造一种艺术形式，它的发展离不开"公共"两字，利用公共资源也为"公共"服务和努力。

Bakaboza is a fictional puppet created by Nini Ayach and based on the desires of the Egyptian citizens of Ard el Lewa, an informal settlement in Cairo, during the first round of presidential elections in 2011. Bakaboza was a utopic symbol and served as an alternative voice to a new and imagined alternate Egypt.

The medium through which Ayach bravely decided to show the bittersweet nature of political utopia and inevitable disappointment in practice was very deliberate; puppetry has traditionally been used globally to provide a satire of political figures. In Egypt, puppetry is a commonplace term to represent all figures in power, but it is noticeable that this particular puppet was born of the people of Ard el Lewa, a population living in one of the many "unofficial" urban developments in Cairo.

Bakaboza promised mango trees and jukeboxes on every street, higher salaries, free healthcare, an end to corruption and illiteracy, even public Aikido lessons by Steven Seagal. Through direct social engagement, the project was effective in its aim of providing participants with the chance to reflect, hope, challenge, and map out whatever figments had lain dormant in their imaginations of their homeland.

Additionally, the effectiveness of the puppetry, in its movements and human representation, provided a very believable aesthetic that allowed children and adults to invest themselves into the artwork, simultaneously knowing that it is utopian, impractical, and surreal.

Bakaboza marched down the streets, backed by children and adults alike, chanting and celebrating him, generating a large following, but reaffirming that he was fictional and based on their desires for a better Egypt. Bakaboza inspired voices and an awareness and a sense of ownership amongst his followers, but also highlighted a flickering moment of real hope.

The exponential take-up of the project is also seen in the full campaign created in the Artellewa art space, where posters, speeches, chants and a large puppet were created to manifest the artwork in its entirety. As defiant as the message of the campaign seemed in the message of its satire, it was also a very politically benign, and astute, manner in which to highlight to Egyptians the real possibilities and impossibilities that would inevitably come from the political upheaval of the time.

As a work of art, the deliberacy with which Ayach portrays the message, through work which may portray a simplicity in its materials yet great sensitivity in the manner in which it evoked public reaction, and one which had a public element to it from inception through to production, is shown through its impact. The form taken by many inspired to create public art which resulted from the "Arab Spring" has usually been one-way, murals or graffiti or slogans for example. But Ayach has managed to create a work of art that developed through, with and for the public, whilst being inspired by politics.

[解读]

偶像

在一个国家或者民族处于精神迷惘或者空白的时候，人民会希望出现一位救世的英雄，他能够力挽狂澜，拯救人民于水火，当时的"茉莉花革命"使埃及穆巴拉克总统黯然下台，此时埃及处于政治的空白期，经济也萎靡衰退，也许动荡的时期人们需要精神的支柱和安慰。而这个玩偶 Bakaboza 作为埃及人民想象中的候选人，也许是未知的一个代名词，是埃及人民给予乌托邦世界的一种幻想。Bakaboza 接受人们的欢呼簇拥、尊敬爱戴，在人们将这个虚幻的木偶作为偶像来希冀和崇拜的时候，是否有那么一瞬间人们的心里充满欢欣和希望。这个活动其实具有政治讽刺意味，我们会反思什么是虚假什么是真实，当艺术触碰到政治，他巧妙而微妙地触动了公众的神经，严肃与荒诞就在这里结合与上演。（李田）

安西娅·莫瓦斯 vs 格雷厄姆斯敦
Anthea Moys VS The City of Grahamstown

艺术家：安西娅·莫瓦斯　　Artist：Anthea Moys
地点：南非格雷厄姆斯敦　　Location：Grahamstown，South African
形式和材料：表演艺术　　　Media/Type：Performance
时间：2013 年　　　　　　Date：2013
推荐人：沃恩·赛迪　　　　Researcher：Vaughn Sadie

在南非，大多数大型行为艺术都是借助主要城市的节日庆祝活动开展的，这些活动项目基本都需要遵循一定的计划条规，并在付费观众那里取得一些基本报酬。国家艺术节每年都在东开普省的格雷厄姆斯敦大学城举办，而东开普省在历史上看时南非经济状况最落后的地区。观众的看法则是了解这一艺术作品重要性的关键。

莫瓦斯作为 2013 年度南非渣打银行行为艺术青年艺术家的获奖者，被委任负责开展格雷厄姆斯敦的国家艺术节。她借此机会将她自己的艺术概念包括表演、游戏、演奏、参与、公共空间以及些许冒险用一种简洁而又主题明确地方式传达给格雷厄姆斯敦这座城市，这种概念无论在艺术活动场地还是活动氛围都有体现。

为了演出 Moys 为自己安排了与雷厄姆斯敦当地居民竞赛的活动，尽管她对这些竞赛活动都不精通。这些竞赛科目包括：描述 1745 年苏格兰与南非的反抗战争重现活动；与来自格雷厄姆斯敦的 Rhodes

DanceSport 和来自博福特港的 Mtshizz 在舞厅比赛跳舞；作为一个独唱者与维多利亚女子高中唱诗班以及 Pro Carmine 唱诗班举行歌唱比赛；和来自 Rhodes 象棋俱乐部的五名棋手同时进行象棋比赛；自己一个人抢断 MARU 足球俱乐部成员的足球；和来自东开普省 Shotokan-Ryu 空手道会馆的六名黑带空手道选手进行较量。

通过三个月的培训，莫瓦斯与各个团队进行了训练以取得必要的竞赛技能，并发展人际关系，让人们理解使这座城市居民作为文化实践者参与进这个实际节日的意义，这不仅仅只是一个社会舞台背景，主办方或者当时的服务提供者。Moys 邀请他们在这个国家平台上与她进行平等竞赛。

在筹备这个锦标赛过程中，莫瓦斯尽可能地号召各种各样的"当地人"以及来自各种分类的群体参与进来，他们包括不同的年龄段、种族、性别以及不同的经济地位（这在一定程度上暂时掩盖了这座小城的等级种族划分）。这些群体的参与便可使莫瓦斯可以创建一个具有多样性和综合性的观众群体，他们的兴趣会随着主观角度而转变。当他们看到一个人荒诞地在与不同的团队进行不同项目的竞赛，所有人会在信仰的带动下变得团结，这或许会激发他们对不被看好的一方加油打气，或许就会意识到自己的这座小城就是在与一个学生城的身份抗争，这是一座全国的中产阶级大学生城。

尽管长期影响并不够明显，这个项目置身于一个复杂的情形下，在这个环境里，这个城市的身份会在外来中产阶级人群以及节日参与者的影响下不断明确和重塑。不如说，他们的关系基本是由当代艺术的或者创造性的文化和高等教育规定的，这在南非国家（其大部分的贫困人口数是参与不到节日庆祝的）是主要的特权领域。因此 Moys 努力将不同的群体带进这个关联空间，通过活动创建一个新的社会环境，在这里他们可以作为一个充分体现并带有感官的机制相互了解。

In a South African context, most large-scale performance art is commissioned through major city-based festivals, most of the projects are temporary and are subject to programme constraints, with a primary consideration for a paying audience. In the context of the National Arts Festival, which is held annually in the university town of Grahamstown in the Eastern Cape, historically the most economically depressed region in South Africa. The idea of audiences becomes vital to understanding the importance of the work.

Moys was awarded a commission to develop a work for the Grahamstown National Arts Festival, as the inaugural winner for The

Standard Bank Young Artist for Performance Art in 2013. She took this opportunity to bring her vocabulary of performance, gaming, play, participation, public space and risk, together in a succinct and conceptually cohesive manner that explored Grahamstown both as site and context.

For her performance Moys was scheduled to compete against residents of Grahamstown, though she herself was not a specialist in any of the disciplines she would compete against. The disciplines included: Depicting the 1745 Scottish rebellion with the South African Battle Re-enactments (SABRE) group; to dance in a ballroom competition against Rhodes DanceSport from Grahamstown and Ta Mtshizz from Fort Beaufort; compete as a soloist against the Victoria Girls High School Choir and the Pro Carmine Choir; take on five chess players simultaneously from the Rhodes Chess Club; tackle the MARU Football Club on her own and finally; challenge six Karate black belts from the East Cape Shotokan-Ryu Karate in combat.

Over a three-month period Moys trained with each the respective teams to develop the skills necessary to compete, developing relationships and understanding of what it would mean to bring the residents of the town into the actual festival as cultural practitioners, and not just a social backdrop, hosts, or service providers of the period. Moys invited them into the festival as participants to compete against her as equals on a national platform.

In creating this tournament, Moys gathered together the widest possible variety of "locals" and the participants across all categories of age, race, sex and economic status: which in some way temporally hides the class and race divides in the town. Add the Festival audience to this and Moys creates a heterogeneous collective audience whose own interests' shift according to their own subjective position. Everybody might be united in their belief in the absurdity of an individual competing against various teams in multiple events, yet they might be galvanised in their support for the underdog or perhaps support the town which in itself struggles with its own identity as a student town for middleclass tertiary students from all over the country.

Though the long-term impact is not clearly evident, the project inserts itself into complex situations, where a towns' identity is continually defined and reshaped by a migrant middleclass population of tertiary learners and festival-goers. The assumption would be that neither is really concerned nor has a vested interest in a sustained and continually developing understanding of place. Rather, their relationship to place is primarily defined by contemporary artistic/creative culture and tertiary education, which in a South African context is mainly the realm of the privileged, with the poorer economic demographic for the most part excluded from the festival. Thus Moys attempts to draw in some of these groups into a relational space through performance that creates new social contexts in which these identities can be reflexively explored as an fully embodied and with its own sense of agency. One of the participating teams were happy to finally be included in the history of the festival.

[解读]
为共同的信仰

这是一场精致的表演，是一次视觉的盛宴，是一次欢乐的庆典，还是一场观众自己选择阵营的比赛，艺术家将行为艺术融入国家艺术节（National Arts Festival），策划了这样一场有着众多观众的盛大活动。艺术家 Moys 将不同阶层、不同职业、不同年龄、不同性别的人组织在一起，形成一个综合性的观众群，艺术家这样在一定程度上掩盖了小城的等级种族划分，他们能够暂时忘却自己的身份以平等的姿态去参与和审视这场活动。他们的兴趣会随着主观的角度而转换，不同生活背景的人们会因为比赛的进行、相同的信仰而变得团结，他们共同支持着某一方，为他们共同的信仰而呐喊而欣喜或者悲伤。这样的行为艺术唤起了观众内心所共存的属于人类的情感，艺术家就是要将这样的一个活动带入到这样一座全国的中产阶级大学生城，这样的活动短期影响可能是小的，但是活动在人们心里所激起的涟漪是恒久的。（李田）

发挥你的作用
Dlala Indima

艺术家：班图·费哈拉
地点：南非帕卡密撒
形式和材料：涂鸦艺术
时间：2011 年
推荐人：沃恩·赛迪

Artist：Buntu Fihla
Location：Phakamisa, South Africa
Media/Type：Graffiti
Date：2011
Researcher：Vaughn Sadie

Dlala Indima 是一个创造性的集合词，柯萨语翻译为"发挥你的作用"。这个集合词是在一个较大型项目——2011 年 2010 个生活在小城市的理由过程中，被班图·费哈拉、Kwanele Mboso 提出使用的。Dlala Indima 以及 mak1one（开普敦的一个杰出涂鸦艺术家）打算通过涂鸦来美化 Phakamisa，并使年轻人通过这一创造性的行为和直观反射性、环境特有性的场所营造方法来参与到发展自我价值的活动中来。

Phakamisa 是位于东开普省的 King Williams 镇和 former Ciskei 交界处。Ciskei 是"班图斯坦"，又称"黑人家园"，是在由 1913 年"土地法"引起的种族隔离政策下，南非为黑人划分的居民区。这一法律带来的经济排斥所遗留下的问题有着深远持久的影响，一些区域的官方统计平均失业率为 40%，但是在某些区域失业率高达 90%。这对年轻人接受高等教育方面的影响尤为严重，有大批年轻人在接受完义务教育后开始无所事事。

土长在 Phakamisa 的 Dlala Indima 意识到他们应当关注年轻人，他们需要找到一个接近和动员这些年轻人的方式。最初，他们将据点设立在废弃的屠场和邻近的商店里，年轻人常常把这里当做隐蔽场所来吸毒、喝酒。他们对涂鸦艺术和嘻哈文化本有的兴趣，使他们得以融进早有当地年轻人参与的活动和运动中。在所有的这些活动中，这些年轻雅皮士会负责邀请人们来参加 Dlala Indima Centre 的发布会开幕式。这些活动是很有意义的，因为这些年轻人可以引起周边其他居民对环境处理问题上的关注。尽管市政度当局对此持怀疑态度，但社区支持以及较少经费预算的合理使用使这一项目得以实现。在使这一艺术中心逐渐走向概念化的过程中，Dlala Indima 不仅仅是要"复制一个本土化的嘻哈文化或者涂鸦艺术，而是在本地挑战城乡差距氛围的启发下尝试通过新颖的视觉和口头表达方式进行表达"(Sitas, 未发表，2014)。

借鉴他们在项目早期参与活动的一个相似战略，他们在屠场外面布置了一个完善的系统吸引人们加入。这变成了当地居民的一个日常活动，他们可以来这里放松或者来闲暇里帮忙解决问题。这是这个项目的关键，因为这可以给居民与项目建立联系的时间，也可以给艺术家们反思年轻人身上表现出的主题和兴趣。这就随后告知了他们年轻人对最后作品中用到的图片和短语有怎样的想法。通过这样的过程，Dlala Indima 得以从 40 个志愿者组成的核心团队发展到 200 个成员，他们在这个项目的不同领域发挥着各自的作用。这些互动行为进一步帮助他们设立其他据点，比如汽车站和托儿所，这都将得益于这类积极性消息。

Dlala Indima 的成功在于他们可以在农村社区本地化一些关于城市亚文化群的概念，以开启关于环境和审美标准的公共讨论模式。他们成功地通过延伸的过程——"强调了在他们自己环境下得集体合作可以为其涵盖体提供独一无二的机会"(ibid)。这两种元素合二为一，使居民有了对空间标准重新定义和怀疑的动力。他们积极参与到这个项目的多个部分来确认和塑造这一项目视觉识别的各个方面。Dlama Indima 艺术中心当前进行着多种项目，在这里来自 Phakamisa 的青年专家会对年轻人发表演讲，会定期放映纪录片，会围绕最初项目的各种主题来召开研习会。这个团队得到了第二次授意拨款，并正在研究适用于在整个东开普省发展社区中心网络的方法。

Dlala Indima, is a creative collective that means "play your part" in isiXhosa. The collective formalized by Buntu Fihla, Kwanele Mboso, during a larger project—Two Thousand and Ten Reason to Live in a Small Town 2011. Dlala Indima along with mak1one a prominent Cape Town graffiti artist, had the intention of using graffiti as a tool to beautify Phakamisa and engage the youth in developing self-worth through creative action and an intuitive, reflexive and context specific approach to place-making.

Phakamisa, is a township on the edge of King Williams Town in Eastern Cape and the former Ciskei. The Ciskei was a Bantustan or homeland, which was a territory set aside for black inhabitants of South Africa as part of the policy of Apartheid, made possible by the Land Act of 1913. The legacy of this economic exclusion brought about by this legislation has had a lasting impact, with the regions' official unemployment averaging at 40%, though in certain areas it is as high as 90%. This too has had a significant impact on the youth's ability to access resources and tertiary education, leaving a majority idle after completing formal schooling.

Having grown up in Phakamisa, Dlala Indima was aware that they should focus on youth and they would need to find a way for accessing and mobilizing them. Initially they had identified an abandoned butchery and the adjoining shop as a site, which was being used by the youth as a secluded site to take drugs and drink. With their own interests in graffiti and hip-hop they partnered with exiting events and movements that were already working with the youth of the area. Throughout these events flyers were distributed to invite people to opening launch event of the Dlala Indima Centre. These events were valuable because as the youth were raising concerns around how fellow residents were treating the environment. Despite scepticism from the municipality, drawing from community support and a smart use of a relative small budget they were able to realise the project. In conceptualising the centre, Dlala Indima was not an attempt at
"replicating an existing Hip-hop or graffiti vernacular, but were also experimenting across new ways of visual and verbal representation inspired by the local context that challenge the rural-urban divide."
(Sitas, unpublished, 2014)

Appropriating a similar strategy to the events they participated in earlier in their process, they placed a sound system outside of the butchery, to draw people in. This became a daily event for residents, with people either coming to relax or help out if they had spare time. This was vital to the project, as it gave time for residents to develop their own relationship to the project and give the artist the opportunity to reflect on the themes and interests expressed by the youth. This would later inform their thinking around the images and phrases that would be used in the final pieces. Through this process Dlala Indima were able to expand the core group of forty volunteers that they worked with, to about two hundred people who moved in out of the projects at various stages. These interactions further assisted in identify other sites, such as bus stops and crèches that would benefit from this type of positive messaging.

The success of Dlala Indima is that they localised a vocabulary associated with urban subculture in a rural community to open up public discussion about the environment and beauty norms. They successfully—through their extended process— "highlighted that working collectively within ones own context offers unique opportunities for inclusion" (ibid). These two elements combined, created the momentum for residents to redefine and question their own notion of place. They actively participated in several components of the project and determined and shaped aspects of the visual identity of the project. The Dlama Indima Centre is currently operational with a varied programme, where young professionals from Phakamisa come and speak to the youth, regular documentary screenings take place and workshops convened and structured around various themes that came out of the initial project. The team have received a second grant to reflect on this process and are developing an adaptive methodology that could be used to develop a network of community centre across the Eastern Cape.

[解读]

自发的色彩

艺术只是属于艺术家的创造还是可以属于一种群体行为？艺术必须是严谨的还是可以被更为轻松地创造和对待？艺术家 Buntu Fihla，Kwanele Mboso 的项目用涂鸦的形式来美化 Phakamisa，这个曾经被种族隔离，直到如今依然在就业和教育方面存在巨大的空缺的地方，艺术家企图用这种更为生活和轻松的方式引导和发展年轻人的自我价值。他们设立的涂鸦据点本是年轻人吸毒酗酒的堕落之地，而嘻哈文化却在此形成，年轻人的想法可以在此得到交流和表现，让这些曾经肮脏不堪的地方充满色彩并且萌发新的思想。艺术曾经是艺术家的行为和作品，而如今艺术更成为了也许表面上和艺术毫无关系的人们的共同创造，他可以影响大众，改变生活的环境和文化，让观者愉悦，让思想交流，让色彩流溢。这便是艺术带给我们的，也许不止是视觉可感知的美。（李田）

燃烧的博物馆
Burning Museum

艺术家：多位艺术家	Artists：Multi Artists
地点：南非开普敦	Location：Cape Town, South Africa
形式：贴画	Media/Type：Wheatpaste
时间：2013 年	Date：2013
推荐人：沃恩·赛迪	Researcher：Vaughn Sadie

第六区位于开普敦中心商业区东部边缘的临近地带，历史上的种族混合群体社区。在 1966 年的 2 月 11 日，这一地区在 1950 颁布的组织区域法的号召下被宣布成为 White Group。1966 年—19 世纪 80 年代之间，66 000 多人被安置到位于开普平原新的组织区域。这一过程产生的可视影响就是随着房屋、学校、教堂以及电影院的拆除以及限定社区关系的空间法规的实施，可视的文化遗产发生系统性的消除。

《燃烧的博物馆》是一个植根于南非开普敦地区的各学科交互性的集体。这一集体由 Justin Davy, Jarrett Erasmus, Tazneem Wentzel, Grant Jurius 以及 Scott Williams 组成。作为一个集体，他们对历史、身份、空间以及结构等主题法都感兴趣。这源于他们在开普敦自己独有的环境，这种环境可描述为一个："随着暴力排斥的历史发展变得伤痕累累，枯萎焦灼。这些历史构成了难以理解和有时具有无限能力的民主的基础，而这种民主会时不时地在所谓依法规定的公共空间内施展暴力手段"。

在 1980 年代，这里按规定正发展为白人的建筑除了部分大学外，这已经使人们开始关注这一地区的回忆就要流失掉了。在 2005 年，在土地归还过程中，这一地区重建为返回型前期居民区。每个家庭通过已被侵蚀掉记忆的情况下，弄清楚他们的土地，然而被学生和商业占据后的土地已与以往没有半点联系。那是第一次抓住乡绅化浪潮的一个明确标志。如今在城市安家的炒作使第六区成为这些投机项目的牺牲品，抱着转变城市的希望，推动设计成为社会，文化经济发展的工具。在此情况下，关于资本主义式的设计的重要历史的消失成为一种必要。作为回应，燃烧的博物馆将小麦馅饼引进这个区域—来自 Van Kalker Archive 的图片。这份档案包括 300 000 份底片，这些底片记录的是在一个叫做 Gerhardt van Kalker, 的荷兰人在搬来西开普省时，在一个 1937 年建立于 Woodstock Main Road 的经典肖像工作室拍摄的全家福的图片。这个集体关注的是 1950 年代和 1970 年代的图片，以期望回来的居民或者与这里历史相关的参观者也许能认出一些人，或者路过的人可以回忆起他们那已成为这个过程和体系牺牲品的朋友和家庭。

这个项目的关联性在于在那么短的时间框架里就可以引起大家对开普敦试图侵蚀这段重大历史重要性的举动。District Six 的文化遗产在一定程度上可以促进旅游业，但是几乎不能够考虑进开普敦 City Bowl 的城郊重建计划中。在将本项目发展到公共美术馆的过程中，本项目将家庭和政治史作为转变居民区的会话中心，而场景营造也是对关于公共空间这一复杂困难历史的回忆和复述。这一行为也是一些街头艺术策略的挑战，这些发展者一心想美化和优化居民区（这在本地已经十分普遍），他们往往是用感觉舒服的图片涂抹在墙上，而不是利用符合当时环境或者具有社会相关性的图片，而这样的图片才可以开启人们对于关系是怎样在复杂的社会、政治历史的条件下进行调停这一问题的讨论和交流。

District Six was a historical mixed race neighbourhood on the eastern edge of the City of Cape Town's, Central Business District. On 11 February 1966, the area was declared a White Group Area, under the Groups Area act of 1950. Over 66 000 people where displaced and resettled to new group areas situated in the Cape Flats, between 1966 and the 1980s. What this process made visible was the impact of the systematic erasure of visible heritage and culture through the demolishing of homes, schools, churches and cinemas, on the spatial practices that define the community's relationship to place.

The Burning Museum is a collaborative interdisciplinary collective rooted in Cape Town, South Africa. The group is made of Justin Davy, Jarrett Erasmus, Tazneem Wentzel, Grant Jurius and Scott Williams.

As a collective they are interested in thematic approaches such as history, identity, space, and structures. This stems from their own context in Cape Town, which they describe as a : "(space) that has been scarred and seared by a historical trajectory of violent exclusions and silences. These histories form the foundation of an elusive and at times omnipotent democracy that occasionally reveals its muscle in the form of laws and by-laws in public space".

During the 1980s the area was subject to development with the building of a whites only university on a portion of the site, this already evoked concerns that memory of the site would be lost. In 2005 the area was redeveloped for returning ex-residents under the land restitution process, this saw several homes being occupied by returning families. The family had to make sense of an eroding edge condition, occupied by students and business that had no connection to the area. This was a clear sign that the first wave of gentrification was taking hold. Currently with the hype around the city hosting The World Design Capital, District Six has fallen victim to speculative projects that push design as a tool for social, cultural and economic development in the hope of transforming the city. In this instance, the erasure of a significant history for capitalistic design imperatives. In response, Burning Museum have been inserting wheatpasties into the area—images from the Van Kalker Archive. The archive consists of 300,000 negatives that contain the family portraits that were recorded in the classic portrait studio in Woodstock Main Road, established in 1937 upon the arrival of Gerhardt van Kalker, a Dutch immigrant to the Western Cape. The collective focused on images from the 1950s to 1970s, in the hope that returning residents and visitors who have a history with the area might recognize somebody or that passers-by would be reminded of their own friends and family, who were victims of this process and system .

The relevance of the project, is that in such a short timeframe attention is drawn towards Cape Town's attempt to erode the historical importance of a relevant and significant history. The legacy of District Six is leveraged for tourism but barely considered in the planning for the redevelopment of the outskirts of the City Bowl. In turning this edge into a public gallery the project places family and political history at the forefront of the conversation of a shifting neighbourhood and that place-making is also about the act of remembering and retelling of complex and difficult histories in public space. The intervention also challenges the strategy of the appropriation of street art by developers to beautify or gentrify neighbourhoods (which is pervasive in the area), often plastering walls with feel good images, rather than working with contextually or social relevant images that open up conversations and debate about how relationships to places are mediated through complex social and political histories.

[解读]
记忆之城

人类生活和建造过的建筑成为人们存在过的痕迹，有些建筑被风沙和岁月掩盖，有些建筑被战争摧毁，有些建筑被人们遗弃，但是有些地区却因为政治和一项法律的颁布，使一切环境翻天地府，让一切建筑成为历史。开普敦是欧裔白人在南非建立的第一座城市，这座南非白人心中的母城三百年来数易其主，充满了多元欧洲殖民地文化色彩。开普敦地区成为种族隔离的牺牲品，上万家庭的被迫迁徙让这个地区成为永恒的回忆，人们只能从零星的记忆碎片中去寻找，而大多数都随风而逝。300 000份底片所记录下来的这个地区在那时候的人们生活的影像，帮人们回忆起那些被历史洪流席卷而去的建筑，和生活其间的普通人民。开普敦这片有着复杂历史的土地上，人们追忆和忏悔，反思和重建。（李田）

火车上的布道
Sermon on the Train

艺术家：MADE YOU LOOK
　　艺术团体
地点：南非豪登省
形式和材料：表演艺术
时间：2009 年
推荐人：沃恩·赛迪

Artist：MADE YOU LOOK
Location：Gauteng, South Africa
Media/Type：Performance
Date：2009
Researcher：Vaughn Sadie

MADE YOU LOOK 是约翰内斯堡的一个艺术团体，由 Nare Mokgotho 和 Molemo Moiloa 组成。这个团体在 2009 年开始发起了一项名叫《火车上的布道》的项目，这个项目致力于通过在通往索韦托的火车上举办南非金山大学的演讲来推进公有制、互动以及学术这些概念走向新的发展道路。MADEYOULOOK 艺术团体并不把这个项目看作是日常的研究而是作为探究人类互动以及获得知识的实验。

火车是约翰内斯堡及其周边最低廉的公共运输模式，主要的使用者是工薪阶级以及穷人。这列火车在以前种族隔离的政治制度下曾用于从城市中心的边缘地带运来大批的黑人劳动力，他们在 1950 年颁布的《族群住区法》的号召下不得不重新定居。这一法规将种族群体分配到不同的住宅区和城市不同的商业区，以对他们进行隔离。尽管 1994 年的第一次民主选举示意隔离不再强制执行，但是空间隔离的遗留问题仍然会反映在如今交通设施的使用上，因为这一历史性的隔离依旧由于经济因素而存在。

这一团体认为他们之所以对火车感兴趣是因为他们将火车比作动态的公共空间，这里有多种使用者通过它们在约翰内斯堡和索韦托之间往返——这种短暂无常性的状态持续性地影响了他们对于空间理念的定义和形成。MADE YOU LOOK 参与者的空间理念因下面场景产生的影响而不同。

背着笨重，可垂到膝盖的双肩包的孩童……带着撑满各类顾客所需小商品的篮子，从一个车厢走到另一个车厢的小商贩……时不时地会有表演、打鼓、跳舞、ibeshu 或者其他演出。当然这也会是一个教堂。你并不知道这会是个教堂，直到在你上车后的几站，看到手持《圣经》的两个年轻人跳进了你的车厢。也许还会有颂歌，火车上的其他人也许会参与进来，"阿门，哈利路亚"，这些传道声音也许会一直持续到你到达目的地。

《火车上的布道》将一系列正规的演讲带到了火车上，正如所说"观点都出自平常事物，分享启蒙运动，公共的了解而又恰恰在隔离的中心"。这些演讲由三个不同的学者来作，他们是：Anitra Nettleton, Kirsten Doermann 和 Isabel Hofmeyer。而这些演讲的宗旨是"把知识带给大家"，而不是"人们需要它"，是要号召大家将学术界的隔离打破，并鼓励大家对获取和分享知识新方法的探究。

对于《火车上的布道》奖品授予与否的考虑，是因为它属于一个小的、有涵盖意义的项目，而不是有一个变革性的议程。这个项目理解了它所处的环境并谨慎地将自己归纳进一个明确的范围。这个项目有自己的两个公共性：学术和公共交通工具通勤者，并将他们合二为一。这也是对于公共艺术标准的挑战，通常意义上的公共艺术是利用一个宏大的平台来来展示"每天那里发生了什么"的概念，并强调在如约翰尼斯堡这样的城市还存在怎样大得社会差距。火车上的布道就如火车本身体现的一样，既相互连贯也彼此隔离，这种矛盾冲突一直存在，应该去着手解决而不是改变，更可以说是突出体现。它强调的是着眼于当下短暂的瞬间作为可能性——一种新的场所营造方式。

这个项目的成功之处在于，它通过挑战嵌入公共机构的学术来使知识公共化，以此论述了公共的标准；不是通过文章发表和同行业内互评，而是通过公共传递的方式。利用一个简单的方式将日复一日的生活方式转化为公共艺术。

MADE YOU LOOK is a Johannesburg based art collective, consisting of Nare Mokgotho and Molemo Moiloa. The collective ran Sermon on the Train between 2009—2011, a project that looked to push notions of public ownership, interaction and academia into a new direction by facilitating University of the Witwatersrand lectures on the train to Soweto. MADE YOU LOOK saw the project as not being a study of the quotidian but rather as an experiment exploring human interaction and access to knowledge.

The train in Johannesburg and its surrounds is the cheapest mode of public transport, and is predominantly used by the working class and the poor. The train service under the previous political system of apartheid was used as a means of bringing in a large black workforce from the edge of the urban centers, where they had been resettled under the Group Areas Act in 1950. The act assigned racial groups to different residential and business sections in urban areas as a means of keeping them segregated. Despite the first democratic election in 1994 this state induced segregation (no longer enforced), the legacy of spatial segregation still informs contemporary usage of public transport, as this historical separation still exists mainly through economic determinants.

The Collective cite their own interest in trains as dynamic public space with multiple users shaping and defining it daily, with their own notion of place constantly being negotiated as they find themselves in a state of transience that moves between Johannesburg and Soweto. According to MADE YOU LOOK the players that impact on this negation of place vary from:

Children whose massive backpacks reach their knees... small-scale entrepreneurs (who) walk from coach to coach with overloaded shopping baskets of absolutely everything you need... Every now and again there might be a performance, drums, dancing, ibeshu and all. And then of course there is church. You don't always know there is one on the coach you get on and two young men might leap onto your coach, Bibles in hands, a station or two after you get on. There might be singing, other people in the train might take part, preaching might take the form of an elongated relay of pacing and 'amen, hallelujahs' till you get to your destination. (MADE YOU LOOK, 2011, 67)

Sermon on the Train, brought a formal lecture series on the Metro Rail, that "took as its point of departure the ordinariness of train preaching, of sharing enlightenment, of communal understanding and its position slap bang in the centre of isolation" (ibid). The lectures were given by three different academics: Anitra Nettleton, Kirsten Doermann and Isabel Hofmeyer about academic subjects, to wide variety of commuters. These lectures were intended to "take knowledge to the people" (ibid), not because "the people needed it" but because those who produced it did—as a way of calling to account the isolation of academia and encouraging the exploration of possibilities for new ways of making and sharing knowledge.

The consideration for the award is that Sermons on a Train is small and contained, and it is not prefaced with a transformative agenda. The project understands the context that it works in, and carefully places itself between clearly defined conceptual dichotomies. The project understands both its intended publics: academics and public transport commuter and considers the way they come together. It also challenges the notion of public art as a grand spectacle by showing an appreciation of "what's there, everyday" and highlighting the vast social distances that still exist in a city like Johannesburg.

Sermon on the Train took on the fraught clash of connectedness and separation that is embodied in the train, engaged it, yet did not change it, but rather highlighted it. It explores place-making as performative, located within the important moment of exchange and interaction between two diverse and transient publics. The emphasis is placed on the ephemeral through temporary interventions as a possibility—a new form of place-making.

[解读]

流动空间

火车是底层人民的聚集地，火车是封闭的，也是流动的，是分隔开来的，也是彼此连接的，这样一个复杂而有趣的公共空间，让这项公共艺术活动有了更独特的影响和氛围。《火车上的布道》其实是针对这样一个处于文化荒漠的人群，虽然种族隔离已然成为过去式，但是贫富和阶级的差距让这种隔离依旧存在，而这样的布道更是企图将学术界的隔离打破，让知识真正与最缺乏学术知识的那群人互动。也许真正的平等还远没有到来，但是火车上的这种布道成为了沉沦里的一点光明和奋进的力量，艺术家和我们都期待这种形式的知识传播能够改变一些什么，这样的实验让我们能够看到人与知识互动的过程中所产生的一些新的东西。总有那么一天，所有人都能够平等地接受历史给我们留下的知识和遗产，而这种火车上的布道，也成为在那一天来临之前的努力和探索。（李田）

感悟
PERSONAL UNDERSTADING

时至今日，艺术与社会及日常生活的关系愈加密切，如何将艺术从博物馆、美术馆中解放出来，融入我们的城市生活和公共空间，以艺术的力量来推动我们生活的进步，成为人们日益关注的焦点，公共艺术也由之前的边缘走向了当代艺术的中心。

公共艺术的核心即公共性，公共性不仅包括公众的参与，更为关键的是其中蕴含以艺术的手段解决我们的公共问题。随着经济的快速发展，我国城镇化建设取得了显著成绩，但在此过程中，伴随着城乡发展不平衡、人与人之间关系冷漠、人与自然环境失衡等诸多问题，因此，如何在城镇化建设中提高人们的生活品质成为一个值得关注的问题。公共艺术的引入恰恰成为解决诸多问题、净化人们心灵的有效方式之一。公共艺术抛弃之前的"艺术精英论"，将艺术面向公众，引导公众参与其中，针对社区建设、环境保护、能源短缺、人与环境之间的失衡等各种公众问题发出自己的声音，觉醒自己的公众意识。

公共艺术不是一种艺术风格或艺术潮流，它不仅仅停留于以雕塑、绘画等形式在某个地区或公共空间创作几件艺术作品，它更为关注的是以艺术为手段提高公众生活品质，实现公共福祉的最大化，重塑我们日常的生活美学，以此构建更为美好、开放的公共生活环境。

（闫丽祥）

随着人们对精神生活要求的不断提高，公共艺术成为许多专家和学者关注的焦点之一，公共艺术所具有的开放性和参与性，也使得越来越多的人能够理解与接受艺术。由于我国艺术资源在空间上分布不均，有些人可能一生中都没有机会走进美术馆或博物馆看一次展览。一二线城市的艺术博览会与艺术节此起彼伏，三四线城市就连美术馆与画廊都没有，艺术教育在我国还没有普及。艺术不应该是神坛上的圣物，它的目的应该是让更多人感受到美与思想的存在，而公共艺术就是在公共空间中对人们进行的一种艺术教育。

公共艺术应该能够使人们看到艺术是如何去质问诸如城市化与现代化、贫富差距、环境污染、种族平等、政治权利与市民社会等人类社会中最根本的几个问题的。我觉得艺术家应该是生活中的哲学家，把高深莫测的思想用简单易懂的形式告诉大家，而不是在共公共空间中追求那些神乎其神的抽象概念。相对而言，我们国家的公共艺术做得好还的很少，形式还比较单一，也没有引起很多人的共鸣。有些地方的所谓"公共艺术"反倒成为一种"视觉污染"。我想这就是我们大力推进公共艺术发展的意义所在吧。

（于奇赫）

公共艺术作为一种新的艺术门类，逐步打破艺术的边界，让艺术走向大众，成为一个更具综合意义的学科。它不再被供奉在高高的神坛之上，而是成为提出和解决一些社会问题的方法，面对一个有待完善的世界，不管是暴力、贫困、饥饿、污染、歧视，还是很多历史遗留的复杂问题，艺术似乎成为了一种更加美丽和温情的方式，它让人们感动和反省，激发出属于"人"的内心的精神感受。

公共艺术的呈现方式让冷酷的现实和惨烈的过去不那么令人绝望，它慢慢渗入到社会的各个角落和人群里去，它的影响力是潜移默化的。其实公共艺术极大地延展了艺术的领域，它与宗教、哲学、社会学、心理学都有关联，当我看到艺术家们的公共艺术作品的时候，会感叹艺术不再是贵族的玩物、宗教的附庸、个人的宣泄，而是成为拯救人们生活与精神的工具，很多时候，它走在敏感问题尖锐的刀锋上，唤起人们对一些事物的共同反思，也许是针对一个群体、一个地区的共同记忆，或者是所有民众的精神共鸣。在整理这些案例的过程中，我发现虽然处于同一个时代中，但是世界各个地区所面临的问题是极不一样的，有很多贫穷的地区仍面临着生存问题，比如食品短缺、暴力横生、无家可归等，而发达的地区则拥有更高的精神和文化方面的建设。公共艺术的出现在我看来是一个趋势，它使艺术以这样的方式参与和影响民众的生活，让沉沦得以解救，让幸福得以延伸，让伤痛得以缓解，艺术会更加多元化，也会更多地参与到生活中去。

（李田）

我心目中的公共艺术指的是存在于公共空间的艺术形式，形式表达多种多样，涉及雕塑、壁画、装置、景观设施等，公共艺术多出现于公共场合，除了对普通大众具有基本的美育意义，也必然有其自身的性质、意义，譬如：纪念性、象征性、标志性、陈列性、装饰性、趣味性、商业性、寓言性等。

在我生活周边的商业广场，有数尊岳敏君版的"大笑"雕塑立在广场上，偶尔有家长会让孩子站在雕塑旁合影留恋，市民们也会驻足停留观看，这样的公共艺术不仅提高了市民们的审美情趣，也标志着地方政府对公共教育有所建树。

众所周知，公共艺术早已不是拘泥于唯美形式的雕塑作品，艺术创作加入了多媒体艺术、装置艺术、行为艺术等形式，包括材质使用的多样性及对互动的重视。而艺术的施行主题及其创作范围也扩展至灯光设计、空间整体识别系统、街道家具及活动记录等形式，也越来越多地出现在大众视野中，对此感兴趣的朋友如若走上街头、走上公共场所，相信你会有很多关于公共艺术的收获。所以在我看来，在当下社会，公共艺术创作形式的发挥几乎不受限制。

（祁雪峰）

后记
AFTERWORD

本书是上海大学美术学院与国际公共艺术协会（IPA）在2014年11月共同主办的第一届国际公共艺术研究员会议的总结梳理。来自全球的近30位国际公共艺术研究员带来了精彩案例，各抒己见，为建立国际公共艺术学科的研究发展提供了交流平台，也逐步发挥了上海大学美术学院国际公共艺术智库网络的作用。

从会议的策划组织、顺利召开、后期案例的整理和评论，以及本书的编辑、排版、校对和出版，都与大家的努力密不可分。首先感谢上海大学美术学院院长汪大伟教授发起与组织了本次会议，为国际公共艺术领域的研究和交流填补了空白；感谢学校领导和相关部门的大力支持。其次感谢国际公共艺术协会（IPA）组委会主席 Lewis Biggs 和《公共艺术评论》杂志主编 Jack Becker 先生不远万里参与组织会议，带来了国际公共艺术宝贵资源和学术力量；感谢 IPA 的副主席和发起人金江波教授在会议组织中发挥积极主导的作用和活跃思维观点。同时感谢我的同事周娴、陈文佳、张羽洁等在会议中的贡献；感谢美院的博士和硕士研究生团队在本书编辑过程中做出的一切努力，在这里特别感谢博士生高浅在编辑工作中的突出贡献，以及硕士生梁鑫鑫在排版工作中的严谨态度。最后感谢上海大学出版社与柯国富老师为本书的出版发行给予的积极帮助与支持。

感谢所有为本书编辑、出版、发行而付出努力并给予过帮助的朋友们。

章莉莉
2016年12月

撰稿名单
Editor Group

主编：章莉莉

编撰工作组组长：高　浅

编撰工作组成员：闫丽祥　李　田　于奇赫　祁雪峰

排版设计：梁鑫鑫

排版助理：李晓翠　颜　含

图书在版编目（CIP）数据

地方重塑：首届国际公共艺术研究员会议文集/章莉莉主编. —上海：上海大学出版社，2017.7
ISBN 978-7-5671-2866-8

I. ①地… II. ①章… III. ①公共艺术—环境设计—世界—文集 IV. ① TU-856

中国版本图书馆 CIP 数据核字（2017）第 139824 号

责任编辑：柯国富
美术编辑：谷　夫
技术编辑：章　斐

地方重塑——首届国际公共艺术研究员会议文集
章莉莉　主编

出版发行	上海大学出版社
社　　址	上海市上大路 99 号
邮政编码	200444
网　　址	www.press.shu.edu.cn
发行热线	021-66135112
出 版 人	戴骏豪
印　　刷	江阴金马印刷有限公司
经　　销	各地新华书店
开　　本	787×1092 1/24
印　　张	18 １/３
字　　数	367 千字
版　　次	2017 年 7 月第 1 版
印　　次	2017 年 7 月第 1 次
书　　号	ISBN 978-7-5671-2866-8/TU・013
定　　价	120.00 元